Social Sciences Teaching in the Face of the Global Challenges of the 21st Century

Social Sciences Teaching in the Face of the Global Challenges of the 21st Century

Editor

Álvaro-Francisco Morote

MDPI

Basel • Beijing • Wuhan • Barcelona • Belgrade • Novi Sad • Cluj • Manchester

Editor
Álvaro-Francisco Morote
University of Valencia
València, Spain

Editorial Office
MDPI
St. Alban-Anlage 66
4052 Basel, Switzerland

This is a reprint of articles from the Special Issue published online in the open access journal *Social Sciences* (ISSN 2076-0760) (available at: https://www.mdpi.com/journal/socsci/special_issues/Social_Sciences_Teaching).

For citation purposes, cite each article independently as indicated on the article page online and as indicated below:

Lastname, Firstname, Firstname Lastname, and Firstname Lastname. Article Title. *Journal Name* **Year**, *Volume Number*, Page Range.

ISBN 978-3-0365-9350-0 (Hbk)
ISBN 978-3-0365-9351-7 (PDF)
doi.org/10.3390/books978-3-0365-9351-7

Contents

About the Editor vii

Preface ix

Jesús Granados-Sánchez
Sustainable Global Citizenship: A Critical Realist Approach
Reprinted from: *Soc. Sci.* **2023**, *12*, 171, doi:10.3390/socsci12030171 1

Teresa Nogueiro and Margarida Saraiva
TQM and SDGs for Erasmus+ Programme—Quality Education, Reducing Inequalities, Climate Change, Peace and Justice
Reprinted from: *Soc. Sci.* **2023**, *12*, 123, doi:10.3390/socsci12030123 23

Jordan Correa-González, Abel López-Díez, Jaime Díaz-Pacheco and Nerea Martín-Raya
Climate Change and Sustainability in Spanish Classrooms: State of the Art and Didactic Proposal
Reprinted from: *Soc. Sci.* **2023**, *12*, 108, doi:10.3390/socsci12020108 39

Esther Sánchez-Almodóvar, Isabel María Gómez-Trigueros and Jorge Olcina-Cantos
Climate Change and Extreme Weather Events in the Education of the Citizens of the Twenty-First Century: The Perception of Secondary Education Students
Reprinted from: *Soc. Sci.* **2023**, *12*, 27, doi:10.3390/socsci12010027 71

Diego Luna and José Antonio Pineda-Alfonso
Conflicting Knowledge Paradigms: Competence Discourse and Disciplinary Reality in Social Sciences Teaching
Reprinted from: *Soc. Sci.* **2022**, *11*, 553, doi:10.3390/socsci11120553 89

Roberto García-Morís and Ramón Martínez-Medina
Trainee Teachers' Perceptions of Socio-Environmental Problems for Curriculum Development
Reprinted from: *Soc. Sci.* **2022**, *11*, 445, doi:10.3390/socsci11100445 105

Fadil Latifi and Endrit Latifi
Did the Characteristics of Kosovar Teachers Influence the Results of Students in TIMSS 2019? Findings from the Performance of Kosovar Students in TIMSS 2019
Reprinted from: *Soc. Sci.* **2022**, *11*, 344, doi:10.3390/socsci11080344 119

Yamilé Pérez-Guilarte, Francisco Xosé Armas-Quintá and Xosé Carlos Macía-Arce
Social Sciences Teaching: Building a Holistic Approach from Student Teachers' Social Representations
Reprinted from: *Soc. Sci.* **2022**, *11*, 307, doi:10.3390/socsci11070307 133

Cristina Yáñez de Aldecoa and Isabel María Gómez-Trigueros
Challenges with Complex Situations in the Teaching and Learning of Social Sciences in Initial Teacher Education
Reprinted from: *Soc. Sci.* **2022**, *11*, 295, doi:10.3390/socsci11070295 151

José-Manuel González-González, Jesús-Gerardo Franco-Calvo and Darío Español-Solana
Educating in History: Thinking Historically through Historical Reenactment
Reprinted from: *Soc. Sci.* **2022**, *11*, 256, doi:10.3390/socsci11060256 171

Álvaro-Francisco Morote and María Hernández
What Do School Children Know about Climate Change? A Social Sciences Approach
Reprinted from: *Soc. Sci.* **2022**, *11*, 179, doi:10.3390/socsci11040179 189

Pilar Rivero, Iñaki Navarro-Neri and Borja Aso
Who Are the Protagonists of History? Exploratory Study on Historical Relevance after Completing Compulsory Secondary Education in Spain
Reprinted from: *Soc. Sci.* **2022**, *11*, 175, doi:10.3390/socsci11040175 . . . . **207**

Nena Vukelić, Nena Rončević and Sven Toljan
Student Teachers' Willingness to Act in the Climate Change Context
Reprinted from: *Soc. Sci.* **2022**, *11*, 47, doi:10.3390/socsci11020047 . . . . **223**

About the Editor

Álvaro-Francisco Morote

Álvaro-Francisco Morote is a lecturer at the Department of Experimental and Social Sciences Education (Faculty of Teaching Training, University of Valencia, Spain). His research work in relation to the didactics of the Social Sciences is focused on the teaching–learning of climatology, climate change, natural risks (floods, droughts, etc.), natural resources (water), landscapes, and field trips. These lines of work are carried out at school level (primary and secondary education and Baccalaureate) as well as at the university level (undergraduate and postgraduate), and from three points of view: (1) social representations; (2) school textbooks; (3) didactic proposals.

Preface

1. Preface to "Social Sciences Teaching in the Face of the Global Challenges of the 21st Century"

This reprint is based on a Special Issue on the same topic, published by the Social Sciences journal in 2022–2023. The objective of this reprint is to present different studies regarding the challenges encountered in Social Sciences teaching during the 21st century. The world is experiencing global changes that are affecting the social, economic, political, and environmental fields, among others. Climate change, natural hazards, social inequalities, immigration, refugees, and the spread of diseases (COVID-19, etc.) are just some examples of topics in the realm of Social Sciences. All of these themes can be addressed in the teaching field and across all educational stages (including university). Today, we are facing significant problems that must be addressed in Social Sciences classes (mainly those of Geography and History) and that can be linked to the so-called Sustainable Development Goals (SDGs).

This Special Issue aims to answer the following research questions through theoretical and empirical studies: (1) What are the purposes of teaching History, Geography, and Social Sciences in schools today? (2) What roles do social thought formation and social problems have in Social Sciences learning/teaching?

The reprint consists of three parts:

1. Introduction
2. Section dealing with the teaching of Social Sciences content, consisting of chapters on the following topics:
   a. Sustainable Global Citizenship: A Critical Realist Approach.
   b. TQM and SDGs for Erasmus+ Programme—Quality Education, Reducing Inequalities, Climate Change, Peace and Justice.
   c. Climate Change and Sustainability in Spanish Classrooms: State of the Art and Didactic Proposal.
   d. Climate Change and Extreme Weather Events in the Education of the Citizens of the Twenty-First Century: The Perception of Secondary Education Students.
   e. Conflicting Knowledge Paradigms: Competence Discourse and Disciplinary Reality in Social Sciences Teaching.
   f. Trainee Teachers' Perceptions of Socio-Environmental Problems for Curriculum Development.
   g. Did the Characteristics of Kosovar Teachers Influence the Results of Students in TIMSS 2019? Findings from the Performance of Kosovar Students in TIMSS 2019.
   h. Social Sciences Teaching: Building a Holistic Approach from Student Teachers' Social Representations.
   i. Challenges with Complex Situations in the Teaching and Learning of Social Sciences in Initial Teacher Education.
   j. Educating in History: Thinking Historically through Historical Reenactment.
   k. What Do School Children Know about Climate Change? A Social Sciences Approach.
   l. Who Are the Protagonists of History? Exploratory Study on Historical Rele-vance after Completing Compulsory Secondary Education in Spain.

m. Student Teachers' Willingness to Act in the Climate Change Context.

3. Conclusions.

2. Social Sciences Teaching in the Face of the Global Challenges of the 21st Century: Conclusions of the Special Issue

This reprint aims to present, to the international audience, different studies and experiences regarding the challenges of Social Sciences teaching in the 21st century. Currently, the world is experiencing global changes that are affecting the social, economic, political, and environmental fields, among others. Climate change, natural hazards, social inequalities, immigration, refugees, and the spread of diseases (COVID-19, etc.) are just some examples of topics in the realm of Social Sciences. All of these themes can be addressed in the teaching field and across all educational stages (including university). Today, we are facing significant problems that must be addressed in Social Sciences classes (mainly those of Geography and History) and that can be linked to the so-called Sustainable Development Goals (SDGs).

This reprint encompasses 13 articles that are related to different themes, including global citizenship, the SDGs, climate change, paradigms of Social Sciences teaching, the Social Sciences curriculum, social representations, challenges related to Social Sciences teaching and learning, and History education. For example, in relation to the first topic, the article titled "Sustainable Global Citizenship: A Critical Realist Approach" presents a theoretical reflection on sustainable global citizenship from a critical realist perspective, with the aim of relating realism and pragmatism to the personal and so-cial transformations that are necessary to achieve sustainability. The contribution of this work is the proposal of a conceptual framework that is structured using the following five key dimensions of citizenship: governance, status, social–ecological systems, social conscience, and engagement. The author of this article explains that agency–structure dualism requires more comprehensive approaches that integrate the self-awareness of all components that intervene in the autonomous decision to act, and which include personal capabilities, the desire and motivation to engage, and the real possibility of participation, which is determined by the social context and the individual situation.

In relation to the SDGs, the article "TQM and SDGs for Erasmus + Programme—Quality Education, Reducing Inequalities, Climate Change, Peace and Justice" aims to explore the potential relationships and synergies between HE sustainability and Total Quality Management (TQM) issues through the SDGs. The methodological approach involved a qualitative study of academic papers on TQM, sustainability, and the SDGs in HE, as well as on the analysis of (EU) 2021/817, which was established by Erasmus+. The authors concluded that TQM and sustainability have interactions related to the SDGs, and that the Erasmus+ Programme can support the sustainability of HEIs by promoting these SDGs. Leadership, education and training, the participation of staff members, measurement, evaluation, and control, and other stakeholders are essential factors of the effective implementation of TQM and sustainability in HEIs.

With regard to climate change, there are up to four articles that are directly related to this topic. For example, the research paper titled "Climate Change and Sustainability in Spanish Classrooms: State of the Art and Didactic Proposal" presents different proposals, because the 2023–2024 academic year is the first in which the curricular content developed from the Organic Law 3/2020 of December 29, which modifies the Organic Law 2/2006 of May 3 on Education, popularly known as LOMLOE, will be fully implemented. The authors designed a learning approach to sustainability and climate change that can be implemented in the Spanish and European contexts while responding to Rosenshine's principles of instruction, a circumstance that gives it enormous flexibility and makes it an interesting resource focused on helping Geography teachers in facing current challenges from an innovative, scientific, and inclusive perspective.

Another article related to global warming is "Climate Change and Extreme Weather Events in the Education of the Citizens of the Twenty-First Century: The Perception of Secondary Education Students". The objectives of this study, which was focused on third and fourth year ESO (Obligatory Secondary Education) state school students in a town in the province of Alicante (Region of Valencia, Spain), seek to reveal the perceptions of students regarding climate change and extreme weather events, in accordance with the subject through which these contents are taught, and to analyse whether the students have acquired a basic knowledge of the topic by the end of their secondary education. The students considered climate change as a threat to human beings and believed that anthropogenic action was the main cause of this threat. They also perceived an increase in extreme weather events, although it is necessary to qualify this aspect.

In the article titled "What Do School Children Know about Climate Change? A Social Sciences Approach", focused on school children (Primary Education—10 to 12 years old; third cycle, Secondary Education—12–16 years old; and pre-university, Baccalaureate—17–18 years old) in the Region of Valencia (Spain), the aims were to analyse the following: the main information channels through which children receive information on climate change; the causes and consequences that they recognize with respect to this phenomenon; and the main greenhouse gas that they believe is in the atmosphere. The results indicate that the three main forms of information media are digital (TV—82.8%; internet—56.2%; social networks—49.4%). With respect to the causes of the phenomenon identified by the students, particularly noteworthy was pollution (70.1%), and, in terms of effects, the increases and changes in temperature (61.7%) were of particular note. Finally, with regard to greenhouse gases, the majority responded with CO2 (63.5%; "incorrect answers").

The last study related to global warming is "Student Teachers' Willingness to Act in the Climate Change Context". The purpose of this article was to measure the willingness to act in the climate change mitigation and adaptation context, attitudes towards climate change, the perception of action possibilities in the climate change mitigation and adaptation context, interests in climate change, and concerns about ecological problems. A total of 201 student teachers from the University of Rijeka (Croatia) participated in the study. It was determined that (I) attitudes towards climate change, (II) the perception of action possibilities in the climate change mitigation and adaptation context, and (III) interest in climate change represented significant predictors of willingness to act in climate-change direction and mitigation contexts.

In relation to the paradigms of Social Sciences teaching, the article "Conflicting Knowledge Paradigms: Competence Discourse and Disciplinary Reality in Social Sciences Teaching" aimed to offer an inside look at the coexistence of different conception and organisation methods regarding Social Sciences teaching in schools. To this end, the methods of autoethnography and critical discourse analysis were used in a case study focused on the teaching experience of one of the authors. A system of categories constructed allowed us to empirically verify the existence of a conflict between two major epistemological paradigms, competency and disciplinary, in clear correspond-ence with a gap between innovative educational discourse and traditional school practices. In the case analysed, this conflict led to curricular over-dimensioning, which was used in order to expand on elements for teachers and their students to improve on, both qualitatively and quantitatively. The causes of this phenomenon are related to neo-liberal pedagogical parameters, and thus allowed us to conclude that we should not conceive competency-based learning as an educational change by itself, nor as a strategy that guarantees change.

The subject of the curriculum is analysed in the work "Trainee Teachers' Perceptions of Socio-Environmental Problems for Curriculum Development". The purpose of this paper was to ascertain how future primary and secondary education teachers in the area of Social Sciences perceive the socio-environmental problems that affect today's society, and how this impacts their subsequent inclusion in the curriculum. To this end, the study was carried out by means of a questionnaire, which showed that trainee primary and secondary teachers have a high consideration of these environmental problems, with hardly any differences noted based on sex or the type of degree studied. The authors explain that this perception is positive in terms of the subsequent inclusion of socio-environmental problems in the development of the curriculum performed by the participating trainee teachers, who largely preferred an autonomous model for curricular imple-mentation.

There were two articles related to social representation. The first, titled "Did the Characteristics of Kosovar Teachers Influence the Results of Students in TIMSS 2019? Findings from the Performance of Kosovar Students in TIMSS 2019" aimed to explore the impact of Kosovar teachers' characteristics on the performance of fourth-grade students in TIMSS 2019. This paper addressed the characteristics of teachers, both demographic and academic, which are closely related to the overall performance of the students. Data from the teacher survey show that the characteristics of the teachers, such as age, professional development, pre-service education, contemporary approach to teaching, etc., also played an important role in the low-scoring performance of Kosovar students in TIMSS 2019. Students who had a teacher with an average age of 30–60 years had a better performance on the test, achieving a difference of 20 points more, compared to students who had younger or older teachers (under 30 years and over 60 years). The lack of professional development of teachers also had a major impact on student performance. On average, 25% of teachers had attended training in certain areas in the last 2 years, and, on average, 80% of them were willing to attend training programs that would improve their overall competencies.

The second article related to this topic (social representation) is "Social Sciences Teaching: Building a Holistic Approach from Student Teachers' Social Representations". This work aimed to investigate the social representations that future primary school teachers have about socio-environmental problems, sustainability, the purposes of socio-environmental education, and training strategies for global citizenship educa-tion and education for sustainable development in Galicia (Spain). The results showed that students focus on problems such as climate change or environmental pollution, leaving aside issues such as social inequalities, poverty, or gender issues. In addition, they show difficulties in recognizing the strategies through which teachers can provide their students with critical thinking that leads to social transformation. Significant differences were observed in some of the aspects researched, based on the university of origin, the grade, and the gender of the students, with gender being the aspect that in-fluenced the greatest number of questions.

With regard to the topic of challenges in Social Sciences teaching and learning, the article "Challenges with Complex Situations in the Teaching and Learning of Social Sciences in Initial Teacher Education" presents various educational innovations in Social Sciences teaching and cultural heritage education based on the use of CBL during initial teacher training at the University of Andorra. These methodological proposals take into account the TPACK model (technological pedagogical content knowledge) based on the interrelation of three types of knowledge: pedagogical content knowledge (PCK), technological content knowledge (TCK), and technological pedagogical knowledge (TPK). Students work on the various dimensions of the SSCC, thereby de-veloping spatial competency, teaching competency, and competency in democratic citizenship. ICTs are included throughout in order to develop students' digital compe-tency. As a result, students feel empowered through having acquired different competencies and having developed an awareness of the value of cultural heritage as a cornerstone of democratic citizenship.

Finally, there were two articles related to History teaching. The first study, "Educating in History: Thinking Historically through Historical Reenactment", aimed to identify trends in the scientific literature that link the following two concepts: historical thinking and historical reenactment. The definitions of both concepts and their commonalities were examined. Local and global identity and heritage, emotions, the reproduction of objects, the use of sources, relevance, empathy, multiple perspectives, causation, communication, the relationship between past and present, and the sus-tainable economy proposed by the 2030 Agenda are all aspects that should be central to History education, both currently and in the future. Finally, the article "Who Are the Protagonists of History? Exploratory Study on Historical Relevance after Completing Compulsory Secondary Education in Spain" had the objective of discovering which four historical figures students identify as being the most important in human history after completing their compulsory schooling, and the types of reasons they give to justify their answers. By means of a mixed study, this research involved asking 165 students aged 16 to 17 which four figures they think are the most relevant in history and why. The results show a clear tendency in students to choose key figures in Spain's history, such as Christopher Columbus or Francisco Franco, or a prominent person, such as Adolf Hitler, as the most relevant historical figures that appeared in the curriculum in their last year of compulsory secondary education.

This Special Issue also demonstrates that Social Sciences teaching is imperative in younger cohorts, as it is one of the few subjects in which contents related to daily social problems are dealt with. It is, without a doubt, a subject that raises awareness and informs the citizens of the 21st century on the different challenges facing humanity, such as climate change, migration, pandemics, sustainability, pollution, economic and energy problems, social inequalities, etc.

All in all, this Special Issue offers some examples and international experiences related to the abovementioned phenomena, and how to deal with them more effectively in Geography and History classes. Ultimately, this Special Issue has been written with the aim of achieving a better and more resilient world in the midst of these problems.

Álvaro-Francisco Morote
Editor

social sciences

MDPI

Article

Sustainable Global Citizenship: A Critical Realist Approach

Jesús Granados-Sánchez

Faculty of Education and Psychology, University of Girona, 17004 Girona, Spain; jesus.granados@udg.edu

Abstract: The current crisis of unsustainability has renewed academic interest in sustainable global citizenship. Classical approaches to this type of citizenship have turned out to be quite abstract, utopian, and naive. This article is a theoretical reflection on sustainable global citizenship from a critical realist perspective, with the aim of bringing realism and pragmatism to the personal and social transformations necessary to achieve sustainability. The contribution of this work consists of the proposal of a conceptual framework that is structured by the following five key dimensions of citizenship: governance, status, social-ecological systems, social conscience, and engagement. These dimensions have been interpreted and described from two core ideas of critical realism: the position-practice system and the seven-scalar laminated system. The main conclusions are that agency-structure dualism requires more comprehensive approaches that integrate self-awareness of all the components that intervene in the autonomous decision to act, and that include personal capabilities, the desire and motivation to get engaged, and the real possibility of participating determined by the social context and the personal situation. It is also necessary to increase the number of types of agencies, especially with the recognition of the group as a key entity. The resolution of the dichotomy on state-global scale relationships can be articulated by differentiating between government and governance, and the role of social innovation in the latter.

Keywords: citizenship; sustainability; sustainable global citizenship; critical realism; agency; position-practice system; citizenship education

1. Introduction

Bhaskar (2002) considered that human society has entered a planetary polycrisis manifested in all four planes of social being: an ecological crisis due to our relations with nature; a crisis of morality because of how some humans treat other humans; a crisis in our social structures; and an identity crisis (who we are and whom we think we are). "*On each of these four levels we are profoundly alienated, and we are alienated by things that we cause to ourselves. They are not natural causes, they are causes which are mediated by human agency, as a result of which human beings are profoundly alienated*" (Buch-Hansen 2005, p. 63). The ecological crisis could be measured through humanity's ecological footprint (Galli et al. 2015), and it is now 73% higher than the world's ecosystems' carrying capacity. From 1 January to 29 July 2021, humanity used as much from nature as the planet could renew in the entire year (GFN & Schneider Electric 2021). Our impact is manifested as climate change, biodiversity loss, stress on freshwater, and deforestation, among many other processes and effects. Hartwig (2015) states that the roots of the planetary polycrisis are found in capitalism and that we have to move on to a post-capitalist way of doing things. For him, critical realism is above all the philosophy and social theory of transition to a planetary post-capitalist sociosphere, in which social transformation will enable universal human flourishing.

In recent years, and because of these current crises, together with the expansion and impact of globalization, there has been a renewed scholarly interest in citizenship and citizenship education with a focus on making the global agenda the citizens' agenda (UNU-CS 2018), and on assigning citizens a decisive role in the achievement of sustainable development (Granados-Sánchez 2021). If citizenship represents the normative guide to leading an

Citation: Granados-Sánchez, Jesús. 2023. Sustainable Global Citizenship: A Critical Realist Approach. *Social Sciences* 12: 171. https://doi.org/10.3390/socsci12030171

Academic Editor: Alvaro Morote

Received: 31 December 2022
Revised: 6 March 2023
Accepted: 9 March 2023
Published: 13 March 2023

active, committed, and meaningful life, sustainability has profound implications for 21st century citizenship because it forms the founding ethical basis for rethinking the requirements, rights, and responsibilities of citizenship in a global context (Kurian et al. 2014).

Citizenship is a contested concept. Its meaning has been broadened and has become increasingly complex. The general conventional definition of citizenship includes two main meanings: membership in a political community and a form of active, responsible, and good behaviour toward the community (Gosewinkel 2010). According to the theory of citizenship (Peterson 2020), there is a distinction between different traditions of citizenship, being the liberal and the republican the main ones. In both traditions, citizenship is associated with belonging to a defined political territory and with the development of shared identity, which constitutes a crucial social link. The Liberal tradition focuses citizenship on the individual and on the rights and guarantees that the state grants and must protect and assure to the person as a citizen. On the other hand, the Republican tradition places its emphasis on the community and on the responsibility and obligations that every citizen has for the common good. For Barry (2016), the interesting and valuable thing about republican citizenship is the importance of action and contribution to the common good. Consequently, citizenship should be taught, learned, and stimulated, and the necessary conditions should be created so that it could be exercised.

Historical records tell us that when social conditions change, some aspects of citizenship issues change with them. The approach of citizenship linked to sustainable development is relatively recent and it has been theorized mainly through concepts such as environmental citizenship (Connelly 2015; Dobson and Bell 2006; Hadjichambis et al. 2020), ecological citizenship (Dobson 2006; Nash and Lewis 2006; Wolf 2007), sustainability citizenship (Horne et al. 2016), and sustainable citizenship (Barry 2006; Granados-Sánchez 2008, 2021). The sustainable development goal (SDG) 4, target 4.7 (UN 2015) states that it must be ensured that all citizens (as learners) acquire the knowledge and skills needed to promote sustainable development, a culture of peace and the practice of global citizenship (amongst other things). According to Agbedahin and Lotz-Sisitka, "*there is an absence of guidance as to how such processes can be engaged and conceptualized*" (2019, p. 104). Hadjichambis and Reis also state that "*the conceptualisation of environmental citizenship in educational context remains an imperative need*" (2020, p. 2).

This paper aims to fill this gap in the literature with a conceptualization of sustainable global citizenship from a critical realist perspective. Critical realism is a meta-theory that offers a way forward for theoretical and methodological innovation within the field of global learning (Khazem 2018). It contributes towards the understanding and achievement of transformation towards sustainability and can help to ensure the flourishing of both current and future generations. Critical realism advances the project of human emancipation and involves reclaiming reality for itself and from the ideologies that usurp, deny, and obscure it (Bhaskar and Hartwig 2011). Critical realism is concerned with other forms of explanation, the nature of causation, agency, structure, relations, and the implicit or explicit ontologies we are operating with.

The conceptualization of the link between human agency and social structures is one of the core issues in social theory (Archer 1995). Specifically, the purpose of this paper is to undertake a critical review of current interpretations of sustainable global citizenship, to broaden their scope and provide a more comprehensive, realistic and relational approach to agency and structure. The starting assumptions of this work are the following:

- Most current approaches to sustainable citizenship are based on reductionism. For example, individualism is "*a position that privileges agency over structure, considering structures as the intentional product of the activities of actors*" (Buch-Hansen and Nielsen 2020, p. 51). Humanism also places agency at the core of the explanation (Mateus and Resende 2015), while structuralism privileges structures. The agency-structure dualism needs a more holistic, integrative and relational approach (Elder-Vass 2010).
- Agency is focused primarily on an individual level of analysis (or personal agency). The conceptualization of agency should go beyond the individual to include group,

collective, and other agencies, and it should be expanded to consider all the aspects that enable or constrain each of these types of agencies in a variety of contexts.

- Agency is considered an abstract and pre-existing value based on intentionality and desire (Thorne 2005). Instead, it should be seen as a culturally informed development shaped by self-awareness of one's capacities and contextual determinations for participation in specific real practices.

This paper is a theoretical reflection on how critical realist concepts, methods and explanation models can enrich and enhance the understanding and practice of sustainable global citizenship, with a special focus on agency and structure. Its methodological processes consisted of the following three different actions and stages:

- An initial non-systematic literature review was carried out to define and conceptualise sustainable global citizenship and to identify its main dimensions.
- A second action consisted of a selection of key principles and concepts of critical realism that are useful to broaden the conception of sustainable global citizenship, in such a way that it brings realism.
- The final stage was a process of reflection on how the dimensions of sustainable global citizenship are interpreted and enriched through the lens of the principles and concepts of critical realism.

The two first processes involved a document analysis (Bowen 2009). This qualitative research methodology is a procedure for reviewing or evaluating documents. The first phase consisted of a review of the literature in the field of sustainable citizenship. Although the literature review was not a systematic one and it is not reported according to the PRISMA 2020 statement (Page et al. 2021), several documents were analysed to track change and development in the field of sustainable global citizenship and to delimit key dimensions, aspects, and appropriate questions regarding this type of citizenship. The second phase consisted of critical reading and interpretation of some works of Roy Bhaskar and other relevant authors of critical realism, with the aim of selecting those ideas and concepts of this philosophical approach that, from the point of view of the author of this work, offer powerful forms of analysis and explanation of reality.

The article is structured in three theoretical sections according to the three stages mentioned previously. The first section provides a definition of sustainable global citizenship and a selection of its main dimensions. The following section presents a brief synthesis of social realism and the description of key ideas and principles that will underpin our interpretation. In the third theoretical section, the selected key conceptual dimensions of sustainable global citizenship are interpreted following the main critical realist principles and concepts. The article ends with a conclusion section where the main arguments of this theoretical reflection are briefly summarized, and their educational implications are discussed.

2. Sustainable Global Citizenship

Sustainability citizenship is unclear and controversial (Nelson 2016). Currently, some authors use the concepts of environmental, ecological, and sustainable citizenship as equivalents to escape the debate of differentiation of terminologies and they choose one of them as an umbrella concept that includes them all at once. This is the case of the European Network for Environmental Citizenship (ENEC 2018; Hadjichambis et al. 2020), which considers that, from an educational point of view, it is important that there is a single concept and that the most appropriate one is that of environmental citizenship. For their part, Van Poeck and Vandenabeele (2013) choose sustainable citizenship as a uniting concept. In my opinion, the umbrella concept is used to avoid the terminological debate, but it is important to emphasize that conceptualization is not banal or neutral, it is intentional, and the nuance is important.

Environmental citizenship has placed its emphasis on claiming environmental rights, that is, the right that all human beings have to a healthy and adequate environment to develop a life with health and well-being, and this requires society and governments to protect and ensure a quality environment. For Barry (2006) it is a very narrow and limited concept of citizenship, since it focuses mainly on environmental aspects and forgets key dimensions of sustainability such as society, politics, culture and the economy. In addition, it contains a rather passive liberal conception of citizenship and is mainly circumscribed at local and state levels. Kurian et al. (2014) see environmental citizenship as a continuation of the status quo, as citizens carry out symbolic actions such as planting a tree or consuming in an environmentally friendly way. Neoliberal environmental citizenship theorists shifted the focus from politics to economics, replacing the citizen as a political being with the consumer (an economic being), the state with the corporation, and politics with markets (Cao 2015). In this sense, sustainability depends on consumer choice.

Ecological citizenship, as described by Dobson (2003, 2011), differs from environmental citizenship in four essential characteristics:

- Citizens' obligations and responsibilities are understood as being non-reciprocal and must come first. The main obligation of each citizen is to ensure that their individual impact does not impair the possibilities of others to have opportunities and to meet their needs. Dobson (2004) uses the ecological footprint as a measure of this impact.
- Virtue is a very important concept in citizenship. Justice and equity stand out as first-rate virtues. Justice has to be applied spatially and temporally, that is, among all citizens of today and of the future and regardless of their origins (intragenerational and intergenerational justice). A second order of virtues includes care and compassion.
- Both the public and private spheres of people are considered since private individual and group acts often have implications for the public sphere. The private sphere is closely linked to standards and lifestyles.
- Ecological citizenship extends beyond the state because it needs to be global (it must include the whole Earth). For Wolf (2007) it is a citizenship without territoriality since national borders are nothing more than an obstacle to effective action on problems such as climate change, which is caused by, affects, and has to be remedied with the participation of all terrestrial citizens, governments, and global actors. Thus, ecological citizenship needs to be global by definition.

These four characteristics are the basis of what Dobson calls a post-cosmopolitan citizenship. The main criticisms of Dobson's theorizing about ecological citizenship highlight the fact that the author insists a great deal on the individual agency of citizens and little on other types of agencies, in addition to ignoring the social, economic, political, and cultural structures that restrict people's ability to act (Sáiz 2005).

Barry (2006) prefers to talk about sustainable citizenship, as he considers it more ambitious, multifaceted, and challenging because it cares more about social justice, equity, and democratic governance. Sustainable citizenship is well aware of the structural causes of socio-environmental degradation and, above all, deepens the social and economic aspects of sustainable development, such as respect for human rights, social inclusion, otherness, solidarity, equality and equity, quality of life, deliberative democracy and participation for good governance. Bullen and Whitehead (2005) argue that the notion of sustainable citizenship destabilizes the spatial, temporal, and material parameters on which forms of modern citizenship are based. The norms of sustainable citizenship go beyond state boundaries and the present to include individuals who do not know each other, either because they live in distant parts of the world or because they are people who are yet to be born. It is not reciprocal and assumes that individuals do not act on the basis of personal gain or by claiming certain rights, but because they feel responsible for others and are motivated to act to achieve social justice and equity (Atkinson 2014). For Van Poeck and Vandenabeele (2013) a sustainable citizen is an active, critical and independent citizen who is able and willing to play an active role in solving problems and issues related to sustainable development. Dobson (2011) also conceives sustainable citizenship as a pro-sustainability behaviour, both in the public and private spheres, based on equity in the distribution of environmental goods and on the co-creation of policies for sustainability. Precisely, Kurian et al. (2014) affirm that deliberative processes constitute the key element of sustainable citizenship because they go beyond mere routine dialogues and allow the confrontation of ideas that can bring out shared values and carry out sustainable actions that combat, for example, inequality as a result of political and economic power. Barry (2016) goes one step further and proposes that citizens have the obligation to carry out a service to the sustainability of the community, in the spirit of contributing to the common good.

In this paper, the author uses sustainable global citizenship because the concepts of sustainability and sustainable development are more holistic and intentional than that of the environment and, moreover, direct us towards a horizon. I, therefore, believe that it is more appropriate as an object of citizenship. In addition, using sustainable global citizenship makes it easier to link the field of citizenship with that of education for sustainability, two fields that until now have evolved in parallel and without a clear and determined connection that must be solved. Last, but not least, the ascription of the global scale to citizenship is important for highlighting the importance of our role in the current and future health of our planet.

There is no widely agreed definition of global citizenship. As stated by UNESCO (2017, p. 2), "global citizenship does not entail a legal status. It refers more to a sense of belonging to the global community and a common sense of humanity, with it presumed members experiencing solidarity and collective identity among themselves and collective responsibility at the global level". Political approaches to sustainable global citizenship include globalist (Beck 2010) and pluralist theories (Cao 2015) to emphasize the importance of interconnection and interdependence, universalism, plurality, and the need for inclusion of diversity, and difference. Abdi (2015) advocates decolonising global citizenship by problematizing the basic meanings and assumptions coming from Western countries and moving to an epistemic pluralism. For him, "the current mono-epistemicalizations of global citizenship education which are disempowering and de-culturing people in more ways that we can count here, should be redesigned and reconstructed with multi-locational knowledge and cultural pluralisms that can effectively and inclusively respond to the realities of lived citizenship contexts that are not fixed or static but are active and dynamically shifting as demanded by the contexts and relational categories that sustain them" (Abdi 2015, p. 23).

As Nelson (2016) argues, despite the fact that there is no consensus when it comes to defining sustainable citizenship, we can identify the main characteristics that make it distinct from other forms of citizenship. Table 1 contains three synthetic lists with the main dimensions that could structure the approach to sustainable citizenship, according to the opinion and analysis reported by the selected authors. Most of these dimensions are represented by concepts that constitute dualisms or dichotomies.

Table 1. Examples of key dimensions of sustainable citizenship.

Dobson (2010)	Kurian et al. (2014)	Nelson (2016)
Territorial and Non-territorial Citizenship	State and non-state	Pragmatic Glocal Citizenship
Rights and Obligations	Rights and Responsibilities	Collective Responsibilities and Obligations
Active and Passive Citizenship	Democracy and Capitalism	Participatory Democracy and Shared Governance
Public and Private spheres	Public and Private	Socioecological approach
Individual and Community	Universal and Particular	Individual, People and Environment (one, the other and otherness)
Virtue	Human and Non-human	"Being" rather than "Having"

Synthesis of the works of Dobson (2010); Kurian et al. (2014) and Nelson (2016).

The first dimension of sustainable citizenship confronts state-based sovereignty and how it is linked to political territories, with other non-territorial forms of citizenship. Secondly, the three cited authors coincide in pointing out the importance of rights, responsibilities, and obligations. Regarding this, Nelson (2016) suggests going beyond the individual scale and emphasizes the importance of the collective in relation to responsibilities, obligations, and rights. In third place, the authors address citizen participation in decision-making from different angles: Dobson (2010) pays attention to the active or passive attitude of citizens, while Kurian et al. (2014) and Nelson (2016) focus on the political system, the latter stressing the need for participatory democracy and shared governance in order to achieve sustainability. Although agency and structure play a decisive role in this dimension, the authors do not emphasize these two elements beyond political participation. The next dimension differentiates the public and private spheres, or how to overcome this dichotomy with a more inclusive approach. Another characteristic of sustainable citizenship highlighted by these three authors is the opposition to individualistic and collective positions. In their analysis, they do not consider the group as a category that is decisive, as well as the diversity of complex relationships that occur between individuals, groups, and the collectivivity. The last key dimension for Dobson (2010) is virtue. The sustainable citizen does the right thing, not because of the incentives, but because it is the right thing to do (Dobson 2003). People often choose to do good for reasons other than fear of punishment or loss, or a desire for financial reward or social status. People sometimes do good because they want to be virtuous (Beckman 2001). According to Dobson (2010), justice is the primary virtue, with empathy, compassion, and caring being secondary virtues. Nelson (2016) stresses the need to understand self-realization outside of materialism and to move from "having" to "being".

3. Critical Realism Key Principles and Concepts Related to Citizenship

Critical Realism is a philosophical approach to the functioning of society proposed by Roy Bhaskar (1978). As a meta-theory, critical realism is concerned with the investigation, discussion, and structure of other theories to improve our ability to understand the world and, therefore, reality. To Bhaskar (1998) there is a world independent of our knowledge about it and this world both pre-exists as a condition and is reproduced or transformed by human action.

According to Archer et al. (2016) defining critical realism is not an easy task because there is not a unitary framework and social realist scholars may have a heterogeneous series of positions, but the feature which unites them is a commitment to the following critical realist principles:

- *Ontological realism*: the world is real, structured, and complex. It exists independently of our knowledge or awareness of it.
- *Epistemological relativism*: knowledge is a continuous process that is socially produced under specific social, historical, and cultural conditions and is, therefore, changeable, and fallible. As knowledge is contextual, conceptual, and activity dependent it must embrace a form of epistemic relativism.
- *Judgemental rationality*: Bhaskar and Hartwig (2016) state that, despite the fact that knowledge is fallible, it is possible to arrive at decisions between competing beliefs of theories because not all interpretations are epistemically or morally equal and there can be rational grounds for preferring one to another (or judging which accounts about the world are better or worse).
- *Cautious ethical naturalism*: this principle is an attempt to reconnect facts and values. Facts are value-laden, and values are fact-laden and, therefore, values are open to empirical investigation and critique.

These principles relate to three domains of reality (or strata of knowledge): the real, the actual and the empirical (Mateus and Resende 2015). The real domain refers to whatever exists, be it natural or social ("*everything there is*"). The real cannot be observed and exists independent from human perceptions, theories, and constructions. It includes the power-generating structures and the event-generating mechanisms. The actual domain refers to what happens when these powers and mechanisms are activated and produce change (*"everything we can grasp"*). The empirical domain is what we know about reality and that we gain through experience and with the perception of the effects of actualities (*"everything we can observe"*). Therefore, critical realism makes a clear distinction between the real world and the observable world and represents a shift from epistemology to ontology and from events to the mechanisms that produce them (Agbedahin and Lotz-Sisitka 2019).

The *Transformational model of social activity* (TMSA) (Bhaskar 1998) is a key theoretical contribution that makes a distinction between individuals acting and the society that enables and constrains their actions. For Bhaskar (1998) the relationship between structure and action is transformational and not dialectical because they don't constitute two moments of the same process. Societies are the material condition and the result of human action and, therefore, it is both the ever-present condition and the continually reproduced outcome or result of human agency. On the other hand, praxis is the reproduction or transformation of society. The TMSA offers two central ideas for the conceptualization of sustainable global citizenship: the *position-practice system* and the *seven-scalar laminated system*.

The TMSA suggests that agency emerges from the structure via the *position-practice system*, a point of contact or mediating system between human agency and social structures. This mediating system is key to understanding the evolving, dynamic and unpredictable open system of reality. To better understand the concept of the position-practice system, Bhaskar (1998) points out that positions should be understood in a more broad and elaborate way, as the positions that a person occupies and the practices in which is engaged because of these positions. The post or position involves functions, rights, duties and tasks and certain degrees of assumption and enactment. The practices are the activities within the system in which individuals are involved. For this reason, the position-practice binomial is conceived relationally and as part of a system. States, corporations, intergovernmental organizations, NGOs, and other group and collective partnerships also have an agency and a position-practice system as entities in social systems and structures and global governance.

The seven-scalar laminated system is another core idea of the TMSA that is useful for analysing and examining emergent and complex issues. Table 2 shows the seven scales or levels and their descriptions according to their potential in agential transformative praxis.

Table 2. The seven-scalar laminated system.

Levels and Scales	Description
Level 1. The sub-individual psychological scale.	It is concerned with the intrinsic personality of the individual. It includes the individual's nature, identity, character, and psychology, as well as the individual's motivation, aptitude, confidence, intentions, interests, desires, and concerns.
Level 2. The individual person or biographical scale.	Describes the person that is studied and their capacity to determine the impact on social events and actions. The individual agency and the position-practice system are influenced by the state of mind and body of individuals (such as being healthy or sick, being capable and skilled, having access to training and resources, and so on).
Level 3. Small group micro scale.	The micro level represents the studied group of the population, especially when individuals interact. Human interaction denotes the relationship between the individual, the group, and the collective, and can develop an emergence of group and collective agency as a result.
Level 4. The meso scale: structures and functional roles.	This level of the laminated system explores structural factors that give rise to individual and collective experiences (focusing, for example, on relations between functional roles). Structural factors include mechanisms and/or powers which may be known or unknown, constructive, or destructive, and pleasant or unpleasant.
Level 5. The macro scale: societies and territories.	This layer of reality is concerned with the functioning and operation of societies and/or their territories/regions. The understanding of societies as a whole includes knowledge about its composition, constitution and configuration, and how these elements influence individuals, groups, collectives, and structures.
Level 6. The mega scale.	It is the analysis of civilizations and their traditions, which are the result of different geo-historical trajectories.
Level 7. The planetary scale (or cosmological level).	It is the superior level, and it understands the planet as a whole (as a planetary system). Global bodies and agencies such as the United Nations have a key role in developing, implementing, and monitoring global policies.

Author's synthesis from the following sources: Bhaskar (2010); Price (2014) and Agbedahin and Lotz-Sisitka (2019).

4. Sustainable Global Citizenship Key Dimensions through the Lenses of Critical Realism

Dobson (2010) and Kurian et al. (2014) argue that sustainable citizenship requires a deliberative dialectic on dichotomies, that is, it deals with deliberation based on the discussion of pairs of opposing or contradictory ideas and the confrontation of these in a variety of contexts so that shared values are found. Dialectics allows for interpretative flexibility and achieves synthesis, poses options and aids decision-making. A critical realist notion of sustainable global citizenship goes beyond reductionist and deconstructionist approaches to dichotomies and dualisms because reality is complex. "*Critical realism constitutes an ambitious attempt to transcend all the dichotomies/dualisms (...) by offering a nuanced both/and perspective as opposed to an uncompromising either/or perspective.*" (Buch-Hansen and Nielsen 2020, p. 49). Table 3 presents the key dimensions that, from my personal interpretation of critical realism and sustainable global citizenship, should structure the conceptualization of this type of citizenship. These dimensions are governance, status, social-ecological systems, social consciences, and engagement. The dimensions are described in the following sections of the paper, and they are characterized by unique concepts that represent citizenship's main areas of attention. In turn, each dimension contains a series of threshold concepts arranged in the form of dualisms, but which also integrate different nuances and possibilities. The intention behind this approach is to focus on the relationships, interactions, and dynamics between the two poles of each dualism, as well as the relationships between dualisms in the same dimension and with other dualisms in other dimensions, with a special focus on agency and structure.

Table 3. Key dimensions and threshold concepts of sustainable global citizenship, from a critical realist interpretation.

Dimensions	Threshold Concepts and Dichotomies	
Governance	Sovereignty, state sovereignty, national governments, legality, power, statal authority. Territoriality.	Global governance, international relations, multi-level politics, cosmopolitanism, planetarity, diffusion of authority. Non-territoriality, aterritoriality.
Status	Responsibilities, duties, obligations. Belonging, membership, identity.	Rights, guarantees. Exclusion, multiple identities.
Social-ecological systems	Individual. Personal. One. Public. Society, social processes.	Group(s), communities, collective/collectivity. The other, otherness. Private. Nature, ecosystems, natural processes.
Social conscience	Agency ("agencies"). Self-consciousness, self-efficacy, locus of control.	Structures (social, political, economic), culture. Unconsciousness. Social innovation.
Engagement	Commitment. Participation, empowerment, action, activism. Self-determination, codetermination.	Inhibition. Inaction, passivity. Hesitance.

Author's own proposal.

4.1. *Governance*

The concept of citizenship has a long history in political science, but in recent decades, it has been reinterpreted in many ways and has acquired multiple meanings beyond the political. One of the current main challenges in citizenship is how we combine our agency related to the state and the agency beyond the state and sovereignty.

Agnew (2018) repudiates the false dichotomy of state sovereignty and the global as antithetical political realities. I also disagree with the conventional vision in which it is believed that globalization and states must compete with each other. On the contrary, my conception is of complementarity, which leads to an approach of dual reality, with the first component that is national citizenship (which maintains the political sovereignty of citizens circumscribed in the nation-state and their role in the election and control of a government), and a second component that is global citizenship understood as an alternative construction of "the global" and globalization, from governance and social innovation. Government is a formally centralized political authority that administers a territory or state and executes policies in a hierarchical relationship and within a legality. On the other hand, governance is characterised by a diffusion of authority and power, where non-state actors, such as transnational corporations and social movements, engage in processes of decision-making (Rüland and Carrapatoso 2022).

In the following subsections, the paper presents a critical realistic vision of citizenship that integrates its political, social, cultural, and economic variables, in situations of territorial sovereignty (what is known as a nation-state), in global governance (through multi-level politics in different territorial scales and situations), as well as in other non-territorial life spheres. The dualism state-global could be analysed through the seven-scalar laminated system model. The state and its government are represented by the meso and macro scales (levels 4 and 5), while the global constitutes the planetary scale (level 7). Individual and group agencies of citizens (levels 2 and 3) are affected by the subindividual psychological scale (level 1), and it is of a different nature depending on the levels where action is taken or affected. In this sense, as we move vertically or horizontally within levels, we pass from state to global realities, and from sovereignty to social innovation.

4.1.1. Citizenship and Territorial Sovereignty

Traditionally, citizenship has been related to belonging to a geographically delimited community with which we identify and share an identity, a political space, sovereignty, and a legality. Nation-states are the territories that have played a decisive role in the determination of citizenship because it is where we have the greatest capacity for political decisions or sovereignty in all its scales, from local to state. Sovereignty is a question of the real status of people, which is why the legal system of the country from which we originate applies to us and from which we benefit from civil, political, and social rights, as well as taking on obligations and responsibilities. This conception of citizenship is one of the fundamental components of the socio-legal framework of the Westphalian world of state political communities (or modern states) (Cortés and Piedrahita 2011). For Dalton (2008), citizenship in democratic states is characterized by a series of rules established and regulated by the community itself and which determine what is understood as a good citizen. This expectation revolves around four basic principles:

- Public participation is necessary to make sense of and legitimize the democratic process.
- Autonomy presupposes that citizens develop their opinions separately from each other, through information and dialogue, enabling them to understand various points of view and to form their own.
- Citizens accept the legitimacy of the state and of obey the law.
- The relationship with others means that citizenry includes ethical and moral responsibility towards others based on justice and solidarity.

At present, citizenship as an idea and as a practice is multidimensional and its link to a territory or country is today insufficient (Anderson 2008) for several reasons. First, it does not respond to the complex realities of states. Second, as Dobson (2010) points out, territoriality is a discriminatory characteristic of citizenship since it is a condition or requirement of belonging that, when fulfilled, means privilege and when not fulfilled, such citizenship is denied and, therefore, it excludes. Third, states alone do not provide solutions to the needs and challenges imposed by globalization, climate change and other major problems of global unsustainability are included in the 2030 Agenda (UN 2015). Fourth, there are sovereignties of an integrationist type (Agnew 2018), such as the European Union, which transcend the nation-state and are causing the centrality of states to diminish in favour of territorially augmented sovereignties, with a recognition of supra-state citizenships (Malatesta and Granados-Sánchez 2013).

Although sustainability transcends the borders of countries, I do not believe that the idea of sovereign states should be discarded as proposed by Kurian et al. (2014) because it is not realistic. Instead, states should be considered as a reality with which we live, and which can be an object of transformation or susceptible to being transformed through collective agency. Nor do I believe that states should have such a central and coercive position as Barry (2016) advocates.

4.1.2. Global Citizenship

The Anthropocene suggests to us that we belong to a unique human community that, with its form of development, is influencing the destiny of ourselves and the planet. This notion of human community has led to various conceptualizations such as global citizenship (UNESCO 2018), planetary citizenship (Gadotti 2017), and terrestrial citizenship (Morin 2004), to mention just a few. For Morin, "*we humans have a common identity: not only the same genetic code, the same brain capacity, but the same capacities of emotion, of sympathy, of friendship and, therefore, of hatred. Likewise, among us we have a community of destiny* " (Morin 2004, p. 73). This common destiny also underlies the approach of Gadotti (2017), which supports planetary citizenship in a unifying vision of the planet and of a single-world society that practices "planetarity" and considers the planet as an intelligent evolving being. For the author, planetary citizenship is, in essence, active, full, and just in social, political, cultural, institutional, and economic terms and implies a planetary democracy.

According to Delanty (2015), cosmopolitanism is a normative idea about the world that is located in the broader totality or world community (kosmopolis) and that gives relevance to the perspective of the other and to compassion for the rest of humanity. Diogenes is attributed to be the first to pronounce himself a citizen of the world (in Greek kosmospolitês) and to originate cosmopolitan thought in classical Greece. However, cosmopolitanism has been developed as a theory throughout history: in his essay on perpetual peace, Kant (1795) called for the creation of a global federation of states that would enter into a cosmopolitan order where all people would be treated as equals, forming a universal community. Kantian cosmopolitanism recognized an ethic of hospitality that would diminish the meaning of state borders and that should abandon the idea of hostility between countries and people. Cosmopolitan citizenship also promotes inclusion, impartiality, and non-discrimination (Bullen and Whitehead 2005). It is fundamentally different from nation-state-based citizenship because it does not involve making distinctions about who is a citizen and who is not and who is inside and who is left out. Cosmopolitanism seeks global dialogue and allows everyone to have a voice and speak from their perspective, although in the end, the best argument is the one that should prevail.

In recent decades, globalization has posed a challenge to the world order for many reasons: economic activities have expanded beyond state borders; networks and trade flows have increased and intensified; interactions and the dissemination of information and knowledge have accelerated; people's mobility has increased and expanded; and all these global processes are increasing the impact on the planet and have altered the functioning of states (Held 2002). Globalization is an asymmetrical and stratified process because not all citizens of the world are in the same position to "be global" and, as a result, the benefits generated by globalization are unevenly distributed (Dobson 2010) and benefit some elites and a small group of people. Or put another way, not all of us influence and contribute to the polycrisis mentioned earlier in the same way. We all can be affected by the consequences of globalization, and by the risks posed by climate change, but "*there are sustainability issues and risks that are peculiar and more prevalent in some continents and their respective member states than in others*" (Agbedahin and Lotz-Sisitka 2019, p. 108). For example, Africa has a relatively small carbon footprint compared to other continents in the world but is particularly affected by the effects of climate change and other unsustainable problems highlighted by the Agenda 2030 (UN 2015), such as hunger, food insecurity, malnutrition and poverty. In short, individuals, groups, states, and regions influence and are affected by global processes unevenly. At the same time, *"we need to appreciate the complex lived contexts of people"* (Abdi 2015, p. 20). Thus, there is a principle of differential responsibility towards global issues that must be assumed by individuals, groups, and states, depending on their global positions over time.

Cortina (2004) proposes not leaving globalization to chance and orienting it towards voluntarily desired goals. To that end, she believed that international institutions must be reformed and new ones created to ensure transnational communities that join through agreements. According to Fraser (2007), this hypothetical new transnational public power should allow and guarantee the legitimacy of public opinion through the communication process that includes all those affected, regardless of their political citizenship. These contributions of political cosmopolitanism and global citizenship are being criticized for a lack of realism and because they arise from a Western, elitist perspective that focuses on globalization (which has a negative connotation for the rest of the world for its social and ecological consequences) and on the construction of utopian global political systems that do not exist today and that are more than questionable (Boni and Calabuig 2015). So, it would seem more appropriate and realistic to flee from the pretension of global political organizations with their unknown consequences and which compete with states.

Dingwerth and Pattberg (2022) identify two cross-cutting themes that global governance needs to engage: non-state agency and the complexity of global governance structures. The first feature of global governance is that we need to talk about governance in the global rather than governance of the global. There is a multicentric world made up of constellations of diverse actors and governed by hybrid institutional complexes. Transnational non-state actors are key in global governance and include intergovernmental organizations, international non-governmental organizations, multinational corporations, social movements, transnational professional communities, and globally influential individuals. Some of these actors play a role in governance through orchestration or through influence. Some pursue a lucrative interest while others do not. The second key issue in global governance is the risk of overemphasizing the global scale and the Western vision. Global governance is a multi-level system that includes political processes and governance systems at all levels of human activity, from the family to the international organization, and all are inseparably linked. Scale interrelations (both vertical and horizontal) should be considered as part of global governance, and there is an overgeneralization of Western experiences that hinders the visibility and development of the agency of the South.

It seems that there is a move towards another type of political activity that is already materializing and being channelled through social movements and transnational networks (Kartal 2012), which additionally have the function of acting as a counterpower. Global civil society is uniting for causes related to human rights, environmental conservation and many great causes that affect humanity as a whole and are carrying out collective actions that are generating changes. This new global citizenship nears the idea that Dobson (2011) called post-cosmopolitanism, which contemplates both relations between citizens and citizen-state relations and posits that citizens' obligations have a socioecological nature and must be guided by virtue. Post-cosmopolitism focuses on developing the sense of interconnection and interconnectivity to feel that we are part of multiple collectives and groups at different territorial or aterritorial scales. In addition, taking account of interconnection involves looking at the present and the future, as well as the implications and consequences of the actions carried out at any given time.

Finally, it is important to bear in mind that global citizenship can also be understood through life facets different and away from politics. This way of conceiving and living life as a common vital project (Gimeno 2003) could not be attached to a specific territory and be based on certain values that bring together people from different backgrounds (and without universalist pretensions) who share the same way of seeing the world and a horizon to which they tend. The Internet facilitates digital environments where individuals can create communities and exercise new types of citizenship. Digital citizenship as described by

Ribble (2015) lacks elements of digital action and community building and engagement. These elements are critical for the development of what Elder-Vass (2022) calls digital utopias (utopian projects that emerge in the digital space).

4.2. Status

Human beings belong to political communities that confer them citizenship status that grants benefits, as well as ascribes duties. Just as countries have different political realities, the status of citizenship and its characteristics will also differ from country to country. The classic discourses of citizenship have focused on the recognition of the rights, responsibilities and obligations that exist within the functioning of the political structure of a community or a state, which arise as a result of social relations. Rights have evolved and expanded over time but with spatial variations and singularities. Marshall (1992) distinguished an initial threefold typology of rights (civil, political, and social) that has been increased up to four in recent years (Granados-Sánchez 2008). The first set of rights achieved were civil rights, including the right to freedom of expression and the right to free association. Political rights came after and, thanks to them, most citizens in many countries obtained the right to vote. Later social rights linked to the welfare state were won and, finally, environmental rights are recognised today in many countries. The tensions in the field of rights are related to whether these should be considered socially or privately based, whether they should be a combination of individual rights and community rights, and how to seek universality and attention to diversity with specific rights for minority groups with particular needs.

The dimensions of responsibility include taking into account the object and subject of responsibility, as well as the territorial scales concerned. As objects of responsibility, we find oneself, the "we," other members of the global community, and future generations. There must also be responsibility for the Earth and all forms of life. Individuals, groups and collectives are the subjects to whom responsibility is assigned. Therefore, there will be individual, group and collective responsibilities and duties. According to Wolf (2007), responsibility is related to the sense of belonging and the perception of interconnection with others that can drive us to the obligation to act. Thus, responsibility arises from the idea that one is part of local, regional, national, and global communities, and other horizontal groupings from which one has or perceives interconnection and to which it associates values of solidarity, security, and sustainability.

The sustainable post-cosmopolitan citizenship of Dobson (2010) is based on the principle of global dialogue and socio-ecological obligations. The obligation goes beyond the immediate reality to extend it to the global, and to lives lived in other parts of the planet and in present and future times. Obligations do not belong equally to all humanity, since the impact that each produce is different. Therefore, the obligations have to be asymmetrical depending on the position and impact of each citizen in the global sphere, in different moments. Dobson (2004) focuses on reducing the ecological footprint, but our obligations must go much further and extend to social, political, and moral aspects. Asymmetric obligations also have to be extended to groups, corporations and nations based on the inherited world order and the impact on the global change generated (such as, for example, the contribution to climate change), highlighting diluted communities and dense communities with historical obligations.

In the context of sustainability, both rights and responsibilities are important, because achieving sustainability is a collective undertaking. Deliberation between rights and obligations requires negotiating ideas of sufficiency, sustaining nature, and fulfilling obligations to present and future generations. Ultimately, the responsibility of sustainable global citizenship is to ensure efficient governance at all levels, from the local, regional, state, supra-state, and global levels.

4.3. Social-Ecological Systems

Nature is an essential part of our lives that must be incorporated into the discourse and exercise of sustainable global citizenship. The concept of social-ecological systems refers to the relationships between nature and society, and it is used to describe, analyse, and model human-nature interactions. The social-ecological systems are defined as hybrid and emergent systems resulting from interactions of various social and natural components over space and time (Liehr et al. 2017). The key components of social-ecological systems are ecosystem functions and social actors. Ecosystem functions are part of natural dynamics, but they can change due to societal actions. Thus, they can provide services and goods, or disservices and harm society. Social actors include individuals and groups of persons who influence ecosystem functions with the practice of their agency. Görg et al. (2017) find that *"what is largely missing in the current transformation debate are analyses that focus in more depth on the interactions between globalized societies and the natural environment, analysing resource use patterns and its social implications in terms of global inequalities as much as its impact on global ecosystems without denying local (including everyday), regional and national scales of problems and action"* (2017, p. 5).

One claim of this article is to include the group as a key category in citizenship and, thus, transcend the individual-collective dualism. Schild (2016) also advocates for the need for an expanded view on citizenship that goes beyond individual-level analysis and the personal duty and lifestyle approach. Wight (2006) points out that not only individuals have emergent properties. The activities of groups of people can also display emergent properties that cannot be reduced to the activities of single individuals. Among these emergencies is the development of identities linked to belonging to various groups, as well as the influence that these close groups exert for the development of a type of agency different from the individual. Hill (2012) argues for self-organized, action-oriented, problem-solving groups enabling people to recognize structures of social control and power relationships, develop oppositional forms of action, generate novel ideas and create new kinds of knowledge that can help us in a better relationship with nature. In sum, both in nature and in society, there are individuals (and phenomena) that can act alone or united in groups or collectivities, and these can be governed by certain particularities. In turn, these groups are subject to laws of universal value. In this way, the individual, the particular, the group, the collectivity and the universal are mutually linked. According to Görg et al. (2017) *"a critical concept of social-ecological transformations points at a better understanding of the social-ecological dimensions of current transformation processes. This includes a better understanding of scale interactions, i.e., global, regional, and local processes, and the systemic processes as much as the actor constellations and power relations involved"*(2017, p. 6).

Another tension in the field of sustainable global citizenship is the different points of view between individualised and profit-seeking self-interest promoted by current hegemonic neoliberalism and the global common good. In other words, the dualism between private and public interests. From liberal political theory, the public sphere is the space of politics, power, and civic engagement. The private sphere, on the other hand, is the intimate family space of care and protection, where all personal things are protected from the power of the state. Republicanism places public interest before private interest, which implies distinguishing that something that may be good for an individual may not be good for the same individual as a member of a community (Dobson 2006). Inherent to this is the explicit commitment to freedom as non-domination and the acceptance of plurality in the way of living. Sustainable global citizenship recognizes that the public and private spheres, although different, are interrelated since private acts can have public implications and vice versa, and both have an impact on natural systems and on the achievement of sustainability. According to Dewey (1954) democracy (and sustainability) must begin at home and has to be built on face-to-face interactions in which human beings work together cooperatively to solve the ongoing problems of life. Small groups and bigger partnerships are the places where relationships and trust are built. These elements, together with the transdisciplinary and transformative competence (Granados-Sánchez 2022), are essential for effective group and collective engagement.

One cannot separate and oppose the public from the private, the individual from the group and the collective, or personal life from political life. To Biesta "*the most important transformation that is at stake in the experiment of democracy is the transformation of private troubles into public issues*" (2015, p. 5), but I would also add that we need to transform the public troubles into private concerns. He continues pointing out that "*what is always at stake in the democratic experiment is the question to what extent and in what form private "wants" (that what is desired by individuals or groups), can be supported as collective needs (that is considered desirable at the level of the collective), given the plurality of individual wants and always limited resources*" (Biesta 2015, p. 5). This leads us to reflect on how to satisfy our needs and desires in simple ways and move from "having" and "pretence and appearance" to "being".

4.4. Social Conscience

Bhaskar states that "*in order to explain what a person does, you must make reference to their reasons, their conscious intentionality*" (in Buch-Hansen 2005, p. 60). Conscience is like a person's moral and emotional compass that provides a sense of what is right and wrong. "*Individual conscience compels us to act morally in our daily lives (...) whereas social conscience compels us to insist on moral action from the wider institutions of society and to seek the transformation of social structures that cause suffering (...) most people experience a gap between the kind of world they see and the kind they want. On a personal level, social conscience is what bridges that gap. If we understand our own social conscience, we can make more conscious choices to help shape society according to our values*" (Goldberg 2009, p. 105).

Bhaskar's (1998) critical realism provides us with good tools for the development of a more accurate social conscience. This theory maintains that reality is formed by different layers and by internal elements united in a multiplicity of complex structures and interrelationships. The ontological principle of critical realism is precisely the stratification of social reality. Social consciousness has to allow us to know reality and our role in it at any moment. Social consciousness is the knowledge that a person possesses and that is imbued with value judgments about things. According to Goldberg (2009), social conscience is composed of three interrelated elements: agency, structure, and consciousness. Agency is defined as the ability of individuals to decide and act independently or in agreement with others and implies a sense of free will, choice, or autonomy (Hay 2002). It is the sense of personal power and constitutes personal responsibility. However, as Hill (2012) points out, human agency not only depends on the wishes of people; to get involved in the community, in addition to the will, it is necessary to have the skill and the opportunity; that is, it is about seeing where it is possible to intervene personally and with others, where we have room for decision and action, and if we are able to do so. Thus, agency is conditioned cognitively, by emotional and affective aspects, character traits, contextual frame of mind, self-efficacy, and performativity. There are disagreements among critical realists as to what aspects of agency are most prominent (Buch-Hansen and Nielsen 2020). Some authors recognise the importance of the unconscious and habits (following Bourdieu), whilst others ascribe great importance to reflexivity and intention of agents, and the interactions between individuals. In addition, sustainable citizenship should emphasize the importance of "agencies", because actions are not always just individual: there are group and collective actions that involve a larger number of people and the community. It should also be remembered that many groups and minorities, due to their particularities and fit and treatment in society, do not have the same capacities and opportunities to engage in political, social, and civic activities and are systematically set apart or excluded (Jones and Gaventa 2002).

Social structures are abstract entities of economic, political, social, and juridical-legal origin that organize the life of the community. Social structures are context-dependent, and they are human activity-dependent in time and space. These structures facilitate and motivate agency and make it possible and, at the same time, limit or constrain it. That is why it is so important to know the structures, their construction and reconstruction mechanisms, in order to participate in public life.

Buch-Hansen (2022) raises the question of to what extent human beings have agency. His reply to it, from a realist perspective, is that humans are always confronted by structures, yet exercise agency. "*While those structures constrain actions, they never determine them. Human beings are understood to have their own unique capabilities that are irreducible to, but not unaffected by, social structures and culture. Such capabilities vary from person to person: for instance, a human being can be more or less strategic, rational, creative, kind,* etc. *On this view, then, human beings are able to exercise agency*" (Buch-Hansen 2022, pp. 10–11).

For critical realism, the relationship between structure and agency is not dialectical but rather recursive and transformational, since society (structures) and human practice (agency) do not constitute moments of the same process: society and its structures are the existing material condition for praxis (which is later) to take place. Agency comes after the structure: "*there was structure; there is now that agency; and there will be the structure that this agency produces*" (Buch-Hansen 2005, p. 63). The result of praxis can reproduce social structures, or it can transform them and lead us to a new reality. Thus, society is both an ever-present condition and a result of continuous transformation. Archer (2017) has advanced in the comprehension of the agency-structure relation through the concept of morphogenesis. Her model focuses on how the interplay between structure and agency can be studied by cycles of structural conditions, social interaction, and structural development. It builds on the notions of stratification and emergence so that structures and agency have emergent properties of different natures. Archer (2010) also separates culture from structures because culture concerns existing relations between ideas. The confrontation of complementary or conflicting ideas takes place in sociocultural interactions, constituting another ontological dimension. In sum, the activities of agents are conditioned both by social structures and a cultural system.

Structures create a system with hierarchies, and people and groups occupy different places in this reality, stratified according to the moment and context. In addition, the resources and facilities of the system are available differentially among social actors. As a result, power and the possibility of participation and action are unevenly distributed among individuals, groups, and institutions. Having social conscience allows us to know the position we occupy in the system and the power and possibilities that this gives us to be able to act and transform reality. The dialectic of critical realism states that as people engage personally and with others, they can generate new ideas and transform structures through collective agency (transformative praxis) (Khazem 2018).

Bhaskar (1998) understands emancipation as being the transition from an unwanted, unnecessary, and oppressive situation to a desired, necessary, and empowering situation that allows people to flourish. Social innovation is a process that allows social emancipation. For Belda Miquel et al. (2019, p. 27) "*social innovation from citizenship for the transition to more sustainable models is a process of innovation through collective action with multiple dimensions: it is a process of individual and collective transformation in the micro, but also a process of experimentation, testing and prefiguration of possible systemic transformations on another scale; it is a learning process, and it is a complex and contradictory process. This innovation does not put the State, but the citizenry, at the centre, but it does so by challenging public policies. It constitutes spaces for mobilization that concretize and make the demands of the neighbourhoods and the general indignation against the system real, since they already draw and practice other possible models.*" Social innovation is the social interaction scenario that allows collective agency.

4.5. Engagement

Civic engagement is one of the leitmotifs of most republican approaches to sustainable citizenship and it is manifested through commitment, participation, and action (and their opposites).

Participation is the force that guides and legitimizes democracy. Banks (2008) classifies citizenship according to the level of participation of citizens into the following four types:

- The most superficial level of citizenship is *legal citizenship*, which applies to citizens that don't participate in politics in a meaningful way.
- *Minimal citizenship* refers to legal citizens that just vote in local and state elections.
- *Active citizenship* involves actions that are designed to support and maintain existing social and political structures.
- *Transformative citizenship* requires the promotion of social justice through the execution of actions that challenges and changes current norms and structures.

This classification is useful for understanding the degree of our commitment to sustainability, from inhibition to transformation. On the other hand, it has some limitations: it is focused on the individual, and it presupposes that a citizen always participates in the same way and degree in all aspects of his/her life, while commitment and engagement depend on the context. Some authors such as Macedo et al. (2005) and Putnam (2000) maintain that the political participation of young people, in most Western societies, is problematic because it is in serious decline. Hibbing and Theiss-Morse (2002) affirm that there is an apathy towards politics and the public and that citizens do not want to be involved in political decisions or know the details that influence and determine decision-making. If we look at participation in the last Spanish elections in 2019 and the American elections in 2020 and 2022, we can see that political participation is revitalized if the context demands it and the electorate is mobilized. For Dalton (2008), the conclusions about the decline in political participation are not entirely correct since they have considered only part of the political activity ("institutionalized" political participation) as if they were the only rules of citizenship. According to him, there is a multitude of rules of citizenship and forms of political participation and action that are simply changing (some are weakening, while others are strengthening). Thus, while citizen obligations such as voting in elections could be eroding in certain periods, other ways of engaging in political action are increasing. Inglehart and Welzel (2005) believe that we are at a time when new generations in advanced countries with better education have changed political party membership by actively participating in specific social movements or interest groups, which allow for direct action and greater influence on outcomes, although this requires greater dedication and commitment. This coincides with the results of research on citizenship and climate change carried out by Wolf (2007), in which the participants stated that they seek to act and develop their agency on issues that are important to them beyond governance and the top-down opportunities that the state provides them. In short, we are facing a renaissance of democratic participation (Dalton 2008) understood in a very personal way and motivated by desire for a self-realization of the individual and groups through having a higher quality life and social relationships. As a result, we are facing a challenge to the traditional political elites, moving from the "politics of loyalties" (or of supporting political institutions and parties) to the "politics of election" (Atkinson 2014), where citizen participation is configured through collective social actions separate from the state (Jones and Gaventa 2002).

One of the main objectives of education for sustainability since its inception has been to provide the appropriate tools for each citizen to develop action competence so that he or she can participate actively and meaningfully in public life and in the private sphere. Some scholars go further and foster activism and even transgression (Lotz-Sisitka et al. 2015). Should we all be activists all the time, or should we be self-conscious critical citizens who engage differentially, considering personal and social necessities and complexities? There are no real emancipatory proposals that train citizens in agency self-awareness and self-regulation of their own individual, group, and collective participation. Our foundation is based on the ideas of Goldberg (2009), who proposes that there are many causes, problems, phenomena, or objects of interest, while the time available to us is limited. *"Individuals cannot care deeply and act effectively on every social and ecological problem they come across, but they can identify problems they feel are important and that they have the agency to act on"* (Goldberg 2009, p. 108). This implies that we first need to decide if we want to act or stay passive and inhibited in each situation. Inaction may be due to health issues or other reasons of a personal nature that require our attention and energy and prevent us from engaging. If we choose to get involved in actions, we need to know which causes we join, know with whom we join, what barriers or structural possibilities we face, what real possibilities of intervention exist and decide our level of involvement and commitment (Granados-Sánchez 2011, 2021). These decisions will not only depend on the social urgency, because our personal context at that moment will also be decisive, which will depend on our training and knowledge of the phenomenon to be treated, along with our motivation, personal values, state of mind, the position we occupy in the group or collective to which we integrate, and the time available. Our sense of self-efficacy and locus of control are also decisive. Self-efficacy is the belief in one's own abilities to succeed in a particular situation. Locus of control is the degree of control a person thinks he or she has over what happens in their life, as opposed to outside forces.

Participatory democracy needs horizontal and bottom-up forms of shared governance, as well as top-down opportunities for civic participation. Thus, governments and public institutions as well as social movements and global networks have to provide opportunities for citizens to participate in the elaboration of policies or actions that encourage sustainability. It is also about the citizenry being able to create spaces, programs and even structures in which social groups and actors can visualize and materialize sustainability from carrying out daily sustainable practices. It is about executing, creating transformation processes, and executing them as a result of collective effort. It is living and breathing the bond with the community (Nelson 2016). It is about creating social learning and building human and social capital.

5. Conclusions and Educational Implications

This paper approached sustainable global citizenship from a critical realist perspective. The position-practice system and the seven-scalar laminated system are two core concepts of critical realism that have helped in the analysis of the selected key dimensions in the field of sustainable global citizenship and how they relate to agency-structure dualism. The main conclusions of this theoretical reflection are the following:

- *Governance*: The practice of citizenship takes place in communities to which we belong and are linked. These communities may have a territorial component and may or may not be linked to sovereignty and legality or be non-territorial or aterritorial (Carpenter 2019). At present, the legal recognition of the sovereignty of the citizens of the world is limiting in terms of territorial scope. In any case, the nation-state does not determine the entire space of participation or the feeling of social, cultural, or other belonging. The challenge of sustainable global citizenship is, precisely, to recognize that there are other territories as well as communities on which we have a vital and moral link beyond sovereignty, and to determine how we articulate that reality outside the nation-state and how we participate in a political, social, cultural, moral and economic way on a global scale, but also in different parts of the world, because the global citizen is a citizen involved and affected by all spatial scales, the territorial interrelationships, and the non-territorial communities to which he or she joins by other facets and objectives that they share. In short, we need to integrate engagement in both government and governance through participation in politics and in social innovation initiatives, to monitor and control policies and to act as counterpower.
- *Status*: global citizenship is stratified and asymmetrical. There is a principle of differential responsibility towards the global that must be assumed by individuals, groups, and states, according to their impact due to the global positions they occupy over time. At the same time, big efforts should be made to achieve symmetry in the status of all people.
- *Social-ecological systems*: Sustainable global citizenship is based on the principle of socio-ecological obligations and global dialogue. Nature and its processes should be integrated in sustainable global citizenship as essential elements and not as external entities. Or put another way, we should consider ourselves as part of the social ecological systems, both as individuals and as part of groups or collectives. The

group is a category that needs more attention because many of our actions occur in connection with close people and this relationship of familiarity produces emergencies that strongly determine our agency as individuals and beyond. Social-ecological systems have distinctive characteristics and threats according to scale and context.

- *Social Conscience*: To explain reality and social outcomes, we have to consider the ways in which agency, structure, and culture are intertwined. This approach is more comprehensive and goes beyond other reductionist proposals that give centrality to the agency and, specifically, to abstract active participation. Social conscience enables citizens to have real power over their agency.
- *Engagement*: There are many factors that determine civic engagement and participation, and citizens should be aware of them in order to decide how to act as autonomous and conscious individuals at all times and thus avoid social apathy and frustration.

This theoretical reflection is an initial stage in the development of the conceptualization of sustainable global citizenship from a critical realist perspective. Its limitations include the lack of detailed specification or development of some of the key components of agency (and their relationships with the structure), that have been described throughout the analysis of the six dimensions.

We could say that there is a general agreement in understanding education for sustainable citizenship as a process that has to do with the acquisition of knowledge, skills, competencies, and values that allow sustainable development, understood as a sign of good citizenship. However, the way in which we understand how this should be carried out is not so clear. To Hadjichambis and Reis (2020) the different political and philosophical approaches make the fostering of citizenship related to the environment and sustainability very complex for educational practice. The latest trends in sustainable citizenship and global citizenship increasingly point towards transformative learning (UNESCO 2018). Mezirow (1995) focuses transformative learning on the cognition of individuals but does not propose how cognitive transformation leads to social action or the development of the individual, group, and collective agencies, especially for the collective transformation of human actions. According to Lotz-Sisitka et al. (2015), all forms of transgressive and transformative learning require pedagogical methodologies that involve a multitude of voices and social actors and that make co-learning, cognitive justice and the development of individual, group, and collective agencies possible. For Biesta (2011), it is about moving from a socializing learning focused on the learning of norms, to a learning based on experiencing democracy and sustainable development. That is, learning to develop sustainably by carrying out sustainable actions that, in addition, make democracy extend, grow, and even lead us to different democratic forms that are more relevant. Education therefore becomes not merely a key method of achieving sustainable development, but part of the development itself. This type of education involves citizenship as practice (Lawy and Biesta 2006) or citizenship in practice (Van Poeck and Vandenabeele 2013). What Biesta (2011) calls "democratic experiment" constitutes a process of transformation of the person and the community and involves exceeding the preconceived standards for sustainable global citizenship and moving from thinking about what skills citizens should acquire to promoting sustainable spaces and practices. Thus, education emerges as a space in which citizens are invited to explore issues that concern and occupy them (Van Poeck and Vandenabeele 2013).

In conclusion, we could say that education for sustainable global citizenship from a critical realist perspective should be based on respect for difference and diversity, relational knowledge, the importance of consensus and disagreement, critical self-reflection, the ability to navigate between the general and the particular, individual and collective responsibility, and the ability to decide how to get involved, act and commit oneself (Kahn and Agnew 2015). It involves the advance in planetary awareness and the culture of sustainability (Gadotti 2017). Future research on education for sustainable global citizenship should focus on the development of multiple identities (Sen 2006), both as citizens of a nation-state, as global citizens, as employees, as consumers and as members of many territorial and non-territorial communities. It should entail the experience of sustainability, which challenges people to manage the complexity of reality at all scales, and which requires the understanding of and working with social actors because sustainability is the result of a collective effort. Sustainable global citizenship should make us reflect on our place in an increasingly connected and network-intensive world, on ourselves and the other, us and them, here and there and that, in the end, we are all one.

Funding: This research received no external funding.

Institutional Review Board Statement: Not applicable.

Conflicts of Interest: The author declares no conflict of interest.

References

Abdi, Ali A. 2015. *Decolonizing Global Citizenship Education*. Edited by Ali A. Abdi, Lynette Shultz and Thashika Pillay. Rotterdam: Sense Publishers, pp. 11–26. [CrossRef]

Agbedahin, Adesuwa, and Heila Lotz-Sisitka. 2019. Mainstreaming education for sustainable development: Elaborating the role of position-practice systems using seven laminations of scale. *Journal of Critical Realism* 18: 103–22.

Agnew, John. 2018. *Globalization and Sovereignty: Beyond the Territorial Trap*. Lanham: Rowman & Littlefield.

Anderson, Jon. 2008. What is geography's contribution to making citizens? *Geography* 93: 34–39.

Archer, Margaret. 1995. *Realist Social Theory: The Morphogenetic Approach*. Cambridge: Cambridge University Press.

Archer, Margaret. 2010. Routine, reflexivity, and realism. *Sociological Theory* 28: 272–303.

Archer, Margaret. 2017. *Morphogenesis and Human Flourishing*. Dordrecht: Springer.

Archer, Margaret, Claire Decoteau, Philip Gorski, Daniel Little, Doug Porpora, Timothy Rutzou, Christian Smith, George Steinmetz, and Frédéric Vandenberghe. 2016. What is Critical Realism? *Perspectives: ASA Theory* 38: 4–9. Available online: http://www.asatheory.org/uploads/4/3/3/7/43371669/perspectives_38_2__fall_2016_final.pdf (accessed on 30 December 2022).

Atkinson, Lucy. 2014. Good Corporate Citizenship: Predictors of support for corporate social justice as an element of sustainable citizenship norms in Europe. *International Journal of Communication* 8: 1–20.

Banks, James A. 2008. Diversity, group identity, and citizenship education in a global age. *Educational Researcher* 37: 129–39.

Barry, John. 2006. Resistance is fertile: From Environmental to sustainability citizenship. In *Environmental Citizenship*. Edited by Andrew Dobson and Derek Bell. Cambridge: The Massachusetts Institute of Technology Press, pp. 21–48.

Barry, John. 2016. Citizenship and (Un)Sustainability: A Green Republican Perspective. In *The Oxford Handbook of Environmental Ethics*. Edited by Stephen M. Gardiner and Allen Thompson. Oxford: Oxford University Press (Oxford Handbooks). [CrossRef]

Beck, Ulrich. 2010. Remapping social inequalities in an age of climate change: For a cosmopolitan renewal of sociology. *Global Networks* 10: 165–81. [CrossRef]

Beckman, Ludvig. 2001. Virtue, sustainability and liberal values. In *Sustaining Liberal Democracy. Ecological Challenges and Opportunities*. Edited by John Barry and Marcel Wissenburg. Houndmills: Palgrave, pp. 179–91.

Belda Miquel, Sergio, Victoria Pellicer Sifres, and Alejandra Boni Aristizábal. 2019. La innovación social desde la ciudadanía para la transición a la sostenibilidad: Enfoques, conceptos y cuestiones. In *Innovación Social Colectiva. Experiencias para la Transición a la Sostenibilidad desde la Ciudadanía Organizada*. Edited by Sergio Belda Miquel and Victoria Pellicer Sifres. Barcelona: Icaria, pp. 11–32.

Bhaskar, Roy. 1978. On the possibility of social scientific knowledge and the limits of naturalism. *Journal for the Theory of Social Behaviour* 8: 1–28. [CrossRef]

Bhaskar, Roy. 1998. Societies. In *Critical Realism: Essential Readings*. Edited by Margaret Acher, Andrew Collier, Tony Lawson and Alan Norrie. London: Routledge, pp. 206–67.

Bhaskar, Roy. 2002. *Reflections on Meta-Reality. Transcendence, Emancipation and Everyday Life*. London: Sage.

Bhaskar, Roy. 2010. Contexts of interdisciplinarity: Interdisciplinarity and climate change. In *Interdisciplinarity and Climate Change: Transforming Knowledge and Practice for Our Global Future*. Edited by Roy Bhaskar, Cheryl Frank, Karl G. Hoyer, Petter Naess and Jenneth Parker. Oxon: Routledge, pp. 1–23.

Bhaskar, Roy, and M. Hartwig. 2011. 'Beyond East and West'. In *Critical Realism and Spirituality*. Edited by Jamie Morgan and Mervyn Hartwig. London: Routledge, pp. 187–202.

Bhaskar, Roy, and Mervyn Hartwig. 2016. *Enlightened Common Sense: The Philosophy of Critical Realism*. New York: Routledge.

Biesta, Gert. 2011. Learning in Public Places: Civic Learning for the 21st Century. Inaugural Lecture on the occasion of the Award of the International Francqui Professorship, Gant 17 February 2011.

Biesta, Gert. 2015. What is education for? On good education, teacher judgement, and educational professionalism. *European Journal of Education* 50: 449–61.

Boni, lejandra, and Carola Calabuig. 2015. Education for global citizenship at universities: Potentialities of formal and informal learning spaces to foster cosmopolitanism. *Journal of Studies in International Education* 21: 22–34. [CrossRef]

Bowen, Glenn. 2009. Document analysis as a qualitative research method. *Qualitative Research Journal* 9: 27–40. [CrossRef]

Buch-Hansen, Hubert. 2005. Critical realism in the social sciences. *Distinktion: Journal of Social Theory* 6: 59–69.
Buch-Hansen, Hubert. 2022. Contending philosophy of social science perspectives: A flexible typology. *Journal for the Theory of Social Behaviour* 1–17. [CrossRef]
Buch-Hansen, Hubert, and Peter Nielsen. 2020. *Critical Realism: Basics and Beyond*. London: Red Globe Press (Macmillan Education Limited).
Bullen, Anna, and Mark Whitehead. 2005. Negotiating the networks of space, time and substance: A geographical perspective on the sustainable citizen. *Citizenship Studies* 9: 499–516.
Cao, Benito. 2015. *Environment and Citizenship*. London: Routledge.
Carpenter, Michael J. 2019. Understanding aterritorial borders through a BIG reading of Agnew's globalization and sovereignty. *Borders in Globalization Review* 1: 123–26. [CrossRef]
Connelly, James. 2015. *Sustainability and the Virtues of Environmental Citizenship*. London: Routledge.
Cortés, Francisco, and Felipe Piedrahita. 2011. *De Westfalia a Cosmopolis: Soberanía, Ciudadanía, Derechos Humanos y Justiciar Económica Global*. Bogotá: Siglo del Hombre.
Cortina, Adela. 2004. Una ética transnacional de la corresponsabilidad. In *Ética y Globalización: Cosmopolitismo, Responsabilidad y Diferencia en un Mundo Global*. Edited by Vicente Serrano. Madrid: Biblioteca nueva, Colección «Razón y Sociedad», vol. 34, pp. 15–32.
Dalton, Russell J. 2008. Citizenship norms and the expansion of political participation. *Political Studies* 56: 76–98. [CrossRef]
Delanty, Gerard. 2015. No todo se pierde en la traducción. Variedades mundiales de cosmopolitismo. *Papers: Revista de Sociología* 100: 281–302. [CrossRef]
Dewey, John. 1954. *The Public and Its Problems*. Denver: Alan Swallow.
Dingwerth, Klaus, and Philipp Pattberg. 2022. Theorizing global governance. In *Handbook on Global Governance and Regionalism*. Edited by Jürgen Rüland and Astrid Carrapatoso. Cheltenham: Edward Elgar Publishing Limited, pp. 21–35. [CrossRef]
Dobson, Andrew. 2003. *Citizenship and the Environment*. Oxford: Oxford University Press.
Dobson, Andrew. 2004. Inclusió social, sostenibilitat ambiental i educació ciutadana. In *Les Dimensions Socials de la Crisi Ecològica*. Edited by Joaquín Valdivielso. Palma de Mallorca: Edicions UIB, pp. 17–43.
Dobson, Andrew. 2006. Ecological citizenship. A defence. *Environmental Politics* 15: 447–51. [CrossRef]
Dobson, Andrew. 2010. *Ciudadanía y Medio Ambiente*. Barcelona: Proteus.
Dobson, Andrew. 2011. *Sustainability Citizenship*. Weymouth: GreenHouse. Available online: https://www.greenhousethinktank.org/uploads/4/8/3/2/48324387/sustainability_citizenship_inside.pdf (accessed on 30 December 2022).
Dobson, Andrew, and Derek Bell, eds. 2006. *Environmental Citizenship*. Cambridge: The Massachusetts Institute of Technology Press.
Elder-Vass, Dave. 2010. *The Causal Power of Social Structures: Emergence, Structure and Agency*. Cambridge: Cambridge University Press.
Elder-Vass, Dave. 2022. Ethics and emancipation in action: Concrete utopias. *Journal of Critical Realism* 21: 539–51.
European Network for Environmental Citizenship (ENEC). 2018. *Environmental Citizenship*. Available online: http://enec-cost.eu/our-approach/enec-environmental-citizenship (accessed on 30 December 2022).
Fraser, N. 2007. Transnationalizing the public sphere: On the legitimacy and efficacy of public opinion in a post-Westphalian world. *Theory Culture and Society* 24: 7–30.
Gadotti, Moacir. 2017. To educate for planetary citizenship: A Latin-American perspective. *Global Commons Review* 1: 39–42.
Galli, Alessandro, David Lin, Mathis Wackernagel, Michel Gressot, and Sebastian Winkler. 2015. Humanity's Growing Ecological Footprint: Sustainable Development Implications. Global Footprint Network. Brief for GSDR 2015. Available online: https://sustainabledevelopment.un.org/content/documents/5686humanitysgrowingecologicalfootprint.pdf (accessed on 30 December 2022).
Gimeno, José. 2003. Tornar a llegir l'educació des de la ciutadania. In *Ciutadania, poder i Educació*. Edited by Jaume Martínez. Barcelona: Graó.
Global Footprint Network (GFN) & Schneider Electric. 2021. *Strategies for One-Planet Prosperity: How to Build Lasting Success on Our Finite Planet*. Ebook. Available online: https://www.se.com/ww/en/download/document(earth_overshoot_day (accessed on 20 November 2022).
Goldberg, Myshele. 2009. Social conscience. The ability to reflect on deeply-held opinions about social justice and sustainability. In *The Handbook of Sustainability Literacy. Skills for a Changing World*. Edited by Arran Stibbe. Devon: Green Books, pp. 105–10.
Görg, Christoph, Ilrich Brand, Helmut Haberl, Diana Hummel, Thomas Jahn, and Stefan Liehr. 2017. Challenges for social-ecological transformations: Contributions from social and political ecology. *Sustainability* 9: 1045. [CrossRef]
Gosewinkel, Dieter. 2010. Citizenship. In *International Encyclopedia of Civil Society*. Edited by Helmut K. Anheier and Stefan Toepler. New York: Springer. [CrossRef]
Granados-Sánchez, Jesús. 2008. Educación para una ciudadanía sostenible en la formación inicial del profesorado. In *Didáctica de las Ciencias Sociales, Currículo Escolar y Formación del Profesorado. La Didáctica de las Ciencias Sociales en los Nuevos Planes de Estudio*. Edited by Rosa María Ávila, Alcázar Cruz and María Consuelo Díez. Grupo de Investigación HUM-167 y Asociación Universitaria del Profesorado de Didáctica de las Ciencias Sociales. Jaen: La Casa del Libro, pp. 217–29.
Granados-Sánchez, Jesús. 2011. La Evaluación de Iniciativas de Aprendizaje Participativo para una Acción Responsable y de Servicio. In *La Evaluación en el Proceso de Enseñanza y Aprendizaje de las Ciencias Sociales*. Edited by Pedro Miralles, Sebastián Molina and Antoni Santisteban. Murcia: Asociación de Profesores Universitarios de Didáctica de las Ciencias Sociales, pp. 343–52.

Granados-Sánchez, Jesús. 2021. Conceptos poderosos para la construcción de una ciudadanía sostenible. In *Enseñar y Aprender Geografía para un Mundo Sostenible*. Edited by Jesús Granados-Sánchez and Rosa María Medir Huerta. Barcelona: Ediciones Octaedro, pp. 177–97.

Granados-Sánchez, Jesús. 2022. La competencia transdisciplinar y transformadora a través de la didáctica de la geografía. In *La Enseñanza de la Geografía en el siglo XXI. Retos, Recursos y Propuestas Docentes ante los Nuevos Desafíos Globales*. Edited by Jorge Olcina Cantos and Álvaro F. Morote-Seguido. Alicante: Universidad de Alicante, pp. 51–62.

Hadjichambis, Andreas C., and Pedro Reis. 2020. Introduction to the conceptualisation of environmental citizenship for twenty-first-century education. In *Conceptualizing Environmental Citizenship for 21st Century Education*. Edited by Andreas Hadjichambis, Pedro Reis, Demetra Paraskeva-Hadjichambi, Jan Činčera, Jelle Boeve-de Pauw, Niklas Gericke and Marie-Christine Knippels. Cham: Springer, pp. 1–14.

Hadjichambis, Andreas C., Pedro Reis, Demetra Pareskeva-Hadjichambi, Jan Cincera, Jelle Boeve-de Pauw, Niklas Gericke, and Marie Christine Knippels, eds. 2020. *Conceptualizing Environmental Citizenship for 21st Century Education*. Cham: Springer.

Hartwig, Mervyn. 2015. MetaRealism. *Journal of Critical Realism* 14: 339–49. [CrossRef]

Hay, Colin. 2002. *Political Analysis. A Critical Introduction*. New York: Palgrave.

Held, David. 2002. Globalization, corporate practice and cosmopolitan social standards. *Contemporary Political Theory* 1: 59–78. [CrossRef]

Hibbing, John, and Elizabeth Theiss-Morse. 2002. *Stealth Democracy: Americans' Beliefs about How Government Should Work*. New York: Cambridge University Press.

Hill, Robert J. 2012. Civic engagement and environmental literacy. *New Directions for Adult and Continuing Education* 2020: 41–50. [CrossRef]

Horne, Ralph, John Fien, Beau B. Beza, and Anitra Nelson, eds. 2016. *Sustainability Citizenship in Cities. Theory and Practice*. London: Earthscan-Routledge.

Inglehart, Ronald, and Christian Welzel. 2005. *Modernization, Cultural Change and Democracy. The Human Development Sequence*. Cambridge: Cambridge University Press.

Jones, Emma, and John Gaventa. 2002. *IDS Development Bibliography, 19. Concepts of Citizenship: A Review*. Brighton: Institute of Development Studies (IDS).

Kahn, Hilary, and Melanie Agnew. 2015. Global learning through difference: Considerations for teaching, learning, and the internationalization of higher education. *Journal of Studies in International Education* 21: 52–64.

Kant, Immanuel. 1795. *Perpetual Peace. A Philosophical Essay*. Translated by M. Campbell Smith. London: George Allen & Unwin Ltd. Available online: https://www.gutenberg.org/files/50922/50922-h/50922-h.htm (accessed on 30 December 2020).

Kartal, Filiz. 2012. Cosmopolitan Citizenship as a Thin Concept: Who Is Willing to Die for Humanity? In *Nationalism and Human Rights*. Edited by Grace Cheng. New York: Palgrave Macmillan, pp. 123–40. [CrossRef]

Khazem, Dima. 2018. Critical realist approaches to global learning: A focus on education for sustainability. *International Journal of Development Education and Global Learning* 10: 125–34. [CrossRef]

Kurian, P., R. Bartlett, and D. Munshi. 2014. Sustainable citizenship for a technological world: Negotiating deliberative dialectics. *Citizenship Studies* 18: 435–51.

Lawy, Robert, and Gert Biesta. 2006. Citizenship-as-practice: The educational implications of an inclusive and relational understanding of citizenship. *British Journal of Educational Studies* 54: 34–50.

Liehr, Stefan, Julia Röhrig, Marion Mehring, and Thomas Kluge. 2017. How the Social-Ecological Systems Concept Can Guide Transdisciplinary Research and Implementation: Addressing Water Challenges in Central Northern Namibia. *Sustainability* 9: 1109. [CrossRef]

Lotz-Sisitka, Heila, Arjen E. J. Wals, David Kronlid, and Dylan McGarry. 2015. Transformative, transgressive social learning: Rethinking higher education pedagogy in times of systemic global dysfunction. *Current Opinion in Environmental Sustainability* 16: 73–80.

Macedo, Stephen, Yvette Alex-Assensoh, Jeffrey M. Berry, Michael Brintnall, David E. Campbell, Luis Ricardo Fraga, Archon Fung, William A. Galston, Christopher F. Karpowitz, Margaret Levi, and et al. 2005. *Democracy at Risk: How Political Choices Undermine Citizen Participation, and What We Can Do About It*. Washington: Brookings Institution Press.

Malatesta, Stefano, and Jesús Granados-Sánchez. 2013. A geographical issue: The contribution of citizenship education to the building of a European citizenship. The case of the VOICEs Comenius network. *J-Reading. Journal of Research and Didactics in Geography* 2: 57–66. [CrossRef]

Marshall, T. H. 1992. Citizenship and Social Class. In *Citizenship and Social Class*. Edited by T. H. Marshall and T. Bottomore. London: Pluto, pp. 3–51.

Mateus, Elaine, and Viviane Resende. 2015. The practice-position system as epistemological category: Contributions to critical discourse analysis. *Alfa, Sao Paulo* 59: 431–54. [CrossRef]

Mezirow, John. 1995. Transformation theory of adult learning. In *Defense of the Lifeworld: Critical Perspectives on Adult Learning*. New York: Welton M. State University of New York Press.

Morin, Edgar. 2004. En el corazón de la crisis planetaria. In *La Violencia del Mundo*. Edited by Jean Baudrillard and Edgard Morin. Barcelona: Paidós, pp. 53–94.

Nash, Nicholas, and Alan Lewis. 2006. Overcoming obstacles to ecological citizenship: The dominant social paradigm and local environmentalism. In *Environmental Citizenship*. Edited by Andrew Dobson and Derek Bell. Cambridge: The Massachusetts Institute of Technology Press, pp. 153–84.

Nelson, Anitra. 2016. The praxis of sustainability citizenship. In *Sustainability Citizenship in Cities. Theory and Practice*. Edited by Ralph Horne, John Fien, Beau Beza and Anitra Nelson. London: Earthscan-Routledge, pp. 17–28.

Page, M. J., Joanne E. McKenzie, Patrick M. Bossuyt, Isabelle Boutron, Tammy C. Hoffmann, Cynthia D. Mulrow, Larissa Shamseer, Jennifer M. Tetzlaff, Elie A. Akl, Sue E. Brennan, and et al. 2021. The PRISMA 2020 statement: An updated guideline for reporting systematic reviews. *Systematic Reviews* 10: 89. [CrossRef] [PubMed]

Peterson, Andrew. 2020. *The Origins of Civic Republican Thought*. London: Palgrave Macmillan. Available online: https://link.springer.com/chapter/10.1057/9780230306752_1 (accessed on 30 December 2022).

Price, Leigh. 2014. Critical realist versus mainstream interdisciplinarity. *Journal of Critical Realism* 13: 52–76.

Putnam, Robert. 2000. *Bowling Alone: The Collapse and Renewal of American Community*. Nueva York: Simon & Schuster.

Ribble, Mike. 2015. *Digital Citizenship in Schools. Nine Elements All Students Should Know*, 3rd ed. Eugene: International Society of Technology in Education.

Rüland, Jürgen, and Astrid Carrapatoso. 2022. Introduction: Issues of governance beyond the nation state. In *Handbook on Global Governance and Regionalism*. Edited by Jürgen Rüland and Astrid Carrapatoso. Cheltenham: Edward Elgar Publishing Limited, pp. 1–19. [CrossRef]

Sáiz, Ángel Valencia. 2005. Globalization, cosmopolitanism and ecological citizenship. *Environmental Politics* 14: 163–78.

Schild, Rebecca. 2016. Environmental citizenship: What can political theory contribute to environmental education practice? *The Journal of Environmental Education* 47: 19–34.

Sen, Amartya. 2006. *Identity and Violence. The Illusion of Destiny*. London: Penguin Books.

Thorne, Steven. 2005. Epistemology, politics, and ethics in sociocultural theory. *Modern Language Journal* 89: 393–409.

UN. 2015. *Agenda de Desarrollo Sostenible 2030*. New York. Available online: http://www.un.org/es/comun/docs/?symbol=A/70/1 (accessed on 20 November 2022).

UNESCO. 2017. *The ABCs of Global Citizenship Education*. París: Unesco. Available online: http://www.unesco.org/new/global-citizenship-education (accessed on 30 December 2022).

UNESCO. 2018. *Preparing Teachers for Global Citizenship Education. A Template*. Paris and Bangkok: Unesco. Available online: https://unesdoc.unesco.org/ark:/48223/pf0000265452 (accessed on 20 November 2022).

United Nations University institute on Computing and Society (UNU-CS). 2018. *Engaging Citizens for Sustainable Development: A Data Perspective. Making the Global Agenda the Citizen's Agenda*. Available online: https://sdghelpdesk.unescap.org/sites/default/files/2019-05/UNU-CS-DataSD-Report-Engaging-Citizens-20181004-compressed.pdf (accessed on 30 November 2022).

Van Poeck, Katrien, and Joke Vandenabeele. 2013. Sustainable citizenship as practice. *Lifelong Learning in Europe* 1–9.

Wight, Colin. 2006. *Agents, Structures and International Relations. Politics as Ontology*. Cambridge: Cambridge University Press.

Wolf, Johanna. 2007. The Ecological Citizen and Climate Change. In European Consortium for the Promotion of Political Science (ECPR) Joint Sessions: «Democracy on the Day after Tomorrow». Helsinki. Available online: https://ecpr.eu/Filestore/paperproposal/c9ae934f-f308-4612-95c7-67c9da8d882f.pdf (accessed on 30 December 2022).

social sciences

MDPI

Article

TQM and SDGs for Erasmus+ Programme—Quality Education, Reducing Inequalities, Climate Change, Peace and Justice

Teresa Nogueiro [1] and Margarida Saraiva [1,2,*]

[1] Management Department, School of Social Sciences, Universidade de Évora, 7004-516 Évora, Portugal
[2] Business Research Unit (BRU-IUL), Instituto Universitário de Lisboa (ISCTE-IUL), University Institute of Lisbon, 1649-026 Lisboa, Portugal
* Correspondence: msaraiva@uevora.pt

Abstract: Any element that enables higher education institutions (HEIs) to set themselves apart in a positive and superior way in terms of their performance would be advantageous given the competitive climate in which they operate. The Erasmus+ Programme provides HEIs with yet another option to become more competitive and to contribute to the Sustainable Development Goals (SDGs) via the improvement of educational quality (SDG 4), reducing inequalities (SDG 10), climate action (SDG 13), and peace and justice (SDG 16). The goal of this work was to explore the potential relationships and synergies between HE sustainability and Total Quality Management (TQM) issues through the SDGs. The methodological approach was concentrated on the qualitative study of academic papers on TQM, sustainability, and the SDGs in HE as well as on the analysis of Regulation (EU) 2021/817, which established Erasmus+. We concluded that TQM and sustainability have synergies related to the SDGs, and the Erasmus+ Programme can support the sustainability of HEIs by promoting these SDGs. Leadership; education and training; the participation of staff members; measurement, evaluation, and control; and other stakeholders are essential factors for the effective implementation of TQM and sustainability in HEIs.

Keywords: Total Quality Management; Sustainable Development Goals; Erasmus+ Programme; quality education; reducing inequalities; climate action; peace and justice

Citation: Nogueiro, Teresa, and Margarida Saraiva. 2023. TQM and SDGs for Erasmus+ Programme—Quality Education, Reducing Inequalities, Climate Change, Peace and Justice. *Social Sciences* 12: 123. https://doi.org/10.3390/socsci12030123

Academic Editor: Alvaro Morote

Received: 2 November 2022
Revised: 14 February 2023
Accepted: 20 February 2023
Published: 24 February 2023

1. Introduction

According to Aquilani et al. (2017) experts such as Crosby, Deming, Feigenbaum, Ishikawa, Juran, and Garvin have discussed how to manage quality to gain a competitive advantage through increased customer happiness and higher performance, which is a commonly lacking competence among different organisations. Additionally, the significance of quality management is acknowledged as being of the utmost relevance in new business paradigms such as value co-creation.

It is unclear where the origin of the expression "Total Quality Management" (TQM) lies. Although there is no universal agreement on how to define the term, TQM is generally understood to refer to organisation-wide initiatives aiming to create a culture in which an organisation continuously improves its capacity to provide consumers with high-quality goods and services (Nasim et al. 2020). According to Yahiaoui et al. (2022), Total Quality Management has been described as a management philosophy, a management strategy, an integrated system, an approach to continuous improvement, and an approach to change in higher education.

Quality management is a holistic, all-embracing, and cogent process that encompasses all employees, managers, and staff in an organisation. It is accepted that quality is a cornerstone of success in worldwide competitiveness. TQM is a source of innovation, a crucial component of corporate culture, and a significant factor in an organisation's ability to outperform competitors (Douglas and Judge 2001). Referring to the study carried out by these authors, the results offered reasonably strong justification for the full and vigorous adoption of TQM.

The service sectors, which included the education sector, were first exposed to TQM concepts and methods in the 1980s. The faster globalisation process, which increased rivalry among educational institutions worldwide, especially higher education institutions (HEIs), was a major factor in the development of TQM in the education sector (Asif et al. 2013).

The assumptions that quality education is essential to succeeding in the global competition around education and that the fundamental principle of TQM applies to education management in the same way that it applies to other industrial sectors underlie the increased interest in TQM in the education sector (Militaru et al. 2013).

TQM is a holistic process that encompasses a wide range of stakeholders from the larger society, in addition to academic staff, students, and management in higher education institutions (Nasim et al. 2020). This corroborates the findings of Ruben (2018), who stated that all stakeholders with an interest in higher education should be involved in quality management in order for it to be effective. This includes not only faculty members, but also university administrators, as well as students.

According to Hackman and Wageman (1995), Total Quality Management, when appropriately applied and integrated with the necessary organisational variables, may be a tool that enables organisations to dynamically maintain a fit with their environments in a competitive and sustainable manner.

TQM is one of the elements that could assist organisations in improving their environmental performance. This management approach has the power to improve both individual and organisational performance (Abbas 2020). Not only does it enable businesses to obtain a competitive edge (Zwain et al. 2017), but it also promotes the creation of competitive products and services with a high quality at reasonable prices and with fast response times (Qasrawi et al. 2017).

Organisations improve their employees' knowledge and abilities regarding the effective use of resources through TQM. Employees are more motivated in such a setting to ensure that their products/services not only present outstanding quality but also safeguard the environment. The results of this study showed that if a company manages its TQM operations well it will increase the skills, capabilities, and motivation of its employees to use resources efficiently (Abbas 2020).

Yeung (2018) identified three levels of sustainable development in higher education: organisational stakeholder involvement, educational goals, and community need realisation; teacher awareness of environmental issues, competency-based concepts, and exposing students to real-world situations; and learner role definition.

The need to develop and apply TQM concepts to all elements of higher education academic units, including teaching, research, community services, and administrative support, was reiterated by Castillo (2020).

Higher education institutions have long been at the forefront of social transformation through scientific research, the development of creative solutions, and the training of intellectuals and change agents. The 2030 Agenda expressly acknowledges that certain goals and objectives can only be met with the cooperation of higher education institutions and research centres (Junyent et al. 2018). According to these same authors, universities can further the implementation of Sustainable Development Goals (SDGs) through:

- Governance—the incorporation of the principles of the SDGs into governance and institutional culture.
- Management—fostering sustainable campus management and university operations.
- Teaching and learning—training students to implement and assess the SDGs.
- Research—promoting responsible research and the creation of alternative pathways for the future.
- Partnerships and community engagement—fostering the well-being of communities and creating new partners for change.

The necessity to provide excellent education, be accountable to society, and internationalise education has boosted the profile of higher education quality evaluation dramatically during the last two decades (Ríos 2015; Ryan 2015). The current tendency in higher education is to foster a shared quality culture and conduct a more holistic assessment of the entire institution (Harris 2017).

A good education is important for the development of a civil society, because it teach-es individuals to think critically and to feel a duty to protect resources and ecosystems, defend the environment, and eradicate hunger and poverty. According to the Rome Communiqué, HEIs should promote SDGs, and internal and external quality assurance systems should evaluate and monitor the SDGs implemented by HEIs (Stukalo and Lytvyn 2021).

HEIs, like any other organisation, require an enormous number of available resources (inputs/outputs) due to the large flow of people, information, and activities created and released. These businesses are left with a huge environmental burden, demanding the implementation of sustainable development strategies (Gazzoni et al. 2018). Therefore, "universities are challenged to include the 17 Sustainable Development Goals (SDG) in the wide range of their training offers, and that higher education is expected to contribute with knowledge and innovation to meet societal, economic and environmental challenges through the training of both academic staff and students" (Chaleta et al. 2021, p. 2).

Despite the appeals of top university managers, only half of all curricular units incorporated an SDG, according to Chaleta et al. (2021). This does not imply that professors from various departments are uninterested in accomplishing the SDGs. Considering the current climate, in which the necessity for global cooperation has grown very apparent, the fact that 2021 was highly unusual due to the COVID-19 pandemic may explain the poor uptake of the SDGs. In light of this, institutions may be able to provide a new opportunity for reflection on the importance of sustainable development and the necessity for each member of the academy to become more involved in attaining the 2030 Agenda for Sustainable Development Goals (Chaleta et al. 2021). The results of this (Chaleta et al. 2021, p. 8):

> "showed that the most notable objectives in the curriculum units as a whole were SDG 5—Gender Equality, and Goal 10—Reduced Inequalities, aspects that teachers in the School of Social Sciences consider to be able to work from the curricular units for which they are scientifically responsible. Also highlighted were Goal 8—Decent Work and Economic Growth and Goal 16—Peace, Justice and Strong Institutions. Less mentioned were Goal 11—Sustainable Cities and Communities and Goal 3—Good Health and Well-Being. All other SGDs were less represented and Goal 6—Clean Water and Sanitation was not identified in any courses."

Open, democratic, fair, and sustainable communities, as well as sustained prosperity, entrepreneurship, and employment, require excellent and inclusive universities. Higher education institutes of various forms are characteristic of our European way of life. This diversity is positive, since it provides opportunities for creativity and synergy through mobility and cooperation.

Universities' contribution to the sustainability challenge is critical because of their function as centres of learning, innovation, and research. On the other hand, they can deal with sustainability issues in a variety of ways, as a result of their diverse functions, which should be accurately specified in their strategic plans. Their job is not confined to teaching and research; it encompasses society as a whole through the dissemination of research findings and through scientific and cultural contributions aiming to raise public awareness about specific challenges. Universities may make a substantial contribution to environmental sustainability in this context, both didactically and scientifically (Sisto et al. 2020).

According to Sisto et al. (2020), these institutions can, for example, offer degree pro-grammes centred on sustainability, encourage research projects on environmental protection with the participation of private companies and public institutions, and organise seminars and conferences on environmental issues, all while building relationships with stakeholders in order to foster future partnerships and synergies. Furthermore, universities can take meaningful steps to lessen their environmental effects as product and service users (e.g., energy, water, paper) and waste producers. As a result, universities may be able to make a substantial contribution to sustainable development by implementing personalised policies that are more effective when shared with stakeholders.

The definition of a strategy in HEIs is relevant. Universities can approach sustainability in a variety of ways, all of which should be clearly recognised in their strategy. Corroborating this statement, the study carried out by Sisto et al. (2020) examined the feasibility of back-casting as a participatory method for involving stakeholders in discussions on the most effective steps to promote sustainability within universities' strategic plans.

Through its iconic Erasmus+ initiative, Europe is currently celebrating 35 years of life-changing experiences for more than 10 million young learners. Analysing the European strategy for universities, we see the relevance that is given to the Erasmus+ Programme and to the Sustainable Development Goals, with the former able to contribute to the latter. The document "A European strategy for universities COM/2022/16 final, Strasbourg, 18.1.2022 COM" (Communication from the Commission to the European Parliament 2022, p. 15) states:

> "This Communication is an invitation for closer cooperation between countries and actors of the higher education sector within the European Education Area (EEA), the European Re-search Area (ERA) and the European Higher Education Area (EHEA, Bologna process). Synergies are needed in areas such as transnational cooperation and the institutional transformation of universities, support for fundamental academic values and scientific freedom, developing academic careers, innovative and interdisciplinary learning, teaching and research, as well as the interconnectedness between these, knowledge circulation, international cooperation with partners beyond the EU and the contribution to the United Nation's SDG's.".

The aim of this work was to analyse the possible links and synergies between the themes of TQM and sustainability in higher education through the SDGs; in this context, to link SDGs 4, 10, 13, and 16 to the Erasmus+ Programme, namely its objectives and actions; and to contribute to the improvement of the Erasmus+ Programme, motivating its beneficiaries to identify and associate to their activities and projects the SDGs to which they are contributing.

This paper is structured as follows: the introduction is followed by sections dedicated to the materials and methods used in the research; the results obtained; and, lastly, the discussion, final considerations, limitations, and future research proposals.

2. Materials and Methods

The methodology was focused on the qualitative analysis of scientific articles on the themes of TQM, sustainability, and SDGs in the context of higher education and also on the analysis of Regulation (EU) 2021/817 of the European Parliament and of the Council of 20 May 2021 establishing Erasmus+: the Union Programme for education and training, youth, and sport. The aim of this methodology was to contribute to the debate on the existence of synergies between TQM and sustainability and to better understand the issues addressed in order to obtain answers as to their relationship with the Erasmus+ Programme and the contribution this programme can make to the sustainability of HEIs in four areas: quality education (SDG 4); reducing inequalities (SDG 10); climate action (SDG 13); and peace, justice, and strong institutions (SDG 16).

The research was conducted in several phases using the following methodology:

Phase 1—The qualitative analysis of the content of the articles on the theme of TQM and sustainability. The selection of articles was based on the use of the following keywords in Google Scholar and Web of Science, among other databases: TQM, Total Quality Management, Sustainability, Sustainable, SDGs, and Sustainable Development Goals. The keywords identified gave rise to categories, and from these, subcategories were identified. To further the goal of the methodology, this information is included in the text of this article.

Phase 2—The qualitative analysis of the content of Regulation (EU) 2021/817 of the European Parliament and of the Council of 20 May 2021 establishing Erasmus+: the Un-ion Programme for education and training, youth, and sport was essential for the establishment of the relationship between the programme and the selected SDGs.

Phase 3—The qualitative assessment of the 17 Sustainable Development Goals, namely their targets, means of implementation, indicators, and goals, was necessary to extract those SDGs related to quality education; reducing inequalities; climate action; and peace, justice, and strong institutions that could have a strong relationship with the Eras-mus+ Programme.

Phase 4—The establishment of the relationship between the selected SDGs and TQM was achieved through the assessment of the common critical success factors of TQM and sustainability and the elements/foundations of the 17 Sustainable Development Goals and the Erasmus+ Programme.

Phase 5—We integrated our research findings with a literature review for the discussion.

The qualitative analysis was carried out manually by the authors.

3. Results

3.1. *Total Quality Management and Sustainability*

Based on the research conducted by Nogueiro et al. (2022a), it was possible to identify leadership; education and training; the involvement of all employees; measurement, evaluation, and control; and other stakeholders as common fundamental elements or critical factors for the successful implementation of TQM and sustainability in HEIs. Table 1 presents a description of each of the critical success factors common to TQM and sustainability.

Table 1. Common critical success factors of TQM and sustainability.

Critical Success Factor	Description/Characterisation
Leadership	The top management, who are responsible for defining the quality; social responsibility; sustainability and environmental policies; and mission, vision, and values of the institution. Commitment to the organisation is crucial for an eventual change in the adoption of practices and communication promoting the empowerment of workers and their involvement, including in decision making.
Education and training	HEIs have a mission very focused on teaching and research, and therefore education and training are part of the core business of the institution. It is only through these factors that people gain skills and knowledge to carry out their activities.
Involvement of all employees	Being involved means actively participating in the institution's activities and decisions. We refer to a level of involvement associated with attitudes, participation, teamwork, and cross-functional interactions, which should be provided by the top management. Only through the involvement of the workers will they know the true value of the products or services they provide.
Evaluation, measure-ment, and control	Evaluation, measurement, and control are fundamental for a HEI to understand whether it is achieving the objectives and targets set. The use of measurement tools and the establishment of adequate key performance indicators are relevant and fundamental for the institution to redirect focus when needed, redefine policies, and adopt preventive and/or corrective measures regarding its performance.
Other stakeholders	The other stakeholders correspond to all those who demonstrate that they have needs, demands, and expectations that the HEIs have to manage and meet.

Source: Adapted from Nogueiro et al. (2022a).

3.2. Erasmus+ Programme's General and Specific Objectives

The main finding was that the Erasmus+ Programme 2021–2027 clearly intends to contribute to Sustainable Development Goals 4, 10, 13, and 16. Figure 1 presents the definitions of these SDGs.

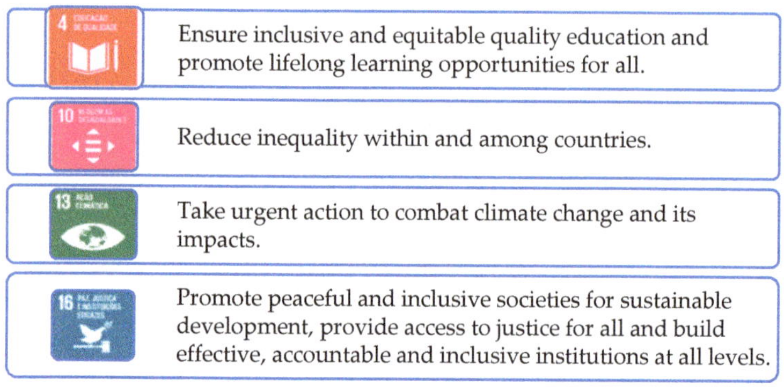

Figure 1. Definitions of SDGs 4, 10, 13, and 16. Source: adapted from https://sdgs.un.org/goals (accessed on 13 February 2023).

According to Regulation (EU) 2021/817 (2021, p. 1), the establishment of Erasmus+ had the following aims, among others:

> "Investing in learning mobility for all, regardless of background and means, and in cooperation and innovative policy development in the fields of education and training, youth and sport is key to building inclusive, cohesive and resilient societies and sustaining the competitiveness of the Union, and is all the more important in the context of rapid and profound change driven by technological revolution and globalisation. Furthermore, such an investment also contributes to strengthening European identity and values and to a more democratic Union."

At the basis of the programme's renewal was the need for it to contribute to the Union's policy objectives and priorities in the fields of education, training, youth, and sport. Lifelong learning is essential for people to manage the different transitions they will face during their lives. The Erasmus+ Programme is a critical element of establishing a European Education Area and continues to play an important role in achieving the goals of quality and inclusive education, training, and lifelong learning, as well as preparing the Union for the digital and green transitions. Erasmus+ shall also assist Member States in achieving the goals of fostering citizenship and the common values of freedom, tolerance, and non-discrimination via education (Informal Meeting 2015). In order to achieve its goals, the Erasmus+ Programme should be made more inclusive by increasing participation among people who have fewer opportunities through a range of measures (Regulation (EU) 2021/817 2021).

Given the challenges to the common values on which the Union was established and which form part of the shared European identity, as well as people's low levels of engagement, fostering a European feeling of belonging and commitment is crucial. The Erasmus+ Programme aims to contribute to the mainstreaming of climate action and the achievement of a global target of 30% of the Union budget expenditure supporting climate goals, and any actions must respect the 'do no harm' principle without changing the fundamental character of the programme. The Financial Regulation's principles of transparency, equal treatment, and non-discrimination should be followed in the programme's performance (Regulation (EU) 2021/817 2021). Table 2 summarises the objectives and European added value of the Erasmus+ Programme.

Table 2. Objectives and European added value of Erasmus+.

General objectives	1. To promote the educational, professional, and personal development of people in the fields of education and training, youth, and sport in Europe and beyond, thereby contributing to sustainable growth, quality jobs, and social cohesion, driving innovation and strengthening European identity and active citizenship through lifelong learning. 2. To function as a crucial instrument for building a European Education Area.
Specific objectives	To promote: (a) The learning mobility of individuals and groups and cooperation, quality, inclusion, equity, excellence, creativity, and innovation at the level of organisations and policies in the field of education and training. (b) Non-formal and informal learning mobility and active participation among young people and cooperation, quality, inclusion, creativity, and innovation at the level of organisations and policies in the field of youth. (c) The learning mobility of sport staff and cooperation, quality, inclusion, creativity, and innovation at the level of sport organisations and sport policies.
European added value	1. Only those actions and activities with potential European added value that contribute to the attainment of the programme's objectives will be supported. 2. The European added value of the programme's actions and activities will be ensured, for example, by their: (a) Transnational character, particularly with regards to learning mobility and cooperation aimed at achieving a sustainable systemic impact; (b) Complementarity and synergies with other programmes and policies at the national, Union, and international level; (c) Contribution to the effective use of Union transparency and recognition tools.

Source: own elaboration based on Regulation (EU) 2021/817 (2021).

The Erasmus+ Programme objectives must be pursued through learning mobility (key action 1), cooperation among organisations and institutions (key action 2), and the supporting of policy development and cooperation (key action 3), which mainly have either a transnational or an international character (Regulation (EU) 2021/817 2021).

3.3. SDGs 4, 10, 13, and 16 vs. Erasmus+ Programme

Analysing the information available on the targets, means of implementation, and indicators of SDGs 4, 10, 13, and 16, our conclusion was that not all are relevant or related to the Erasmus+ Programme's objectives. Therefore, the relevant targets and means of implementation selected were as follows: for SDG 4—4.3, 4.4, 4.5, 4.7, 4.b, and 4.c; for SDG 10—10.3; for SDG 13—13.3; and for SDG 16—16.a (see Figure 2).

The main results for SDGs 4, 10, 13, and 16 are presented below.

3.3.1. SDG 4—Quality Education

Sustainable Development Goal 4 aims to ensure inclusive and quality education and promote lifelong learning opportunities for all (Sustainable Development Goal 4 n.d.). The United Nations defined 10 targets and 11 indicators for SDG 4.

Taking into consideration the basis of the renewal of the Erasmus+ Programme, the objectives outlined, and the definition of the various parameters of SDG 4, it was possible to conclude that the targets to which this European Union programme contributes are as follows (Sustainable Development Goal 4 n.d.):

> "4.3 By 2030, ensure equal access for all women and men to affordable and quality technical, vocational and tertiary education, including university.
>
> 4.4 By 2030, substantially increase the number of youth and adults who have relevant skills, including technical and vocational skills, for employment, decent jobs and entrepreneurship.

4.5 By 2030, eliminate gender disparities in education and ensure equal access to all levels of education and vocational training for the vulnerable, including persons with disabilities, indigenous peoples and children in vulnerable situations.

4.7 By 2030, ensure that all learners acquire the knowledge and skills needed to promote sustainable development, including, among others, through education for sustainable development and sustainable lifestyles, human rights, gender equality, promotion of a culture of peace and non-violence, global citizenship and appreciation of cultural diversity and of culture's contribution to sustainable development.

4.b By 2020, substantially expand globally the number of scholarships available to developing countries, in particular least developed countries, small island developing States and African countries, for enrolment in higher education, including vocational training and information and communications technology, technical, engineering and scientific programmes, in developed countries and other developing countries.

4.c By 2030, substantially increase the supply of qualified teachers, including through international cooperation for teacher training in developing countries, especially least developed countries and small island developing States."

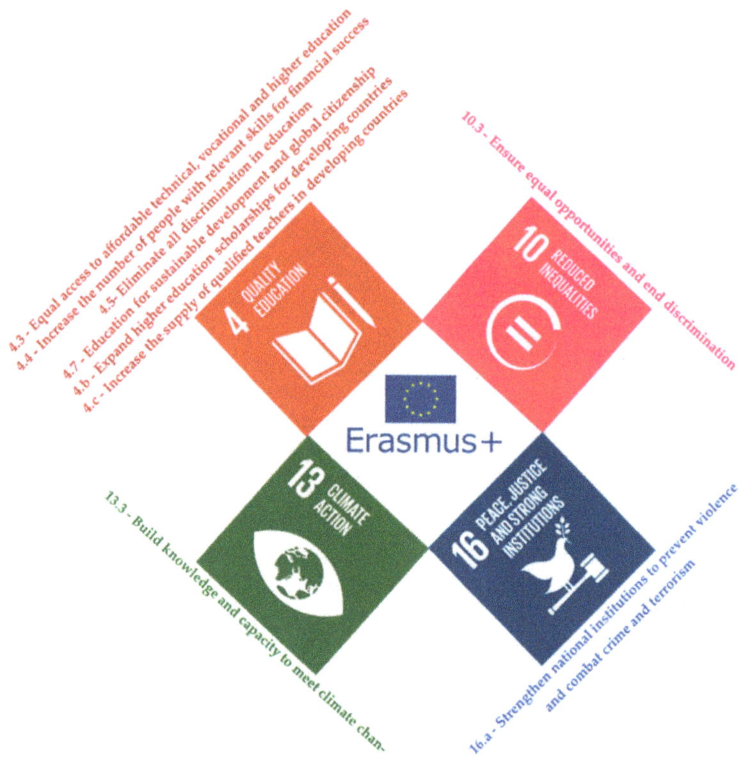

Figure 2. Erasmus+ contributions to Sustainable Development Goals. Source: own elaboration based on https://sdg-tracker.org/ (accessed on 14 February 2023).

There is a perfect match between the programme and SDG 4, which his associated with higher education. In terms of the programme's overall goal, its aim is to support people's educational, professional, and personal development in the fields of education and training (target 4.c), youth, and sport in Europe and beyond through lifelong learning, thereby contributing to sustainable growth, quality jobs (target 4.4), social cohesion, driving innovation, and strengthening European identity and active citizenship (target 4.7). Erasmus+ also aims to promote the learning mobility of individuals and groups, as well as cooperation, quality, inclusion, equity, excellence, creativity, and innovation at the level of organisations and policies in the field of education and training (targets 4.3 and 4.5); non-formal and informal learning mobility and active participation among young people, as well as cooperation, quality, inclusion, creativity, and innovation at the level of organisations and policies in the field of youth; and sport staff learning mobility, as well as cooperation, quality, inclusion, creativity, and innovation at the sport organisation and policy levels. The Erasmus+ Programme will implement three key actions, all of which are primarily transnational or international in nature: (a) learning mobility (key action 1) (target 4.b); (b) cooperation among organisations and institutions (key action 2); and (c) policy development and cooperation support (key action 3) (Regulation (EU) 2021/817 2021).

3.3.2. SDG 10—Reduced Inequalities

Sustainable Development Goal 10 aims to reduce inequality within and among countries (Sustainable Development Goal 10 n.d.). The United Nations defined 10 targets and 11 indicators for SDG 10.

The relationship between the Erasmus+ Programme and SDG 10 is at the level of target 10.3—ensuring equal opportunities and ending discrimination. By 2030, the Union intends to (Sustainable Development Goal 10 n.d.):

> "10.3 Ensure equal opportunity and reduce inequalities of outcome, including by eliminating discriminatory laws, policies and practices and promoting appropriate legislation, policies and action in this regard."

Erasmus+ shall assist Member States in achieving the goals of fostering citizenship and the common values of freedom, tolerance, and non-discrimination via education (Informal Meeting 2015). In order to achieve its goals, the programme should be made more inclusive by increasing participation among people who have fewer opportunities. In other words, inclusion is a major goal of the Erasmus+ Programme, which we can clearly associate with reducing inequalities; promoting actions; and encouraging countries and other stakeholders to put measures in place, develop strategic plans, and expand opportunities for all, without discriminating or leaving anyone behind.

3.3.3. SDG 13—Climate Action

Sustainable Development Goal 13 aims to take urgent action to combat climate change and its impacts. The United Nations defined five targets and eight indicators for SDG 13. The contribution of the Erasmus+ Programme to this SDG relates to target 13.3—building knowledge and the capacity to meet climate change. The goal for this target is to (Sustainable Development Goal 13 n.d.):

> "13.3 By 2030 improve education, awareness-raising and human and institutional capacity on climate change mitigation, adaptation, impact reduction and early warning."

Analysing Regulation (EU) 2021/817 (2021, p. 8) of the European Parliament and of the Council of 20 May 2021, which established the Erasmus+ Programme for the period 2021–2027, it was possible to identify the intent to contribute to this goal from the following passage, wherein it is explicitly mentioned:

> "Reflecting the importance of tackling climate change in line with the Union's commitments to implement the Paris Agreement adopted under the United Nations Frame-work Convention on Climate Change and to achieve the United Nations' Sustainable Development Goals, the Programme is intended to contribute to mainstreaming climate actions and to the achievement of an overall target of 30% of Union budget expenditure supporting climate objectives. In line with the European Green Deal as a blueprint for sustainable growth, the actions

> under this Regulation should respect the 'do no harm' principle without changing the fundamental character of the Programme. During the implementation of the Programme, relevant actions should be identified and put in place and reassessed in the context of the relevant evaluations and review process. It is also appropriate to measure relevant actions that contribute to climate objectives, including those intended to reduce the environmental impact of the Programme."

In article 32, it is also mentioned that (Regulation (EU) 2021/817 2021, p. 28):

> "The Programme shall be implemented so as to ensure its overall consistency and complementarity with other relevant Union policies, programmes and funds, in particular those relating to education and training, culture and the media, youth and solidarity, employment and social inclusion, research and innovation, industry and enterprise, digital policy, agriculture and rural development, environment and climate, cohesion (. . .)"

3.3.4. SDG 16—Peace, Justice, and Strong Institutions

Sustainable Development Goal 16 is committed to promoting peaceful and inclusive societies for sustainable development; ensuring universal access to justice; and establishing strong, accountable institutions at all levels (Sustainable Development Goal 16 n.d.). The United Nations defined 12 targets and 23 indicators for SDG 16.

The Erasmus+ Programme was created with the goal of maintaining peace and justice through actions and projects that are carried out with institutions from all over the world. Therefore, its contribution to this SDG is at the level of the means of implementation, target 16.a—strengthening national institutions to prevent violence and combat crime and terrorism. The aim of 16.2 is as follows (Sustainable Development Goal 16 n.d.):

> "16.2 Strengthen relevant national institutions, including through international co-operation, for building capacity at all levels, in particular in developing countries, to pre-vent violence and combat terrorism and crime." (by 2030)

The mention of peace and justice refers to fundamental rights. Therefore, the Erasmus+ Programme regulation upholds basic rights and adheres to the principles set forth in the European Union's Charter of Fundamental Rights. The programme should also support activities that contribute to citizenship education and participation projects for young people to engage in and learn to participate in civic society, thereby raising awareness of European common values, including fundamental rights, as well as European history and culture (Regulation (EU) 2021/817 2021).

4. Discussion and Final Considerations, Limitations, and Future Research Proposals

The European strategy for universities of the European Commission clearly states that synergies are required in areas such as transnational cooperation and university institutional transformation, support for fundamental academic values and scientific freedom, academic career development, innovative and interdisciplinary learning, and teaching and research, as well as their interconnectedness, knowledge circulation, international cooperation with partners outside the EU, and contribution to the UN's Sustainable Development Goals (Communication from the Commission to the European Parliament 2022).

HEIs must implement CSFs for TQM, since they will help the organisation boost its performance evaluation (Salleh et al. 2018). For the adoption of TQM, these authors named the following CSFs: management commitment and leadership, total customer satisfaction, the involvement of employees, continuous improvement, training, communication, and teamwork.

Identifying CSFs is a crucial step to incorporate them into an organisation's processes, thus providing the organisation with the capability to assess hazards and possibilities in their environment. CSFs also provide a set of criteria for assessing the strengths and weaknesses of organisations (Tambi 2000).

It was possible to determine the alignment between the CSFs for the implementation of TQM and the CSFs for the implementation of sustainability in HEIs, specifically those that are common to both (such as leadership; education and training; the involvement of all employees; measurement, evaluation, and control; and other stakeholders), thanks to the studies conducted by Nogueiro et al. (2022a) and Bayraktar et al. (2008), corroborated/validated by Nadim and Al-Hinai (2016).

Griebeler et al. (2022) defined education quality as the ability to impart information/knowledge to students within a set of requirements established by all those who would need this highly skilled workforce in the future, such as corporations, government agencies, and professional societies. On the other hand, TQM is a way of managing and improving the effectiveness, efficiency, cohesion, flexibility, and competitiveness of an organisation, such as an HEI, as a whole. TQM can be successfully implemented if it includes principles of leadership, commitment, ensuring customer satisfaction, the continuous improvement of products and/or services, total involvement, teamwork, and error prevention (Silva and Mendes 2018).

According to Jermsittiparsert (2020), in order to achieve the SDGs, education quality management is critical. The importance of education in achieving these goals cannot be overstated. An excellent university education has a big impact on community development activities. It raises public awareness and contributes to the welfare of the general population. Education quality management contributes to the achievement of the SDGs by improving society's well-being and reducing inequality (Jermsittiparsert 2020).

The SDGs are presented as a to-do list on behalf of the people and the planet, as well as a plan for success. TQM includes components such as the integrity and promotion of TQM values and principles; equity and openness; a participatory management style; the benefitting of customers, workers, society, and owners and a focus on considering their needs; giving a voice to these parties; and, finally, transparency and openness with regards to wide communication and the sharing of information (Nogueiro et al. 2022b). These are very important elements in higher education institutions. The Erasmus+ Programme will serve as a European Commission tool for education and training that will enable projects to be developed in the most diverse scientific areas and with the most diverse objectives, aligning with both TQM principles and elements of the SDGs (targets, means of implementation, and indicators).

The notion of sustainability has a close association with the Erasmus Programme (Kafarski and Kazak 2022). The authors mentioned that the studies carried out by Nogueiro et al. (2022c) and De La Torre et al. (2022) led to similar results, showing the relevance of certain SDGs for the Erasmus Programme, including SDG 4 (Quality Education).

Aligned with these synergies, it was possible to identify other SDGs, besides SDG 4, to which the Erasmus+ Programme contributes, such as SDG 10, SDG 13, and SDG 16. In total, 37 targets and 53 indicators from the four selected SDGs were analysed, despite not all of them being related to the programme.

In sum, the Erasmus+ Programme aims to support people's educational, professional, and personal development in the fields of education and training (SDG target 4.c), youth, and sport in Europe and beyond through lifelong learning, thereby contributing to growth sustainability, quality jobs (SDG target 4.4), social cohesion, driving innovation, and strengthening European identity and active citizenship (SDG target 4.7). Erasmus+ aims to promote the learning mobility of individuals and groups, as well as cooperation, quality, inclusion, equity, excellence, creativity, and innovation at the level of organisations and policies in the field of education and training (targets 4.3 and 4.5); non-formal and informal learning mobility and active participation among young people, as well as cooperation, quality, inclusion, creativity, and innovation at the level of organisations and policies in the field of youth; and sport staff learning mobility, as well as cooperation, quality, inclusion, creativity, and innovation at the sport organisation and policy levels.

Inclusion is a major goal of the Erasmus+ Programme, which can be clearly associated with reducing inequalities, promoting actions, and encouraging countries and other stakeholders to put measures in place, develop strategic plans, and expand opportunities for all, without discriminating or leaving anyone behind (SDG target 10.3).

The programme is intended to contribute to mainstreaming climate actions and the achievement of an overall target of 30% of Union budget expenditure supporting climate objectives, reflecting the importance of tackling climate change in line with the Union's commitments to implementing the Paris Agreement adopted under the United Nations Framework Convention on Climate Change and to achieving the United Nations' Sustainable Development Goals (SDG target 13.3). The Erasmus+ Programme also has two indicators for climate change: the share of activities addressing climate objectives under key action 1, and the share of projects addressing climate objectives under key action 2 (SDG target 13.3).

The Erasmus+ Programme Regulation (EU) 2021/817 (2021) upholds basic rights and adheres to the principles set in the European Union's Charter of Fundamental Rights. The programme should also support activities that contribute to citizenship education and participation projects for young people to engage in and learn to participate in civic society, thereby raising awareness of European common values, including fundamental rights, as well as European history and culture (SDG target 16.2).

It was concluded that there are synergies between TQM and sustainability, which can be associated with the SDGs, and that the Erasmus+ Programme can, in fact, contribute to the sustainability of HEIs through SDGs 4, 10, 13, and 16. TQM and sustainability, despite the existence of other essential factors for the implementation of each, have elements in common that are equally crucial for their successful implementation in HEIs, such as leadership; education and training; the involvement of all those who work in the institution; measurement, evaluation, and control; and other stakeholders.

It was perceived that the targets to which Erasmus+ contributes are as follows:

- SDG 4—Quality education. Targets 4.3—equal access to affordable technical, vocational, and higher education; 4.4—an increase in the number of people with relevant skills for financial success; 4.5—the elimination of all discrimination in education; 4.7—education for sustainable development and global citizenship; 4.b—the expansion of higher education scholarships for developing countries; and 4.c—an increase in the supply of qualified teachers in developing countries.
- SDG 10—Reducing inequalities. The selected target was 10.3—ensuring equal opportunities and ending discrimination.
- SDG 13—Climate action. The target was 13.3—building knowledge and the capacity to meet climate change.
- SDG 16—Peace, justice, and strong institutions. The selected target was 16.a—strengthening national institutions to prevent violence and combat crime and terrorism.

Making sure that all activities supported by the Erasmus Programme have long-lasting effects is a crucial component (Kafarski and Kazak 2022). To assess the long-term viability of the project results, Alonso De Castro and Peñalvo (2021) conducted a survey among administrative project coordinators. One of the primary conclusions was that the outcomes were successful and could be used long after the grant time had ended, as there were resources available to carry them out.

A limitation to this work was the impossibility of analysing the content of projects already submitted by HEIs for the period 2021–2027 under key actions 1 and 2, i.e., the definition of their partnerships, scientific areas, goals, etc., and the assessment of their probable contribution to sustainable development either under the SDGs selected for this study or under other SDGs. Another limitation to this study was the individual analysis of the SDGs, even though there appear to be relationships, at various levels, between them. A third limitation was the study of TQM in the context of the Erasmus+ Programme and the relationship with the selected SDGs in a higher education environment.

As future research proposals, we suggest the analysis of SDGs based on the realities of each key action of the Erasmus+ Programme; the study of the correlation between the SDGs, the Erasmus+ Programme, and mobility projects; the analysis of the correlation between the SDGs and the projects approved under the Capacity Building for Higher Education action, by scientific area; and the analysis of the contributions of SDGs to each other.

This work will help the European Commission, policymakers, and participants in the Erasmus+ Programme perceive its development from the perspective of continuous improvement and greater sustainability. According to our identification of Sustainable Development Goals and the links between TQM and sustainability, namely those made through the critical success factors, the Erasmus+ Programme, in the context of higher education, is aligned with the 2030 Sustainability Agenda defined by the United Nations.

Following this work, it is recommended that the European Commission integrates in the Erasmus+ Programme and, more explicitly, its contribution to the 2030 Agenda through the SDGs that are considered more relevant and whose contribution is more impactful.

This study is expected to contribute to the continuous improvement of the Erasmus+ Programme by associating it with sustainability through the Sustainable Development Goals and, in the near future, TQM.

Considering the themes investigated and their association with the Erasmus+ Programme, the European Commission has expressed great interest in the continuation of these studies for the improvement and sustainability of the programme.

Author Contributions: Conceptualisation, T.N. and M.S.; methodology, T.N.; validation, T.N. and M.S.; formal analysis, T.N.; investigation, T.N. and M.S.; resources, T.N.; data curation, T.N. and M.S.; writing—original draft preparation, T.N. and M.S.; writing—review and editing, T.N. and M.S.; visualisation, M.S.; supervision, M.S.; project administration, T.N. and M.S.; funding acqui-sition, M.S. All authors have read and agreed to the published version of the manuscript.

Funding: This work was supported by Fundação para a Ciência e a Tecnologia, grant UIDB/00315/2020.

Institutional Review Board Statement: Not applicable.

Informed Consent Statement: Not applicable.

Data Availability Statement: All documents, data, and information used in this work are available to the public.

Conflicts of Interest: The authors declare no conflict of interest.

References

Abbas, Jawad. 2020. Impact of total quality management on corporate green performance through the mediating role of corporate social responsibility. *Journal of Cleaner Production* 242: 118458. [CrossRef]

Alonso De Castro, María Goretti, and Francisco José García Peñalvo. 2021. ICT tools highlighted and their usefulness during the pandemic: Erasmus+ projects related to eLearning. Paper presented at Ninth International Conference on Technological Ecosystems for Enhancing Multiculturality, Barcelona, Spain, October 26–29; New York: Association for Computing Machinery, pp. 219–24.

Aquilani, Barbara, Cecilia Silvestri, Alessandro Ruggieri, and Corrado Gatti. 2017. A systematic literature review on total quality management critical success factors and the identification of new avenues of research. *The TQM Journal* 29: 184–213. [CrossRef]

Asif, Muhammad, Cory Searcy, Ambika Zutshi, and Olaf Fisscher. 2013. An integrated management systems approach to corporate social responsibility. *Journal of Cleaner Production* 56: 7–17. [CrossRef]

Bayraktar, Erkan, Ekrem Tatoglu, and Selim Zaim. 2008. An instrument for measuring the critical factors of TQM in Turkish higher education. *Total Quality Management* 19: 551–74. [CrossRef]

Castillo, Fermin G. 2020. TQM in higher education for sustainable future. In *Sustainable Development and Social Responsibility—Volume 2*. Cham: Springer International Publishing, pp. 373–80.

Chaleta, Elisa, Margarida Saraiva, Fátima Leal, Isabel Fialho, and António Borralho. 2021. Higher Education and Sustainable Development Goals (SDG)—Potential Contribution of the Undergraduate Courses of the School of Social Sciences of the University of Évora. *Sustainability* 13: 1828. [CrossRef]

Communication from the Commission to the European Parliament, the Council, the European Economic and Social Committee and the Committee of the Regions on a European Strategy for Universities. 2022. COM/2022/16 Final, Strasbourg, 18.1.2022 COM. Available online: https://eur-lex.europa.eu/legal-content/EN/TXT/?uri=COM%3A2022%3A16%3AFIN (accessed on 13 February 2023).

De La Torre, Eva, Adriana Perez-Encinas, and Gillian Gomez-Mediavilla. 2022. Fostering Sustainability through Mobility Knowledge, Skills, and Attitudes. *Sustainability* 14: 1349. [CrossRef]

Douglas, Thomas J., and William Q. Judge Jr. 2001. Total quality management implementation and competitive advantage: The role of structural control and exploration. *Academy of Management Journal* 44: 158–69. [CrossRef]

Gazzoni, Fernando, Flavia Luciane Scherer, Ivanete Schneider Hahn, Aletéia de Moura Carpes, and Maríndia Brachak dos Santos. 2018. O papel das IES no desenvolvimento sustentável: Estudo de caso da Universidade Federal de Santa Maria. *Revista Gestão Universitária na América Latina-GUAL* 11: 48–70. [CrossRef]

Griebeler, Juliane Sapper, Luciana Londero Brandli, Amanda Lange Salvia, Walter Leal Filho, and Giovana Reginatto. 2022. Sustainable development goals: A framework for deploying indicators for higher education institutions. *International Journal of Sustainability in Higher Education* 23: 887–914. [CrossRef]

Hackman, J. Richard, and Ruth Wageman. 1995. Total quality management: Empirical, conceptual, and practical issues. *Administrative Science Quarterly* 40: 309–42. [CrossRef]

Harris, N. 2017. El Papel de las Agencias de Evaluación en el Contexto Internacional: Tendencias y Resistencias. ANECA en el Horizonte 2020: Una Agencia de Evaluación al Servicio de la Comunidad Universitaria Española. Santander. September 11–12. Available online: http://www.aneca.es/Sala-de-prensa/Noticias/2017/ANECA-en-la-UIMP-2017-ponencias-y-archivos (accessed on 8 April 2021).

Informal Meeting of European Union Education Ministers. 2015. Paris, Tuesday 17 March 2015—Declaration on Promoting Citizenship and the Common Values of Freedom, Tolerance and Non-Discrimination through Education. Available online: https://eu2015.lv/images/notikumi/2015-3-10_Declaration_EN.pdf (accessed on 11 February 2022).

Jermsittiparsert, Kittisak. 2020. Education Quality Management: A Way Forward to Promote Sustainable Development Goals by Encouraging Wellbeing's and Discouraging Inequality Among the Societies. *Journal of Physics: Conference Series* 1467: 012077. [CrossRef]

Junyent, Mercè, Ingrid Mulà Pons de Vall, and Marta Fonolleda Riberaygua. 2018. The Quality of Higher Education in Andorra and the Sustainable Development Goals: A Proposal for Quality Assessment Standards and Guidelines. Available online: https://www.iau-hesd.net/sites/default/files/documents/7_17-016_estudi_q_ods_maquetat_paper_ang_2_compressed.pdf (accessed on 13 February 2023).

Kafarski, Krzysztof, and Jan K. Kazak. 2022. Erasmus Staff Mobility in the Building of a European Network: The Case of a Central European University. *Sustainability* 14: 4949. [CrossRef]

Militaru, Mădălina, Gabriela Ungureanu, and Alina Ştefania Chenic. 2013. The prospects of implementing the principles of Total Quality Management (TQM) in education. *Procedia-Social and Behavioral Sciences* 93: 1138–41. [CrossRef]

Nadim, Zaynab Shukri, and Ahmed Humaid Al-Hinai. 2016. Critical success factors of TQM in higher education institutions context. *International Journal of Applied Sciences and Management* 1: 147–56.

Nasim, Kanwal, Arif Sikander, and Xiaowen Tian. 2020. Twenty years of research on Total Quality Management in Higher Education: A systematic literature review. *Higher Education Quarterly* 74: 75–97. [CrossRef]

Nogueiro, Teresa, Margarida Saraiva, and António Ramos Pires. 2022a. Critical Success Factors of TQM for Sustainability in Higher Education Institutions—A theoretical contribution. Paper presented at ICQIS 2022, Aveiro, Portugal, May 3–4.

Nogueiro, Teresa, Margarida Saraiva, and Fátima Jorge. 2022b. Total Quality Management and Social Responsibility an Approach Through Their Synergies in Higher Education Institutions. In *Perspectives and Trends in Education and Technology*. Edited by Anabela Mesquita, António Abreu and João Vidal Carvalho. Singapore: Springer, pp. 311–21. [CrossRef]

Nogueiro, Teresa, Margarida Saraiva, Fátima Jorge, and Elisa Chaleta. 2022c. The Erasmus+ Programme and Sustainable Development Goals—Contribution of mobility actions in higher education. *Sustainability* 14: 1628. [CrossRef]

Qasrawi, Bara' Tareq, Soud Mohammad Almahamid, and Shadi Tareq Qasrawi. 2017. The impact of TQM practices and KM processes on organisational performance: An empirical investigation. *International Journal of Quality & Reliability Management* 34: 1034–55.

Regulation (EU) 2021/817 of the European Parliament and of the Council of 20 May 2021 Establishing Erasmus+: The Union Programme for Education and Training, Youth and Sport and Repealing Regulation (EU) No 1288/2013. 2021. *Official Journal of the European Union* 189: 1–33. Available online: https://eur-lex.europa.eu/legal-content/EN/TXT/PDF/?uri=CELEX:32021R0817&qid=1641601177851&from=en (accessed on 18 October 2022).

Ríos, Cristina. 2015. Quality Assurance in Higher Education in Spain: An Overview of the Accreditation System. *International Research and Review: Journal of Phi Beta Delta* 5: 25–45. Available online: https://files.eric.ed.gov/fulltext/EJ1148966.pdf (accessed on 18 October 2022).

Ruben, Brent D. 2018. *Quality in Higher Education*. New York: Routledge.

Ryan, Patricia. 2015. Quality assurance in higher education: A review of literature. *Higher Learning Research Communications* 5. [CrossRef]

Salleh, Norliza Mohd, Norhayati Zakuan, Mohd Shoki Md Ariff, Ahamad Zaidi Bahari, Thoo Ai Chin, Zuraidah Sulaiman, Shafudin Mohd Yatim, Siti Rahmah Awang, and Muhamad Zameri Mat Saman. 2018. Critical success factors of total quality management implementation in higher education institution: UTM case study. *AIP Conference Proceedings* 2044: 020007.

Silva, Rui Jorge, and Luis Mendes. 2018. Mapeamento da aplicação da TQM no ensino superior: Uma revisão bibliográfica utilizando uma abordagem bibliométrica Mapping total quality management research in the higher education context: A literature review using a bibliometric approach. *Revista Capital Científico-Eletrônica* 16: 65–80. [CrossRef]

Sisto, Roberta, Edgardo Sica, and Giulio Mario Cappelletti. 2020. Drafting the Strategy for Sustainability in Universities: A Backcasting Approach. *Sustainability* 12: 4288. [CrossRef]

Stukalo, Nataliia, and Maryna Lytvyn. 2021. Towards Sustainable Development through Higher Education Quality Assurance. *Education Sciences* 11: 664. [CrossRef]

Sustainable Development Goal 4—Ensure Inclusive and Quality Education for All and Promote Lifelong Learning. n.d. Available online: https://sdg-tracker.org/quality-education (accessed on 11 February 2022).

Sustainable Development Goal 10—Reduce Inequality within and among Countries. n.d. Available online: https://sdg-tracker.org/inequality (accessed on 11 February 2022).

Sustainable Development Goal 13—Climate Action. n.d. Available online: https://sdg-tracker.org/climate-change (accessed on 11 February 2022).

Sustainable Development Goal 16—Peace, Justice and Strong Institutions. n.d. Available online: https://sdg-tracker.org/peace-justice (accessed on 11 February 2022).

Tambi, Abdul Malek. 2000. *Total Quality Management in Higher Education: Modelling Critical Success Factors*. Sheffield: Sheffield Hallam University.

Yahiaoui, Fethia, Khalil Chergui, Nesreddine Aissaoui, Said Khalfa Mokhtar Brika, Imane Ahmed Lamari, Adam Ahmed Musa, and Mohmmad Almezher. 2022. The impacts of total quality management practices in Algerian higher education institutions. *Frontiers in Psychology* 13: 874209. [CrossRef]

Yeung, Shirley Mo-Ching. 2018. Linking ISO 9000 (QMS), ISO 26000 (CSR) with accreditation requirements for quality indicators in higher education. *Total Quality Management & Business Excellence* 29: 1594–611. [CrossRef]

Zwain, Ammar Abdulameer Ali, Kong Teong Lim, and Siti Norezam Othman. 2017. TQM and academic performance in Iraqi HEIs: Associations and mediating effect of KM. *The TQM Journal* 29: 357–68. [CrossRef]

social sciences

MDPI

Article

Climate Change and Sustainability in Spanish Classrooms: State of the Art and Didactic Proposal

Jordan Correa-González, Abel López-Díez *, Jaime Díaz-Pacheco and Nerea Martín-Raya

Territorial Management and Risks Group (GEORIESGOS), Disaster Risk Reduction and Resilient Cities Group, Department of Geography, University of La Laguna (ULL), 38200 San Cristóbal de La Laguna, Spain
* Correspondence: alopezd@ull.edu.es

Abstract: Climate change has become a global challenge that must be faced in a cross-cutting manner from multiple fields and involving all citizens. The educational system, as a space that guarantees the training of students and the integral development of the person at the social, intellectual and ethical levels, should be oriented towards increasing the environmental awareness of society, promoting practices and habits that respect the preservation of ecosystems and, in short, education for sustainability. The 2023–2024 academic year is the first in which the curricular content developed from the Organic Law 3/2020, of December 29, which modifies the Organic Law 2/2006, of May 3, on Education, popularly known as LOMLOE, will be fully implemented. This paper designs a learning situation on sustainability and climate change that can be implemented in the Spanish and European contexts, responding to Rosenshine's principles of instruction, a circumstance that gives it enormous flexibility and makes it an interesting resource focused on helping geography teachers to face current challenges from an innovative, scientific, and inclusive perspective.

Keywords: teaching; climate change; sustainability; education; Rosenshine

Citation: Correa-González, Jordan, Abel López-Díez, Jaime Díaz-Pacheco, and Nerea Martín-Raya. 2023. Climate Change and Sustainability in Spanish Classrooms: State of the Art and Didactic Proposal. *Social Sciences* 12: 108. https://doi.org/10.3390/socsci12020108

Academic Editor: Alvaro Morote

Received: 31 January 2023
Revised: 10 February 2023
Accepted: 13 February 2023
Published: 16 February 2023

1. Introduction

It is likely that, apart from the great wars of history, humanity has never faced a threat as pressing as climate change. This phenomenon, whose existence and causes (linked to anthropogenic activity) have been rigorously proven by practically the entire scientific community (Lynas et al. 2021), represents a major problem that has led, in recent decades, to the implementation of numerous regulatory instruments and strategies aimed at promoting the acquisition of environmentally sustainable attitudes and practices (MITECO 2022). Thus, for some time now, public institutions have been trying to raise awareness of the importance of moving towards positions that respect ecosystems and their limits and safeguard natural resources through the promotion of socioeconomic development that does not compromise the right of future generations to meet their own needs and enjoy the Earth responsibly (CMMA 1987).

One of the most noteworthy initiatives in this regard in recent years has been the approval of the Sustainable Development Goals—hereinafter referred to as SDGs—in 2015, which have made it possible to deepen the challenges included in the Millennium Development Goals, promoted in 2000, to embrace a more ambitious and holistic perspective of human development. Within the catalog of seventeen approved SDGs, there are several that are directly linked to the preservation of the environment and the fight against climate change, such as 13—climate action, 15—life on land, or 7—affordable and clean energy (United Nations 2022). At a Spanish scale, in this century both the Central State Administration and the Autonomous Communities have promoted, based on European and international guidelines, the elaboration of numerous strategies and documents of a legislative nature that together make up what could be catalogued as the national regulatory corpus of climate action. Within this compendium, it is worth highlighting, as an example,

Law 7/2021, of May 20, on climate change and energy transition (BOE No. 121, of May 21 2021) or the National Plan for Adaptation to Climate Change 2021–2030 (MITECO 2020).

However, the public is highly skeptical about the effectiveness of these measures. Recent opinion polls show that 92% of the Spanish population agrees that action against climate change and environmental degradation is slow, so that the majority do not believe that Spain will achieve a substantial reduction in its carbon dioxide emissions before 2030 (European Investment Bank 2022). In addition, various interest groups and experts have denounced the measures promoted by the authorities as insufficient to achieve the commitments acquired in climate matters (De Miguel 2020).

The fact is that, according to the Climate Change Performance Index, Spain is in the twenty-third position worldwide, having risen eleven places in this ranking in the last year (Germanwatch 2022). In turn, according to the Environmental Performance Index, the country is the twenty-seventh for the best performance in achieving sustainability in general; although, in terms of climate policies, the Spanish situation can be clearly improved, being placed in a deficient 83rd position at international level (Wolf et al. 2022).

On the other hand, several authors have proven the importance of raising awareness and sensitizing citizens in order to involve them in the fight against climate change through the education system (González-Gaudiano 2007; Anderson 2013; Sánchez-Almodóvar et al. 2022). Thus, most research agrees that educational level is "the strongest predictor of climate change awareness" (Ming-Lee et al. 2015), a circumstance that has encouraged the inclusion of the educational sphere in the main climate action policy documents in most European states.

At the same time, it is worth noting the importance given to education by the Intergovernmental Panel on Climate Change—IPCC—which, in the six assessment reports it has produced since 1990, has included specific proposals to be applied in the field of education (Sánchez-Almodóvar et al. 2022), including the need to create environmental awareness campaigns, encourage the participation of all sectors of the population, and promote collaboration between scientists, policymakers, and all interested parties. Likewise, the Incheon Declaration and the Framework for Action for the implementation of Sustainable Development Goal 4 shows the interest and usefulness of a commitment to quality education that prepares young people and adults for a world in constant change and with enormous challenges to be faced from the perspective of resilience and balance between meeting social needs and caring for the environment (UNESCO 2016).

The European Union, for its part, has on several occasions endorsed international trends in education for sustainability, stating that member states must intensify efforts to ensure that education becomes a space that fully promotes ecological transition and sustainable development, pointing out the importance of implementing interdisciplinary and cross-cutting activities that contribute, directly or indirectly, to encouraging sustainable lifestyles, taking care of nature, etc. (OJEU No. 243 of June 27 2022).

In Spain, Law 7/2021, of May 20, on climate change and energy transition, dedicates its Title VIII to "Education, research and innovation in the fight against climate change and energy transition," indicating that the national education system must promote the involvement of society in the responses designed to combat this phenomenon, expanding knowledge about it and preparing citizens for the development of a resilient and responsible technical and professional activity (BOE No. 121, of May 21 2021). Similarly, Article 35.2 establishes the need to review the treatment of climate change and sustainability in the basic curriculum of the teachings of the educational system in a cross-cutting manner, as well as the obligation to adequately train teachers in these subjects.

The autonomous communities, with broad competences in education (Aragón-Reyes 2013), have also assumed, in their respective legislation on the climate challenge, the relevance of education as an essential instrument to build a culture of sustainability and resilience; examples of this are the climate change laws of the Valencian Community (DOCV No. 9486, of December 9 2022), Andalusia (BOJA No. 199, of October 15 2018), and Catalonia (DOGC No. 7426, of August 3 2017).

However, the notable lack of consensus and educational instability suffered by Spain, which in its democratic period has approved or had in force a total of nine educational laws—LGE, LOECE, LODE, LOGSE, LOPEG, LOCE, LOE, LOMCE, and LOMLOE—is an obstacle to the achievement of the goals set given the short period of effective implementation of the different regulations, which are continually subject to revision and political controversy (Novella and Cloquell 2021). Nevertheless, Spain's prolific legislative activity in the field of education is a resource of undoubted interest that allows us to analyze in detail how the treatment of environmental issues has changed over the last four decades at the regulatory level, a circumstance that has encouraged research activity in this regard (Caballero et al. 2021; Morote and Olcina 2021).

The latest Spanish education law, popularly known as LOMLOE (Organic Law 3/2020, of December 29, amending Organic Law 2/2006, of May 3, on Education), states in its preamble that the education system cannot ignore the challenges posed by climate change, so that schools must become places of "custody and care" of the environment, promoting a culture "of environmental sustainability, social cooperation, developing programs for sustainable lifestyles, and encouraging recycling and contact with green spaces" (BOE No. 340, of December 30 2020). The autonomous communities, for their part, have assumed this commitment in the process of approving the new curricula for Compulsory Secondary Education and Baccalaureate, which insist on the need for students to be aware of the planet's environmental problems, know the causes and consequences of climate change, and acquire sustainable habits in their academic and personal lives (Ministry of Education and Vocational Training 2022a).

What has been mentioned so far justifies the multiple works that have been carried out in recent years on the teaching of climate change and sustainability in the educational system. From different disciplines, the international scientific community has generated abundant and innovative teaching material on the issue, as well as a profound reflection on the objectives to be achieved (Maxwell and Blashki 2016; Siperstein et al. 2017; Murga-Menoyo and Bautista-Cerro 2019)

However, the above mentioned regulatory instability to which teachers are subjected, the continuous remodeling of curricula, the delay of some regional governments in decreeing the organization of subjects—in the Canary Islands, for example, the 2022–2023 academic year has been developed following a draft curriculum—the ambivalence exhibited by the LOMLOE on issues such as sustainability, and the lack of training in geographic content that teachers sometimes show in social sciences, lead teachers to a situation of uncertainty in the face of new content that, in addition, must be addressed in a cross-cutting manner, according to the legislation.

Some or all of the proposed activities can be worked on at different educational levels and from other related subjects, such as biology and geology, not only geography. Consequently, the main objective is to provide teachers with tools, not specific contents—a "how" rather than a "what"—so that this perspective of environmental education can be assumed in multiple educational contexts, from the earliest ages to the university, adapting the level of demand and difficulty to the intellectual maturity of the students.

The principles designed by Barak Rosenshine constitute one of the most effective methods for teaching, proposing a series of flexible steps from which to structure the contents and activities in the classroom so that the student progressively acquires the desired knowledge, consults appropriate sources of information, is able to make critical judgments, and, finally, has the basic skills necessary for the resolution of an independent practice (Rosenshine 2010).

In this context, the main objective of this paper is to present a learning situation that can be totally or partially applied in different European educational contexts, although the 2nd year of baccalaureate is taken as a reference based on the current regulations in force in the autonomous community of the Canary Islands on climate change and sustainability. It is intended, therefore, to provide teachers with specific and updated resources and activities to work on the issue in the classroom following a systematic, rigorous, and problem-solving-oriented methodology.

In the following section, with the purpose of contextualizing the issue at the regulatory level, some guidelines are offered on how the contents related to sustainability fit into the Spanish and European legislative framework.

2. Climate Change and Sustainability in Educational Legislation: Fit, Novelties and Prospects

As expressed in the introductory section, one of the defining characteristics of the Spanish education system is its instability, which is due, in essence, to the absence of consensus among the country's main political forces about what education should be and in relation to various controversial issues such as private and subsidized education or the subject of Religion (Novella and Cloquell 2021; Fernández-Mellizo 2019).

As already mentioned, unlike what happens in other European countries whose educational legislation remains in force for decades, in barely half a century, as many as nine educational laws have been passed (Figure 1), two of which—the LOECE and the LOCE—never came into force, and others, such as the LOMCE, have had a reduced period of validity. Education thus becomes an ideological battlefield, a circumstance that makes it impossible to meet the objectives contained in the regulations and subjects teachers to a situation of continuous uncertainty in the face of the new curriculum.

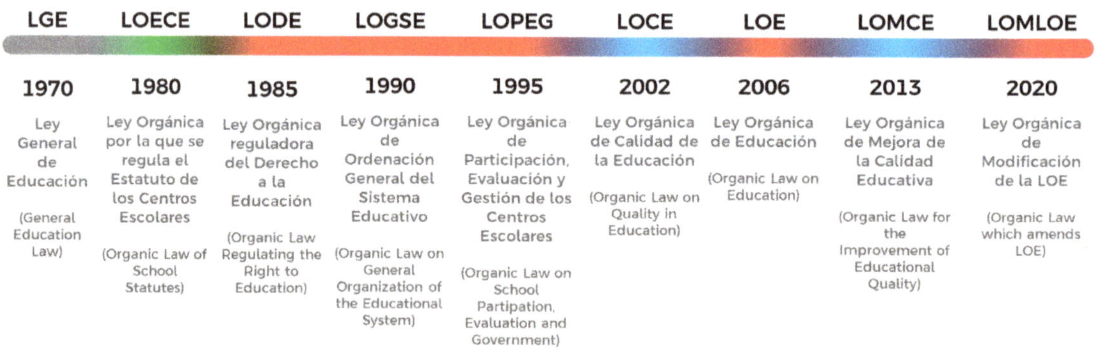

Figure 1. Spanish educational laws in the last half century. Own preparation.

Analyzing the latest educational laws passed in this century—LOCE, LOE, LOMCE and LOMLOE—it should be noted that all of them refer, implicitly or explicitly, to the need to raise awareness among secondary and baccalaureate students of the importance of preserving the planet and contributing to their acquiring adequate training in environmental matters.

Thus, the LOCE, approved in 2002, recognized the need for high school students to have the necessary skills to understand and critically analyze "the contribution of science and technology to change living conditions, as well as to strengthen sensitivity and respect for the environment" (BOE No. 307, of December 24 2002).

Notwithstanding the above, the LOE—2006, the law that fully introduced environmental issues in the educational regulations, states that one of the basic purposes of the Spanish educational system consisted in the acquisition of values that favored "respect for living beings and the rights of animals and the environment, in particular the value of forest areas and sustainable development," and indicated, in addition, that a responsible and committed attitude should be fostered in the battle against climate change and unsustainability (BOE No. 106, of May 4 2006).

The LOMCE—2013, on the other hand, the explicit references to climate change and sustainable development disappeared, although, for example, knowledge and prevention of environmental risks remained among the objectives of vocational training (BOE No. 295, of December 10 2013).

The last educational law, the LOMLOE, was approved in 2020 after a controversial drafting process and a harsh debate and vote in the Congress of Deputies (Torices 2020). In the 2022–2023 academic year, the modifications introduced in the curriculum, the organization, and objectives of the odd courses of primary, secondary, and baccalaureate came into force, while in 2023–2024, those referring to the even courses come into force, including the course with which students are prepared to access university, vocational training, or, directly, the labor market: 2nd year of baccalaureate (Ministry of Education and Vocational Training 2022b).

In any case, the LOMLOE once again places the environmental issue at the center of the educational system. Thus, the promotion of a culture of environmental sustainability and the achievement of sustainable development—referred to on more than twenty occasions in the law—is embedded in the educational regulations with such vigor that they must be addressed in a cross-cutting manner in the different subjects that make up the educational program (BOE No. 340, of December 30 2020).

This is expressed in Royal Decree 217/2022, which establishes the organization and minimum teachings of secondary education, in which teachers are repeatedly invited to encourage students to think critically about the preservation of the environment. It also states that by the end of basic education, students must have analyzed and adopted ideas related to sustainable development. Similarly, it should be noted that the subject of Education in Civic and Ethical Values introduces basic knowledge related to sustainability and environmental ethics, including concepts such as the limits of the planet, climate emergency, circular economy, resilient communities, etc. (BOE No. 76 of March 30 2022). This same philosophy of instruction committed to sustainable development and the environmental challenges facing the Earth is also observed in the Royal Decree establishing the organization and minimum teachings of the baccalaureate (BOE No. 82, of April 6 2022), as well as in the regional decrees that have been approved by the different regional executives to establish with precision the organization and curriculum based on the LOMLOE (BOCM No. 176, of July 26 2022; BOPA No. 169, of September 1 2022; DOE No. 164, of August 25 2022).

3. Methods

By virtue of what has been explained up to this point, it becomes obvious that education in sustainability has ceased to be an objective that is only superficially addressed and has turned, at all levels of the educational system, into an imperative within the curricular development. This fact has encouraged pedagogical research, which, in recent years, has tried to provide teachers with tools to face the new teaching challenges related to the environment with guarantees and resources. Among the most outstanding proposals is Murga-Menoyo's (2015), which identifies four competencies for sustainability: (1) critical analysis, including critical thinking and ethical and intellectual commitment; (2) systemic reflection, which contains relational thinking, the feeling of belonging to the community of life, and holistic thinking; (3) collaborative decision-making, developing argumentative and participatory skills and adopting a clear commitment to democracy and universal human rights; and (4) a sense of responsibility towards present and future generations, which refers to ethical and social commitment, anticipatory, synchronic and diachronic thinking, and, finally, universal responsibility and compassion.

In this context, Barak Rosenshine's didactic proposal constitutes an interesting model from which to work on the intended contents and competencies. Since its publication in 2010 (Table 1), it has become one of the most widely used in the educational field (Sherrington 2019; Buzo 2021), having been designed from research based on the way in which the brain acquires and uses new information, the practices carried out by teachers whose students achieve great goals, and, finally, the results of studies that teach learning strategies to students (Rosenshine 2010).

Table 1. Rosenshine's Principles of Instruction.

No	Principle of Instruction	Objective
1	Daily review	To reinforce connections to prior learning by recalling concepts and relationships
2	Present new information in small steps	To provide new materials progressively, helping students to put the content into practice
3	Ask questions	To help students consolidate new information and connect it to prior learning
4	Provide models	To make models and solved examples that can be applied to solve other problems available to students
5	Guide the student's practice	To consolidate learning by processing and performing trials or practice with the material
6	Check the student's understanding	To examine the degree of understanding at each step to avoid errors and confusion in learning
7	Obtain a high success rate	To achieve a success rate of around 80% among the student body
8	Provide scaffolding for difficult tasks	To allow the student to solve complex tasks through temporary support from the teaching staff
9	Independent practice	To increase the ability and the autonomy of the student to solve tasks or practices in a fluent way with automated procedures
10	Weekly and monthly review	To connect ideas through providing materials, conducting discussions, and application activities

Source: Own preparation from Rosenshine (2010).

The following table reflects the ten principles established by the author, as well as a brief description of their basic objectives:

The learning situation presented below has been designed following these principles. The proposed timing is merely illustrative and should be adapted to the unique characteristics of the students in question. In addition to the activities, the learning objectives, and expected competencies, resources and key concepts to be worked on during the sessions are indicated. It should be noted that multiple online viewers are proposed, whose analysis and interpretation contribute to the acquisition of basic geographic skills.

Regarding the methodologies proposed for the realization of the learning situation, these are the following:

- Collaborative work groups. Group work will be encouraged, promoting the acquisition of skills related to cooperation, coordination, and peer learning.
- Presentation of results and group discussions. The main results of the activities will be presented in groups, and debates will be held to encourage critical thinking.
- Inverted class. In order to achieve that the student works in an autonomous, dynamic, and collaborative way, this technique will be adopted based on the existence of audiovisual resources and other content not specifically explained by the teacher.
- Gamification. The Kahoot tool is integrated in the learning situation as one of the basic resources to favor the assimilation of the contents from a playful perspective.
- Techniques for the construction of concept maps. Students will be invited to elaborate concept maps on the topics addressed.
- Interpretation and elaboration of cartography. Maps analysis is a fundamental geographic technique on which the explanation of phenomena, processes, and relationships can be based.

Regarding the way to evaluate the students' performance in the learning situation, this will depend on the singular characteristics of the group of students with whom we work and the availability of means and time. However, the instruments with which, in principle, it is planned to evaluate the students are the following:

- Participation in group discussions.
- Observation of behaviors and attitudes during the period of development of the activities.
- Elaboration and, if necessary, exposition of the different works and tasks proposed.
- Elaboration of the final review test or exam.

Notwithstanding the fact that each activity developed presents a specific qualification based on the objectives pursued, the following Table 2 presents a proposal of operational objectives and evidences of achievement that allow to evaluate, as a whole, the effectiveness of the training process and, ultimately, to determine whether the student has acquired the competencies and has satisfied the objectives of the learning situation.

Table 2. Operational competences and evidence of achievement.

Operational Competence (Capacity to...)	Evidence of Achievement (The Student...)		
	Level 1 (Approved = Grade C)	**Level 2 (Remarkable = Grade B)**	**Level 3 (Outstanding = Grade A)**
1. Search, use, and analyze data included in official statistical databases and viewers available online	1.1. Is able to describe the available information, identifying general trends by country and region	1.2. Points out the possible explanatory causes of the observed phenomena or process and explains the reason for the territorial differences	1.3. Relates the observed phenomena to other processes and points out the importance of human beings on these processes
2. Analysis, interpretation, and interrelation of a variety of cartographic material	2.1. Analyzes cartographic material in an appropriate and informed manner, using geographic terminology	2.2. Is able to relate several maps to reach general conclusions about a specific phenomenon or process	2.3. Reasons about the underlying variables that can explain the processes represented cartographically
3. Elaboration of simple cartographic compositions	3.1. Produces cartography following the established indications and including the basic geographic elements (scale, key, title, etc.)	3.2. Represents more than one variable on the map, in order to show the relationship between different elements	3.3. Applies a geoprocessing tool, generating new material from the layers provided
4. Analysis of policy documents related to climate action or sustainability	4.1. Is capable of analyzing the content of the regulations related to sustainability, summarizing its most relevant aspects	4.2. Points out the evolution of Spanish regulations in this area in recent decades, especially with regard to the changes and approval of new legislation	4.3. Conducts a comparative analysis of the regulatory content related to climate action at the European, national, and regional levels
5. Preparation of complex documents with a formal structure	5.1. Reproduces the proposed outline, adheres to the established standards, and completes the different headings in a simple but documented manner	5.2. Prepares internally coherent documents, maintaining the same line of argument focused on the resolution of the problem	5.3. Supplements the information required in each task or section with graphics, cartography, examples, audiovisual material, etc. related to the topic
6. Reflection on the interrelation of various concepts (resilience, climate change, sustainable development, adaptation, etc.	6.1. Does not limit him or herself to explaining each of the concepts worked on independently, but links common causes and consequences among different phenomena	6.2. Argues about the bidirectional relationship between sustainable development and climate change	6.3. Identifies proposals and comprehensive solutions to mitigate the malfunctions of the current socioeconomic system that hinder the achievement of the SDGs

Table 2. *Cont.*

Operational Competence (Capacity to...)	Evidence of Achievement (The Student...)		
	Level 1 (Approved = Grade C)	**Level 2 (Remarkable = Grade B)**	**Level 3 (Outstanding = Grade A)**
7. Concretization of theoretical and abstract notions (sustainability, resilience, etc.) into specific and quantifiable actions	7.1. Provides practical examples on resilience, sustainability, mitigation and adaptation in a complementary way to the theoretical side	7.2. Proposes specific measures to contribute to sustainable development and mitigation and adaptation in the face of climate change	7.3. Is capable of establishing simple indicators to evaluate the degree of progress and compliance with the proposed measures
8. Development of scientific reasoning, based on evidence	8.1. Backs up their analyses with rigorous and solid sources of information, verifying their initial hypothesis with scientifically proven reality	8.2. Is able to identify differences and similarities between various databases and sources in relation to the information and data required in the activity	8.3. Critically integrates in his or her reasoning the strengths and weaknesses of his or her own theses, as well as the opposing arguments, pointing out the issues that may generate controversy

Source: Own preparation.

4. Results

The main result of this work is a learning situation resulting from the combination of Rosenshine's methodology with some innovative approaches that have been published in recent years on the need to work in the classroom on the basis of concrete problems that the students are able to solve on their own or with the help of the teacher.

In order to facilitate its understanding, this section is structured in several subsections: learning objectives, competences to be developed, associated SDGs, basic resources for the student, key concepts, and, finally, the training proposal.

4.1. Learning Objectives

- To recover the learning that students had acquired in previous courses on issues related to climate change in order to broaden them and to be able to relate causes and effects.
- To describe and explain the links between anthropogenic action and climate change, paying special attention to the main activities and countries that generate greenhouse gas emissions.
- To describe, explain, and evaluate the most important risks and disasters of climatic origin, mainly in the Spanish and Canary Islands context.
- To evaluate the impact that climate change could have on our lives, health, and economic and social welfare, as well as on the fauna, flora, and waters of Spain and the Canary Islands.
- To describe, explain, and evaluate the main mitigation and adaptation measures that could be adopted, analyzing, in addition, the special vulnerability of island territories to this problem.
- To know the main methods and sources for obtaining and analyzing meteorological and climatic data, as well as to learn how to interpret graphs and perform simple tasks in Geographic Information Systems—hereinafter referred to as GIS.
- To explain what the SDGs are and describe their connection with the fight against climate change.

4.2. Competences to Be Developed

As already indicated, the students' daily work aims to contribute to the acquisition of the four competences for sustainability pointed out by Murga-Menoyo (2015): critical analysis, systemic reflection, collaborative decision-making, and sense of responsibility towards present and future generations. These are broken down into the following seven operational competences:

1. Search, use, and analysis of data included in official statistical databases and viewers available online.

2. Analysis, interpretation, and interrelation of varied cartographic material.
3. Elaboration of simple cartographic compositions.
4. Analysis of policy documents related to climate action or sustainability.
5. Elaboration of complex documents with a formal structure.
6. Reflection on the interrelation of various concepts—resilience, climate change, sustainable development, adaptation, etc.
7. Concretization of theoretical and abstract notions—sustainability, resilience, etc.—into concrete and quantifiable actions.
8. Development of scientific reasoning, based on evidence.

4.3. Associated Sustainable Development Goals (SDGs)

Although Sustainable Development Goal 13 makes explicit reference to climate action, the commitment to and awareness of sustainability and the fight against climate change is inscribed in each and every one of the SDGs (United Nations 2021). Specifically, the contents and activities inserted in the learning situation are directly linked to the goals included in the following Figure 2:

Figure 2. SDGs associated with the learning situation. Own preparation

4.4. Basic Resources

The following are nine basic resources that students should consult in order to adequately solve the proposed tasks (Table 3):

Table 3. Basic resources.

No.	Resource	Access
1	World Bank Climate Change Knowledge Portal (World Bank Group 2021)	https://climateknowledgeportal.worldbank.org/, accessed on 9 February 2023
2	World Bank Climate Change Country Database (World Bank Group 2023)	https://datos.bancomundial.org/tema/cambio-climatico, accessed on 9 February 2023
3	Climate projections for the 21st century (AEMET 2011)	https://www.aemet.es/es/serviciosclimaticos/cambio_climat, accessed on 9 February 2023
4	Global Climate Monitor (Camarillo-Naranjo et al. 2019)	https://www.globalclimatemonitor.org/, accessed on 9 February 2023
5	IPCC interactive atlas (Gutiérrez et al. 2021)	https://interactive-atlas.ipcc.ch/, accessed on 9 February 2023
6	Climate Change Scenario Viewer (Adaptecca.es 2021)	https://escenarios.adaptecca.es/, accessed on 9 February 2023
7	Sustainable Development Report viewer (Sachs et al. 2022)	https://dashboards.sdgindex.org/map, accessed on 9 February 2023
8	Carbon footprint calculator (Cities Footprint Project 2014)	https://huelladeciudades.com/AppHCCali/main.html, accessed on 9 February 2023
9	Illustrated dictionary of meteorology (AEMET 2018)	https://meteoglosario.aemet.es/, accessed on 9 February 2023

Source: Own preparation.

4.5. Key Concepts for Students

The following are the basic concepts with which the students of the 2nd year of baccalaureate will have to become familiar during the development of the learning situation.

4.5.1. Climate

It is the set of atmospheric conditions existing in a given territory during a relatively long period of time. It could also be defined as the average state of the atmosphere in a place on Earth. Conventionally, the minimum period to be able to refer to climate is thirty years; below that figure, one must speak of atmospheric weather, which is the state of the atmosphere at a precise moment or a temporal range of less than three decades (Linés 2010; Torres 2019; WMO 2022).

4.5.2. Climate Change

It is a process of transformation of climatic conditions resulting from the alteration of global atmospheric composition as a consequence, directly or indirectly, of anthropogenic activities, although, at other times, it may have been due to internal factors such as solar cycles or volcanic eruptions, transformations to which it is more convenient to refer to with the expression *climate variability* (United Nations 1992; IPCC 2013).

4.5.3. Sustainable Development Goals

These are seventeen goals adopted by the United Nations in 2015 that seek to move towards improving the well-being of citizens in all countries and human development in all its aspects: education, health, labor, equality, ecology, etc. It is therefore an interrelated framework that seeks to address some of the most pressing issues of our century: inequality, wars, poverty, deteriorating environmental conditions, etc. (United Nations 2015; UNDP 2022).

4.5.4. Climate Emergency or Crisis

This is an expression popularized in recent years that highlights the acceleration of global warming and the other effects of climate change. It is, in short, a term linked to climate change that aims to raise public awareness and draw the attention of institutions to the civilizational threats posed by global change in the making (Amico et al. 2020; Erviti 2020).

4.5.5. Climate Anomaly

It is the deviation recorded in the value of a climate element with respect to its normal values. It is common in scientific analyses of climate change to refer, for example, to thermal or rainfall anomalies—deviations in temperature and precipitation, respectively, caused by climate change (WMO 2017; AEMET 2018).

4.5.6. Sustainability

It is a criterion for action and development based on the fact that the actions of the present should not compromise the right of future generations to enjoy and live healthily on Earth. Therefore, it is a condition in which the use of natural resources and the socioeconomic development of societies does not diminish, deteriorate, or directly suppress the quality of the environment, so that anthropogenic activities can be developed without having a negative impact on the planet (Toro 2007; RAE 2022a).

4.5.7. Vulnerability

The susceptibility of a given population or its assets to be affected by a hazard of any kind. Vulnerability is often assessed taking into account the economic and human resources available to minimize the effects of an event, available infrastructures, political measures related to the issue, etc. (Basque Government n.d.; Biología-Geología.com n.d.; National Association of Safety Directors 2020).

4.5.8. Exposure

The volume of people, assets or systems that may be affected in the case of a disaster of natural origin or of any other kind. Consequently, the level of exposure of an element present in a risk zone will be higher the greater the potential human or economic losses in the event of a disaster (Biología-Geología.com n.d.; Inter-Agency Network for Education in Emergencies 2022).

4.5.9. Prevention

This is a principle consisting of advanced preparation carried out prior to the occurrence of a risk in order to avoid its occurrence or, at least, to reduce its negative impacts on societies or natural and economic resources. Preventive measures are considered the most effective, since they anticipate the occurrence of an accident or disaster by preparing societies or the environment (RAE 2022b; ABC Definition 2022).

4.5.10. Resilience

Resilience is the capacity of living beings—including, of course, human beings—and territories, elements, and systems to adapt to, recover from, and overcome the adverse events in their way and, in addition, to continue their development in spite of the negative circumstances that surround them. In terms of climate, resilience is the capacity of societies and natural environments to cope effectively with the impacts and transformations resulting from climate change (Becoña 2006; FAO 2019).

4.5.11. Mitigation

This is the strategy for combating climate change that seeks to reduce the effects of this phenomenon by reducing or eliminating greenhouse gases present in the atmosphere. For example, actions such as the promotion of renewable energies, the increase in forest mass to enhance CO_2 sinks, or the prohibition against driving polluting vehicles in cities are measures framed within mitigation (IPCC 2018; European Environment Agency 2022).

4.5.12. Adaptation

This is the process of adjusting to the current or potential effects of climate change in order to minimize its damages and take advantage of the opportunities it presents. Some examples of adaptation measures are the provision of shaded areas in cities in the face of rising temperatures, the protection of buildings near the coastal edge in the face of rising sea levels, the approval of emergency plans, or the improvement of the sewage system in the face of possible flooding (IPCC 2018; European Environment Agency 2022).

4.6. Training Proposal

With the purpose of facilitating the understanding of the planned activities and sessions, the training proposal presented below has been divided into seven phases: (1) concepts retrieval; (2) construction of basic schemes; (3) management of information sources; (4) introduction to practice; (5) review of learning; (6) autonomous practice; and (7) presentation and evaluation of results.

4.6.1. Phase 1. Concept Retrieval

Step 1

- Estimated sessions: 2
- Associated Rosenshine's Principle: (1) daily review

First, in order to check the students' prior learning on some concepts related to the topic, each student will be invited to elaborate two mind maps on their initial knowledge about climate change, on the one hand, and the SDGs, on the other hand. In this sense, students will be given total freedom to shape their outlines as they wish, making them more or less complex depending on their knowledge and skills with digital tools. The use of the free portal *GoConqr* (GoConqr 2023) is proposed for its simplicity and visual quality. It will be oriented more closely to students with special educational needs, whose outlines will probably be more synthetic, requiring them to prioritize a few essential ideas.

In any case, it is important that the students do not consult any material during the process of elaborating the outline, since the objective is simply to have a record of the degree of prior knowledge on the subject. This material, once elaborated, should be sent to the teacher, who will then be able to know the ideas that their students have about the basic content of the subject before going on to more advanced steps.

At the end of the second session, a questionnaire will be carried out through *Kahoot* or some other platform with a similar purpose, entitled *What do you know about climate change and the SDGs?*, complementary to the mind maps. With these two tools, the first phase, dedicated to concept retrieval, will be satisfied through a simple gamification strategy. These ten questions that could be asked in the initial questionnaire are proposed in Table 4.

Table 4. Proposed questions for concept retrieval.

No.	Question	Answers
1	Which of the following phenomena is not related to climate change?	a. Increased frequency and spread of large wildfires b. Event of seismic movements that sweep away entire cities ✓ c. Increasing droughts and decreasing availability of water resources d. Sea level rise
2	According to most scientists, climate change is a natural phenomenon that happens every few decades	a. True b. False ✓
3	Which gas is not considered a greenhouse gas?	a. Carbon dioxide b. Methane c. Hydrofluorocarbons d. Hydrogen ✓
4	What is the approximate current concentration of carbon dioxide in the atmosphere?	a. 250 ppm b. 100 ppm c. >400 ppm ✓ d. 350 ppm
5	Which country currently emits the most CO_2 into the atmosphere?	a. China ✓ b. India c. United States d. Germany
6	In which year were the SDGs adopted?	a. 2014 b. 2015 ✓ c. 2017 d. 2020
7	Which organization has been most closely linked to the SDGs since the initiative's creation?	a. European Union b. United Nations ✓ c. OECD d. Council of Europe
8	How many SDGs are there?	a. Ten b. Fifteen c. Eighteen d. Seventeen ✓
9	The SDGs refer only to social issues, such as poverty, resource distribution, or gender equality	a. True b. False ✓
10	There are one or more SDGs that specifically address climate change	a. True ✓ b. False

Source: Own preparation.

Step 2

- Estimated sessions: 3
- Associated Rosenshine's Principle: (6) check the student's understanding

In the third session of the learning situation, the overall results of the completed questionnaire will be analyzed collectively. In addition, the teacher will present a mind map of his own elaboration that will collect the main ideas highlighted by the group of students, even if these were incorrect or inaccurate. Once the results have been objectively presented, the teacher will offer a brief presentation that will clarify the main misconceptions pointed out by the students and will focus on the issues that, in his/her opinion, lend themselves to greater confusion.

The following sessions will begin to work with the following key concepts:

- Climate
- Climate change
- Climate emergency or crisis
- Climate anomaly
- Sustainable Development Goals
- Sustainability

4.6.2. Phase 2. Construction of Basic Outlines

Step 1

- Estimated sessions: 3
- Associated Rosenshine's Principle: (2) present new information in small steps

Given the importance of students learning to use rigorous and primary geographic sources, the class will be invited to carry out the activities included in Appendix A, entitled "Workshop: Sources and tools for the study of climate change," which includes practical exercises with six interactive and updated viewers of maximum interest at different scales.

After having installed the free software *QGIS*, and after having been given a brief reminder on how to use the program—which they should have already used in previous courses—a vector layer on climate change will be distributed for students to load in their projects. The link to download the layer is the following: https://drive.google.com/drive/folders/1ww8TdKH2p5Wah_BaFgNEIXQ2NJ1YrdqZ?usp=share_link (accessed on 9 February 2023).

As can be seen in the layer (Table 5), it includes several variables on the subject at the scale of the autonomous communities:

Table 5. Fields of the proposed vector layer.

No	Field	Description
1	POBL_2021	Population as of January 1, 2021 (INE 2022)
2	DENSITY	Population density (people per km^2)
3	KM_COAST	Coast kilometers (IGN 2022a)
4	EMISSIONS	CO_2 equivalent emissions by autonomous communities in 2019 in kilotons (MITECO 2019a)
5	TMIN	Average of minimum temperatures between 1991 and 2020 in °C (World Bank Group 2021)
6	TMED	Average of average temperatures between 1991 and 2020 in °C (World Bank Group 2021)
7	TMAX	Average of maximum temperatures between 1991 and 2020 in °C (World Bank Group 2021)
8	RAIN	Average annual rainfall between 1991 and 2020 in mm (World Bank Group 2021)
9	ANOMAL_MIN	Estimated minimum temperature anomaly (°C) in 2080–2099 in the *SSP58.5* scenario (World Bank Group 2021)
10	ANOMAL_MED	Estimated average temperature anomaly (°C) in 2080–2099 in the SSP58.5 scenario (World Bank Group 2021)
11	ANOMAL_MAX	Estimated maximum temperature anomaly (°C) in 2080–2099 in the SSP58.5 scenario (World Bank Group 2021)
12	ANOMAL_LLU	Estimated rainfall anomaly (mm) in 2080–2099 in the SSP58.5 scenario (World Bank Group 2021)
13	LEY_CC	Autonomous communities with their own climate change law
14	SUP_FOR	Forest area by Autonomous Community in hectares (MITECO 2019b)
15	DESERT	Approximate proportion of area at significant risk of desertification (Álvarez-Ubría et al. 2007)
16	HIDRICO	Availability of water resources in thousands of cubic meters in 2018. (INE 2018)

Source: own preparation.

Step 2

- Estimated sessions: 2
- Associated Rosenshine's Principles: (2) present new information in small steps and (3) ask questions

After exploring the data provided in the previous session, students should observe the different variables and discuss collectively the following questions relating the data arranged in the layer with the concepts presented in class:

a. Which do you think are the Spanish autonomous communities most affected by climate change?
b. Do you observe differences between regions? Justify your answer.
c. What is the situation of the Canary Islands in the Spanish context?

Next, they will be invited to calculate their own carbon footprint and compare it, according to the information available on the web, with that of citizens from other countries in the world. It is suggested to use the tool created by the *Cities Footprint* project, whose link is the following: https://huelladeciudades.com/AppHCCali/main.html (Cities Footprint Project 2014).

In the two final sessions of this step, students will have to elaborate a small report, in pairs, where they will answer the following questions:

a. Which are the five autonomous communities that emit the most greenhouse gases into the atmosphere? And those that emit the least?
b. Do all the autonomous communities have legislation or strategies focused on mitigation and adaptation to climate change? When did the Canary Islands pass their climate change law?
c. Which autonomous communities will experience a greater increase in maximum temperatures? What is the situation of the Canary Islands compared to the rest of the country?
d. Do the autonomous communities with the highest risk of desertification coincide with those in which rainfall will decrease the most? Explain your answer.
e. Reflect on the relationship between population density and climate change impacts.

Step 4

- Estimated sessions: 2
- Associated Rosenshine's Principles: (3) ask questions and (5) guide the student's practice

Students will be asked to complete the following activity using *QGIS*:

1. Create a folder on your computer or working directory named "Climate_Change_Project" where you will save a *QGIS* project with the name "General_Project." Then, insert the vector layer provided by the teacher.
2. Choose three fields from the layer and represent them in three layouts with all the necessary cartographic elements: title, north arrow, scale, legend, and coordinate system. Students with specific needs will be required only one layout of the variable of their choice.
3. Present the result to the rest of the class, explaining how the product produced can be related to climate change and the SDGs.

Next, a group discussion will be held on the basis of the following questions:

a. Do the Canary Islands share some unique characteristics with the Balearic Islands compared to the mainland?
b. Many experts suggest that the two archipelagos are especially vulnerable to climate change. Why do you think this may be?

4.6.3. Phase 3. Handling of Information Sources

Step 1

- Estimated sessions: 1

- Associated Rosenshine's Principles: (4) provide models

Based on the consultation of statistics and viewers available on the Sustainable Development Report portal (Sachs et al. 2022), students should respond in writing to the following questions:

a. What are the indicators included within each SDG? Select them in the left panel and analyze if there are major differences between them.
b. Observe the differences between continents in the degree of achievement of the SDGs.
c. Compare the Spanish situation with the surrounding countries.
d. Selecting the *trends* button. Do you think that substantial progress is being made in achieving the SDGs?

Step 2

- Estimated sessions: 2
- Associated Rosenshine's Principles: (7) obtain a high success rate, (8) provide scaffolding for difficult tasks, and (9) independent practice

Based on the explanation and subsequent reflection on the concepts of vulnerability, exposure, prevention, and resilience, students will be provided with the following three maps (Figures 3–5):

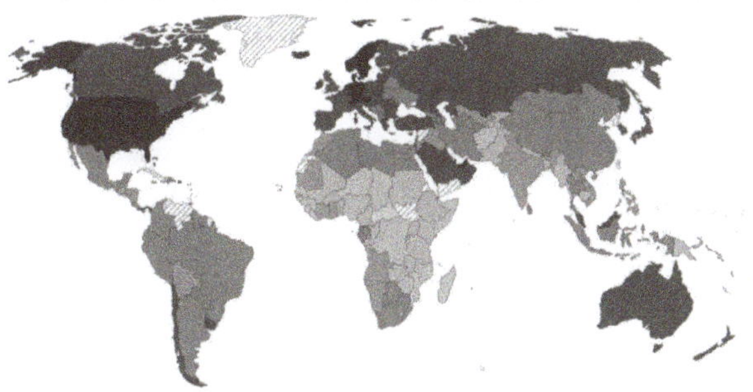

Figure 3. GDP per capita. Source: Our World in Data (2020). Countries with darker colors have higher GDP per capita.

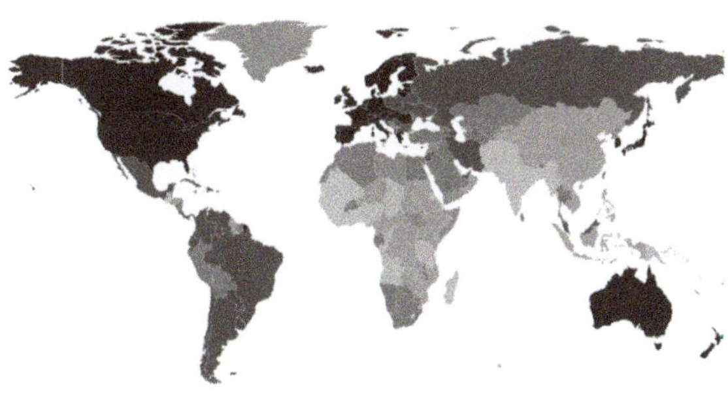

Figure 4. Environmental Performance Index—EPI. Source: Fortang (2021). Countries with darker colors have higher potential for climate resilience.

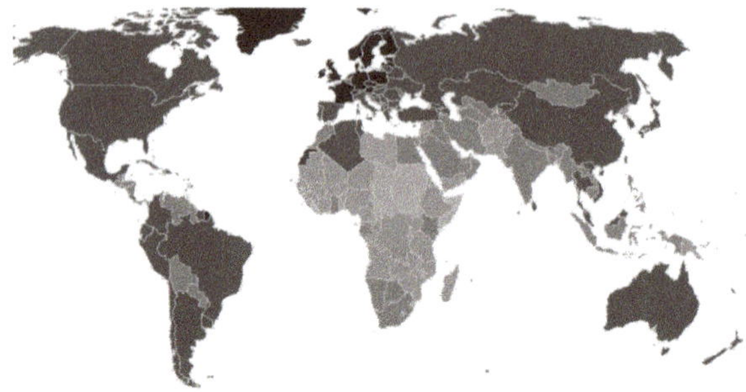

Figure 5. Overall score on total progress toward achieving the SDGs. Source: Sachs et al. (2022). Countries with darker colors have a higher degree of achievement of the SDGs.

Next, a group discussion on international inequalities will be initiated around the following questions:

a. Does anything strike you when comparing the three maps?
b. Do you observe any pattern that is repeated in all of them?
c. Do you think that inequality between countries hinders progress towards the 2030 Agenda Goals?
d. From your point of view, what are the main obstacles to the achievement of the SDGs on climate?

The second session of this step will be devoted to the students' autonomous work. Thus, based on the three map sources (Our World in Data 2020; Fortang 2021; Sachs et al. 2022), students will have to choose, in groups of three members, one state from each continent and analyze it according to the aspects considered—wealth, resilience and achievement of the SDGs. To do so, they will be invited to consult any other reliable source or resource they consider useful, paying special attention to the temporal evolution of economic variables in the selected countries.

In the final session, the results achieved will be analyzed; once the situation of multiple countries belonging to various geographical and economic contexts has been presented, the different groups will be invited to write, in a maximum of 500 words and as a conclusion to this step, a reflection on the need to move jointly towards the achievement of the SDGs, seeking information on whether current efforts and cooperation are sufficient to achieve the goals by 2030.

4.6.4. Phase 4. Introduction to Practice

Step 1

- Estimated sessions: 2
- Associated Rosenshine's Principles: (2) present new information in small steps, (3) ask questions, and (5) guide the student's practice

In this fourth phase, the student should already have internalized most of the key concepts of the learning situation. Thus, after a brief presentation on the notions of mitigation and adaptation, the following activities will be proposed to the groups formed in the previous phase:

a. After consulting the section referring to environmental risks in the *National Atlas of Spain* (IGN 2022b), as well as the one dedicated to environmental quality and impact (IGN 2022c), what are the main environmental risks in Spain as a whole? And in the Canary Islands?
b. Which of the risks included in the atlas do you think are most related to climate change?
c. How do you think the exposed environmental problems hinder the achievement of the SDGs in Spain and the Canary Islands?

d. What do you think would be the mitigation and adaptation measures that could be adopted in the face of these risks?

The fourth question will be the basis for the individual practice included in phase 6, so it will be emphasized that it is important not to go into the question in depth, but only to sketch some elementary ideas. At the same time, it should be noted that in the fifth phase this group exploration work will be reviewed.

Step 2

- Estimated sessions: 2
- Associated Rosenshine's Principles: (2) present new information in small steps, (3) ask questions, and (7) obtain a high success rate

In the second session, the content of Law 6/2022, of December 27, on climate change and energy transition of the Canary Islands (BOC No. 257, of December 31 2022) will be analyzed, encouraging the students' systemic and critical thinking through the following questions:

a. Do you think that the law is aligned with the SDGs?
b. Do you consider that, in general, the measures proposed to address climate change in the islands are sufficient?
c. In the process of drafting the law, an opinion survey was conducted among the Canary Islands population (Consejería de Transición Ecológica 2021—Department of Ecological Transition, Fight against Climate Change and Territorial Planning of the Government of the Canary Islands). Is there anything that strikes you about the public perception?

Finally, based on the work performed by the students and on the main conclusions drawn about the aforementioned law, the results will be discussed in class. In addition, in order for the students to feel prepared for the review that will take place in the fifth phase, a brief overview of the main concepts exposed throughout the learning situation will be carried out.

4.6.5. Phase 5. Learning Review

Step 1

- Estimated sessions: 3
- Associated Rosenshine's Principles: (6) check the student's understanding and (10) weekly and monthly review

In order to check the degree of student understanding and to know if the appropriate outlines have been established, the students will repeat the *Kahoot* quiz proposed at the beginning of the subject. Afterwards, they will have to retrieve the mind maps they created in the first weeks of work, so that they can modify, add, eliminate, clarify or, in short, enrich the previously created outlines. Then, by consensus, the students will establish the common elements to be included in all the outlines—for example, causes and consequences of climate change, SDGs related to this phenomenon, incidence in the Canary Islands, etc.

Once the learners have finished modifying their own mind maps, they will have to send them to the teacher, who will now—and not in the first version—evaluate the result. Likewise, the initial and final mind maps of those students who volunteer will be presented, giving way to a brief discussion in which misconceptions and confusions that existed at the beginning of the topic and that now, after the instruction process, have been clarified, can be identified. In addition, the teacher will dedicate a session to analyze with the students the main elements highlighted by the different groups in the analysis delivered in writing in phase 4.

4.6.6. Phase 6. Autonomous Practice

Step 1

- Estimated sessions: 4

- Associated Rosenshine's Principle: (9) independent practice

The first session will be devoted to the explanation of the independent practice to be carried out by the students. This will consist of the elaboration of a relatively short written work—approximately seven pages, not including cover page, table of contents, introduction and bibliography—entitled Canary Islands strategy for sustainable development, mitigation and adaptation to climate change. The aim is for students to capture in this document the knowledge they have acquired about the SDGs and climate change, applying it to the Canary Islands scale in the form of a strategy where, in addition to clarifying the basic notions linked to the subject, they propose concrete actions at least three levels: public administrations, companies, and citizens. The title of this learning situation—Climate change and sustainability: from science to our daily life—summarizes what is expected from students with this practice: that they are able, starting from an adequate theoretical basis, to provide simple solutions, to be applied in their daily lives and in the territory where they live, to the environmental problems described by the scientific community. The structure, similar to that of a scientific article, should be as follows:

1. Cover page, table of contents, and introduction, which will include an express reference to the objectives of the strategy at the end.
2. Geographical context of the Canary Islands: physical and human elements.
3. Climate change: causes and consequences. The reality of Spain and the Canary Islands.
4. Mitigation: concept and measures to be applied.
5. Adaptation: concept and measures to be applied.
6. The SDGs as an instrument to achieve resilience: concept, relationship with climate change, and link with the proposed measures.
7. Timing and monitoring: time horizon for implementation of the plan and measures or entities in charge of ensuring compliance.
8. Discussion: examples of strategies approved in other territories, possible limitations of the proposed measures, legal framework of the actions at European and national levels, etc.
9. Conclusions and personal assessment.
10. Bibliography.

The importance of consulting various sources and resources will be emphasized, which should be properly cited in the final bibliography section following the APA style (APA 2021). In order to facilitate the preparation of the work, students will be given a list of accessible resources that they can use to support their arguments, such as the National Plan for Adaptation to Climate Change 2021-2030 (MITECO 2020) or all the regulations, strategies, and climate action plans approved by the Government of the Canary Islands (Canarias por la Transición Ecológica 2022).

Thus, starting from a diagnosis of the situation and the reality of the islands, students will have to put into practice the geographic learning they have acquired throughout their academic career in ESO and baccalaureate, since they will have to make a brief description of the geographic context of the islands, including their physical elements—climate, geomorphology, fauna and flora, etc., and human elements—demography, urban configuration, etc. In addition, they will be invited to work with GIS, having to include, at least, a map of their own elaboration referring to any aspect—location of the islands, climate change, proposed measures, etc. Taking into account that the class will have already worked with the *GoConqr* tool, the inclusion of graphs, diagrams, and other types of figures that visually enrich the final product will be positively valued. Regarding the attention to diversity, for students with specific needs, the GIS task will be optional and, instead of seven pages, the extension will be reduced to five.

The assignment, although individual, will be guided at all times in the classroom by the teacher, who will provide materials, resolve doubts, and help students to focus their own strategy. Taking into account that in the sixth step there will be a blind peer review, it is important that students do not write their name inside the document nor include any data that could identify them, in order to guarantee the anonymity of the author and an impartial evaluation.

At the end of each session, fifteen minutes will be allocated to check the students' understanding and verify their progress in writing the report, being able to identify their main difficulties in solving the problem.

4.6.7. Phase 7. Presentation and Evaluation of Results

Step 1

- Estimated sessions: 1
- Associated Rosenshine's Principle: (10) weekly and monthly review

The students will be invited to elaborate a brief presentation where they will indicate the main lines of the work carried out in the previous step.

Taking into account the level of intellectual maturity of a student who is on the threshold of university, vocational training, or the world of work, once this task has been completed, a blind peer review will be carried out. At random, the teacher will assign each student to correct the work of another student, whose name will not be revealed. A simple grid (Table 6) such as the following will be provided for the correction of the strategies:

Table 6. Grid for blind peer review.

Element	Not Completed	Partially Completed	Acceptable	Excellent
1. Writing, spelling, and grammatical expressions denote maturity and are appropriate				
2. Complies with the standards of the work in terms of structure, citation, format, and length				
3. The definitions provided are clear and come from rigorous sources				
4. The proposed measures are realistic and sufficient to meet the indicated objectives and are aimed at the three specified levels: administration, business, and citizenship				
5. The SDGs are well linked to climate change and, in general, are present throughout the strategy				

Source: own preparation.

Each student must send this grid to the teacher, with a brief global evaluation of the strategy analyzed—about half a page—justifying the evaluation made, which, in any case, will be anonymous in the eyes of other students. Obviously, the implementation of this technique requires a detailed explanation of the criteria that should govern the evaluation, including the need to provide a proactive and improving stance—not disqualifying or destructive—and the importance of being rigorous and fair in the evaluation of their peers.

Step 2

- Estimated sessions: 1
- Associated Rosenshine's Principles: (6) check the student's understanding and (10) weekly and monthly review

In the last two sessions of the learning situation, each student, for 5–7 min, will expose his or her presentation on the strategy, having, in addition, to fill in a self-evaluation questionnaire on all the work developed throughout the learning situation. This form should also include questions referring to the degree of satisfaction with the designed activities. Some of the questions could be similar to the following:

a. Do you think that the contents and activities designed have been useful?
b. From 0 to 10, how much have you learned throughout the topic?
c. As a result of the contents raised, has your perception about climate change and sustainability changed?
d. From 0 to 10, how do you rate your personal performance in this topic on climate change and the SDGs?
e. Have you participated in the sessions, contributing ideas, or raising doubts or other questions of interest?
f. Have you handed in all the programmed activities?
g. Have you consulted the resources and bibliographic sources provided in class to solve the problems?
h. Have you modified any daily habits as a result of what has been said about the SDGs and climate change? Do you intend to do so?
i. Taking into account your performance, if you had to give yourself a grade in this topic, what would it be?
j. Which topic was most interesting to you? (Causes of climate change, consequences, climate change in the Canary Islands, international inequalities, the concepts of adaptation, mitigation and resilience, the SDGs, etc.)?

At the end of this activity, the teacher will open a round of discussion on what was raised both in the presentations and in the corrections of the peer evaluation, concluding with a brief mention of the aspects that the teacher considers appropriate to highlight on the topic, the learning process, or the progression of the students.

In order to guarantee the continuity and consolidation of the knowledge worked on, as a suggestion, it is proposed that the next topic to be addressed could be linked to the third group of basic knowledge contemplated in the draft Canarian curriculum, adapted to the LOMLOE, of the subject of geography of the 2nd year of baccalaureate, which refers to "land use planning in the ecosocial approach." In this way, by deepening the issues worked on in this unit, it will be possible to analyze issues such as the sustainability of cities, mobility, urban structure, land occupation in a limited space such as the island, etc., in accordance with the provisions established by the regional authorities of the Canary Islands in the aforementioned draft decree. Likewise, and given the importance of dealing with the SDGs in a cross-cutting manner and common to all subjects, the integrity of the content of the subject should be linked, in one way or another, to the sustainability goals set by international organizations, so that the climate emergency is worked on from multiple perspectives.

5. Discussion

The designed learning situation delves into the pedagogical and geographical research initiated after the approval of the SDGs and the consolidation of the concept of education for sustainability. Meeting the goals set by the international community in the field of education can only be a reality if the teaching staff is capable of integrating contents and activities that encourage students to think critically and commit to the preservation of the planet into their daily classroom routine. As Murga-Menoyo (2021) points out, "education [...] as a transforming social force follows its course between possibilism and utopia," so that, in order to avoid repeating the mistakes and failures committed in other previous initiatives—Education For All movement, Millennium Development Goals, etc. (UNESCO 2016)—practical and updated contents are required to help teachers to place themselves more in the realm of possibilism and less in that of utopia.

All of this acquires great importance given the instability of Spanish education and other realities, such as the fact that, for example, geography is included in the same specialty as history, a circumstance that frequently causes geographic content to be relegated to a secondary position, and that a high proportion of teachers stick to a traditional methodology limited to theoretical issues, without integrating innovative activities or those related to current challenges into their daily practice (Peña-Gallardo et al. 2020).

Moreover, in a context of a significant decrease in the number of students enrolled in university studies of geography (Ministry of Universities 2022), it is urgent to adopt new approaches to bring the discipline closer to society and make it more attractive to students, abandoning traditional methods obsessed with the physical and political description of territories to enter a new approach linked to sustainability, work with updated information sources, and the integration of GIS in the classroom.

In this sense, although since the 1990s it has been claimed that technology has revolutionized geography as a discipline and as a school subject (Lemberg and Stoltman 1999), there is still some reluctance in Spain to fully integrate GIS in the classroom, even though there is no lack of didactic proposals (Boix and Olivella 2007; Martínez et al. 2016). In light of this situation, this proposal seeks to contribute to filling this gap in the Spanish educational system.

In the 1980s, it was asserted from the academic world that there was an inadequacy of geography school programs in relation to the real needs of teaching, the dissemination of ideas associated with the "new geographies," the influence of disciplines such as ecology, the new didactic approaches, and the appearance of new materials and resources of interest (García-Pérez 1987). With no intention of adopting a pessimistic perspective, it is axiomatic that some of the deficiencies identified then are still fully valid, especially when the educational world in general is in a situation of continuous disruption and change, and geography must adapt to what UNESCO defines as the end of a historical cycle that initiates the shaping of new educational patterns (International Commission on the Futures of Education 2021).

As professor Monge states, classrooms must be turned into SDG cultivation fields (Monge 2019). To this end, this learning situation seeks to equip students with the key competencies for sustainable development highlighted by UNESCO: systemic and critical thinking, self-awareness, anticipation, normative competence, problem solving, etc. This explains the diversity of activities proposed in the learning situation: analysis of legal documents, work with GIS, reflection on international inequalities, etc.

Therefore, the learning situation presented above seeks to contribute to the adaptation of teaching–learning processes to the international and ethical requirements demanded by modern challenges, in line with initiatives such as the PRADO Guide, a reference document in the pedagogical field that, in a similar way to the present paper, provides examples of activities associated with the SDGs. This guide pursues, in short and in a way that coincides with this proposal, the ultimate goal of helping teaching practices to shape citizens "actively committed to ecosocial sustainability" (Murga-Menoyo and Bautista-Cerro 2019).

It is worth noting that, in addition to the profuse academic reflection on the need to reorient curricula, methodologies, and teaching approaches towards instruction in competencies for sustainability and the adoption of a biocentric educational approach (Araya 2009; Granados and Medir 2021; Corrales 2021), in recent years there has been an increase in educational innovation projects that, from different approaches and disciplines, encourage the implementation of activities and dynamics that involve the entire educational community in a collective project of respect for the environment and environmental awareness (Consejería de Educación, Universidades, Cultura y Deportes del Gobierno de Canarias 2022; Consejería de Desarrollo Educativo y Formación Profesional de la Junta de Andalucía 2022; Consejería de Educación y Cultura del Gobierno de La Rioja 2022).

The proposed learning situation is based on Rosenshine's method, whose instructional principles are increasingly applied to geographical contents (Buzo 2021; De Lázaro and Puertas 2022). Notwithstanding this, it should be noted that the design and timing of the proposed activities are only intended as a guide that can be freely used by teachers of any subject and educational level in a flexible manner, selecting those aspects considered useful according to the needs of the group, the previous learning and concerns of the students, the legislation of each autonomous community, and the objectives included in the didactic programming documents. Although there are other methodologies that can be used for the configuration of learning situations, Rosenshine offers a way of working based on simple steps that can be applied to different contexts and educational profiles, a circumstance that makes this didactic proposal totally or partially replicable in various contexts. At the same time, it is an approach that facilitates attention to diversity.

Among the limitations of the didactic proposal, it can be pointed out that it may pose overly ambitious objectives considering the reality that the secondary and baccalaureate teacher has to assume: students with specific educational needs, generalized low academic level, lack of interest on the part of large sectors of the student body, scarcity of time and means to develop the pedagogical activity in an adequate manner, etc. However, as mentioned above, the resources and tasks mentioned can—and should—be adapted to the particular characteristics of the students and subject where the learning situation is to be implemented.

Among future lines of research, it would be worth mentioning the interest in developing work that responds to the necessary attention to diversity in geography classrooms, as well as proposals to adapt the instruction of geographic content to the teaching of adults and to integrate geography in the concept, increasingly in fashion, of lifelong learning (Belando-Montoro 2017).

6. Conclusions

The teaching of geography is acquiring an enormous importance nowadays. Contents such as climate change, sustainable development, resilience, and adaptation have been included in educational regulations as transversal elements that should permeate the daily practice of all subjects. Therefore, it is essential to provide teachers of this and other specialties with resources and didactic proposals that facilitate the teaching–learning process and increase the environmental awareness of students who, in turn, demand knowledge on these issues in order to position themselves in a society that is increasingly concerned and informed about the ecological issue. In short, education, as a tool for transformation and social change, is today a field of unquestionable interest for effectively addressing contemporary challenges that cannot be tackled with obsolete methodologies.

To meet these objectives, specific activities are required that, like the learning situation proposed in this paper, place students in a position of active and autonomous work that allows them to acquire proactive competencies and skills to contribute, from their daily lives, to the objectives set by the authorities in terms of sustainability. At the same time, this should be performed without ignoring the importance of instructing students in other elementary geographic knowledge that all students should have at the end of their time in the educational system. Likewise, working with GIS, the analysis of regulatory documents, and the handling of official databases are nowadays basic and priority elements that must be integrated, without delay, in the classrooms of geography, a discipline that must respond, as it has already done in other complex historical moments, to the main social challenges.

Author Contributions: Conceptualization, J.C.-G. and A.L.-D.; data analysis, J.C.-G., A.L.-D. and N.M.-R.; methodology, J.C.-G. and A.L.-D.; visualization, A.L.-D.; writing—original draft, J.C.-G., J.D.-P. and A.L.-D.; writing—review and editing, A.L.-D., J.C.-G., J.D.-P. and N.M.-R. All authors have read and agreed to the published version of the manuscript.

Funding: This research received no external funding.

Institutional Review Board Statement: Not applicable

Informed Consent Statement: Not applicable.

Data Availability Statement: Data are contained within the article.

Conflicts of Interest: The authors declare no conflict of interest.

Appendix A. Workshop: Sources and Tools for the Study of Climate Change

The following is a list of viewers and interactive resources that students should use to reflect on various questions:

1. **IPCC Interactive Atlas**

   Access to the resource: https://interactive-atlas.ipcc.ch/, accessed on 9 February 2023.

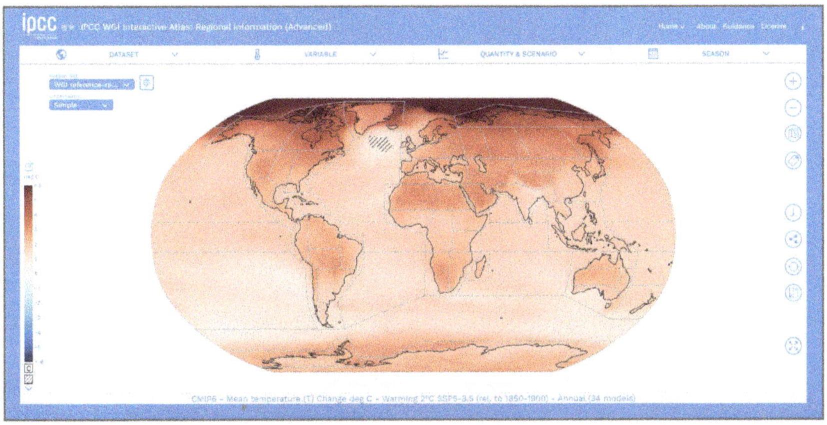

Figure A1. Overview of the resource. Source: Gutiérrez et al. (2021).

Questions for reflection and reinforcement of concepts:

a. According to the various projections, which hemisphere will experience the greatest warming by the end of the century?
b. At what times of the year will the greatest thermal increase be experienced?
c. Observe the differences between ocean warming according to scenarios.
d. Where will sea levels rise the most?
e. Observe the changes in sea pH according to scenarios (variable: ph at surface).
f. Observe the retreat of ice cover by the end of the century according to the worst-case scenario (variable: sea ice concentration).

2. **Climate Change Scenario Viewer (Adaptecca.es)**

   Access to the resource: https://escenarios.adaptecca.es/, accessed on 9 February 2023.
   Questions for reflection and reinforcement of concepts:

a. What is expected to increase more in the Iberian Peninsula, the minimum or maximum temperature (select distant future)? Link it to the evolution of the thermal amplitude variable in degrees.
b. What is the approximate maximum duration of heat waves in the Community of Madrid and the Basque Country according to historical records? And according to the RCP 8.5 scenario?
c. At the end of the century (select distant future), which autonomous communities will see the greatest decrease in precipitation?
d. Regarding the maximum 24 h precipitation variable (select relative anomaly), what do the results suggest to you? What about potential evapotranspiration?

1. **Sea Level Projections Tools (NASA)**

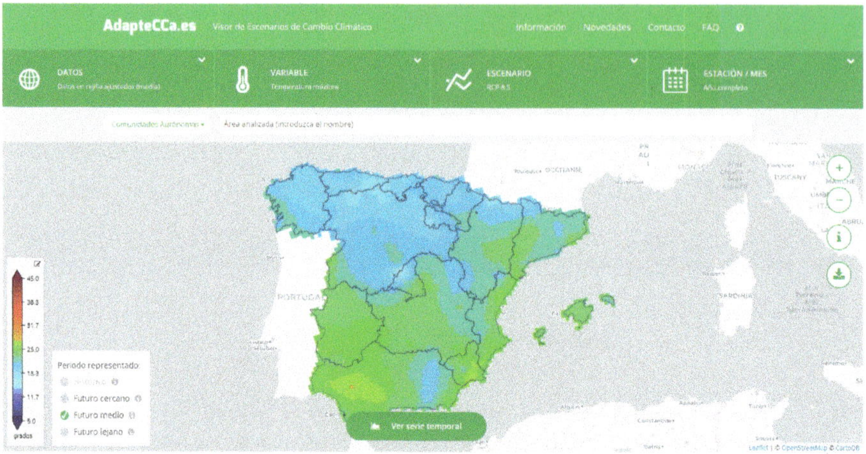

Figure A2. Overview of the resource. Source: Adaptecca.es (2021).

Access to the resource: https://sealevel.nasa.gov/ipcc-ar6-sea-level-projection-tool, accessed on 9 February 2023.

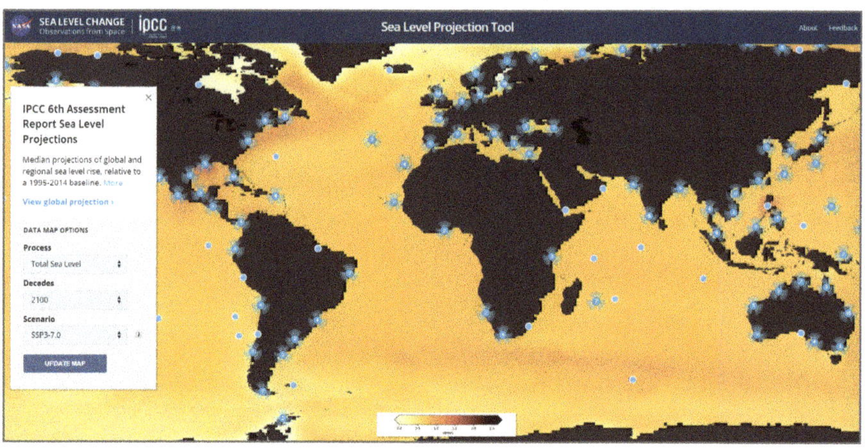

Figure A3. Overview of the resource. Source: NASA (2021).

Questions for reflection and reinforcement of concepts:

a. According to the worst-case scenario, SSP5-8.5, by how many centimeters will sea level rise by the end of the century (2100) in Tenerife, and in other parts of the Canary Islands, and in Barcelona?

b. In which areas of the world will sea level rise the most? Give an example.

2. Risk Zone Map (CLIMATE CENTRAL)

Access to the resource: https://ss2.climatecentral.org, accessed on 9 February 2023.

Figure A4. Overview of the resource. Source: Climate Central (2023).

Questions for reflection and reinforcement of concepts:

a. Observe what would happen in the Canary Islands with an increase of 1, 2 and 5 m in sea level and compare it with the Spanish peninsular coast and, in general, the European coastline. Is there anything that draws your attention?

3. **Coastal Erosion Viewer (GRAFCAN)**

Access to the resource: https://grafcan1.maps.arcgis.com/apps/webappviewer/index.html?id=a1bc45dd09994ac1979479fcff4db989, accessed on 9 February 2023.

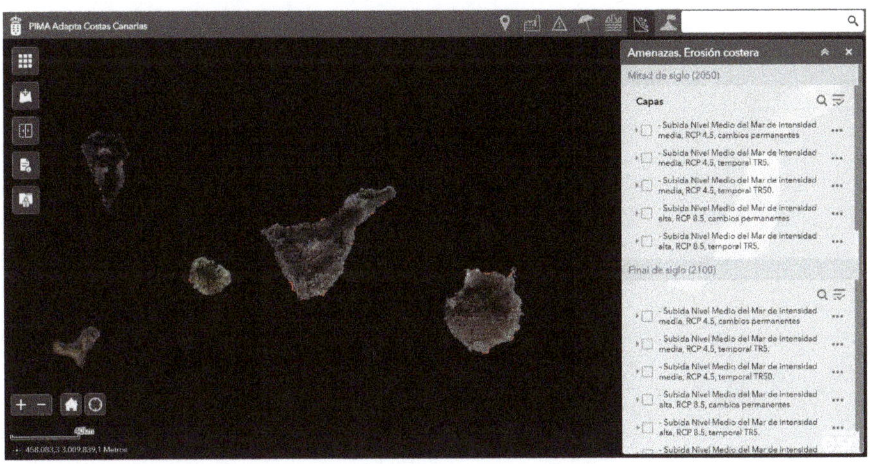

Figure A5. Overview of the resource. Source: GRAFCAN (2022a).

Questions for reflection and reinforcement of concepts:

a. In the Canary Islands, which coastal locations would be most affected by sea level rise? Name at least one place on each island.

b. Reflect on the impact that what is represented in the viewer could have on tourism in the archipelago.

4. **Natural Hazards Viewer (GRAFCAN)**

Access to the resource: https://visor.grafcan.es/visorweb/, accessed on 9 February 2023.

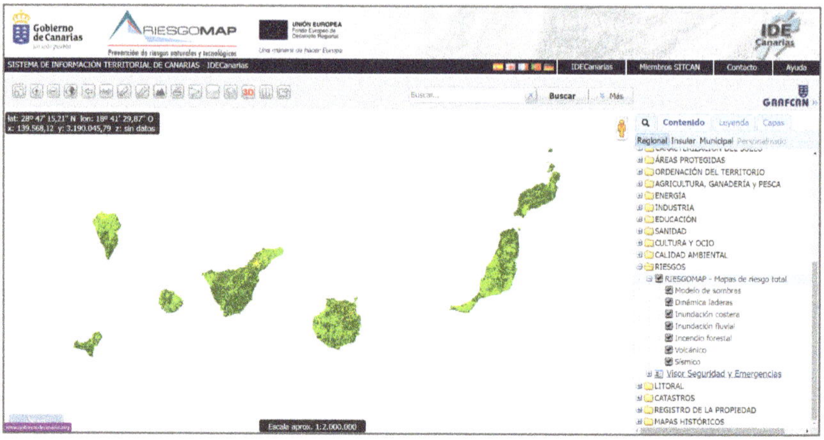

Figure A6. Overview of the resource. Source: GRAFCAN (2022b).

Questions for reflection and reinforcement of concepts:

a. Selecting the risk layer included in the left side menu, click on your area of residence, or any space of interest, and analyze the risk situation.

References

ABC Definition. 2022. Definición de prevención. Available online: https://www.definicionabc.com/general/prevencion.php (accessed on 9 February 2023).

Adaptecca.es. 2021. Visor de escenarios de cambio climático. Available online: https://escenarios.adaptecca.es (accessed on 9 February 2023).

AEMET. 2018. Meteoglosario visual. Diccionario ilustrado de meteorolog ía. Available online: https://meteoglosario.aemet.es/ (accessed on 9 February 2023).

AEMET. 2011. Proyecciones climáticas para el siglo XXI. Available online: https://www.aemet.es/es/serviciosclimaticos/cambio_climat (accessed on 9 February 2023).

Álvarez-Ubría, Pilar P., Leopoldo Rojo, and Zamorano Cristina. 2007. La desertificación en España. *Ambienta*, 60–61. Available online: https://www.mapa.gob.es/ministerio/pags/Biblioteca/Revistas/pdf_AM%2FAM_2007_70_60_61.pdf (accessed on 9 February 2023).

Amico, Bárbara, Andrea Oliveira, and Ileana Zeler. 2020. La comunicación de la emergencia climática en las redes sociales de los activistas ambientales: El caso de Greta Thunberg, Leonardo DiCaprio y Arturo Islas Allende. *Pensar la Publicidad* 14: 281–90. [CrossRef]

Anderson, Allison. 2013. Climate Change Education for Mitigation and Adaptation. *Journal of Education for Mitigation and Adaptation* 6. [CrossRef]

APA. 2021. Guía Normas APA. 7ª Edición. Available online: https://normas-apa.org/wp-content/uploads/Guia-Normas-APA-7ma-edicion.pdf (accessed on 9 February 2023).

Aragón-Reyes, Manuel. 2013. Las competencias del Estado y las Comunidades Autónomas sobre educación. *Revista Española de Derecho Constitucional*, 191–99. Available online: https://dialnet.unirioja.es/servlet/articulo?codigo=4391803 (accessed on 9 February 2023).

Araya, Fabián Rodrigo. 2009. Enseñanza de la Geografía para el Desarrollo Sustentable en Chile. *Uni-pluri/versidad* 9. Available online: https://dialnet.unirioja.es/descarga/articulo/7895949.pdf (accessed on 9 February 2023).

Basque Government. n.d. Anexo: Conceptos de peligro, vulnerabilidad y riesgo. Available online: https://trafikoa.bidehezkuntza.org/primaria/pdfs/c74.pdf (accessed on 9 February 2023).

Becoña, Elisardo. 2006. Resiliencia: Definición, características y utilidad del concepto. *Revista de psicopatología y psicología clínica* 11: 125–46. [CrossRef]

Belando-Montoro, María Remedios. 2017. Aprendizaje a lo largo de la vida: Conceptos y componentes. *Revista Iberoamericana de Educación* 75: 219–34. [CrossRef]

Biología-Geología.com. n.d. Factores de riesgo: Peligrosidad, exposición y vulnerabilidad. Available online: https://biologia-geologia.com/geologia/711_factores_de_riesgo.html (accessed on 9 February 2023).

BOC No. 257, of December 31. 2022. Ley 6/2022, de 27 de diciembre, de cambio climático y transición energética de Canarias. Available online: http://www.gobiernodecanarias.org/boc/2022/257/001.html (accessed on 9 February 2023).

BOCM No. 176, of July 26. 2022. Decreto 65/2022, de 20 de julio, del Consejo de Gobierno, por el que se establecen para la Comunidad de Madrid la ordenación y el currículo de la Educación Secundaria Obligatoria. Available online: https://www.bocm.es/boletin/CM_Orden_BOCM/2022/07/26/BOCM-20220726-2.PDF (accessed on 9 February 2023).

BOE No. 76 of March 30. 2022. Real Decreto 217/2022, de 29 de marzo, por el que se establece la ordenación y las enseñanzas mínimas de la Educación Secundaria Obligatoria. Available online: https://www.boe.es/buscar/act.php?id=BOE-A-2022-4975 (accessed on 9 February 2023).

BOE No. 82, of April 6. 2022. Real Decreto 243/2022, de 5 de abril, por el que se establecen la ordenación y las enseñanzas mínimas del Bachillerato. Available online: https://www.boe.es/buscar/act.php?id=BOE-A-2022-5521 (accessed on 9 February 2023).

BOE No. 106, of May 4. 2006. Ley Orgánica 2/2006, de 3 de mayo, de Educación. Available online: https://www.boe.es/buscar/act.php?id=BOE-A-2006-7899 (accessed on 9 February 2023).

BOE No. 121, of May 21. 2021. Ley 7/2021, de 20 de mayo, de cambio climático y transición energética. Available online: https://www.boe.es/diario_boe/txt.php?id=BOE-A-2021-8447 (accessed on 9 February 2023).

BOE No. 295, of December 10. 2013. Ley Orgánica 8/2013, de 9 de diciembre, para la mejora de la calidad educativa. Available online: https://www.boe.es/buscar/pdf/2013/BOE-A-2013-12886-consolidado.pdf (accessed on 9 February 2023).

BOE No. 307, of December 24. 2002. Ley Orgánica 10/2002, de 23 de diciembre, de Calidad de la Educación. Available online: https://www.boe.es/buscar/doc.php?id=BOE-A-2002-25037 (accessed on 9 February 2023).

BOE No. 340, of December 30. 2020. Ley Orgánica 3/2020, de 29 de diciembre, por la que se modifica la Ley Orgánica 2/2006, de 3 de mayo, de Educación. Available online: https://www.boe.es/buscar/doc.php?id=BOE-A-2020-17264 (accessed on 9 February 2023).

Boix, Gemma, and Rosa Olivella. 2007. Los Sistemas de Información Geográfica (SIG) aplicados a la educación. El proyecto PESIG (Portal Educativo en SIG). In *Las competencias geográficas para la educación ciudadana*. Edited by María Jesús Marrón-Gaite, Julia Salom-Carrasco and Xosé Manuel Souto-González. Grupo de Didáctica de la Geografía de la Asociación de Geógrafos Españoles: pp. 23–32. Available online: http://didacticageografia.age-geografia.es/docs/Publicaciones/2007_comp_ecogeo.pdf (accessed on 9 February 2023).

BOJA No. 199, of October 15. 2018. Ley 8/2018, de 8 de octubre, de medidas frente al cambio climático y para la transición hacia un nuevo modelo energético en Andalucía. Available online: https://www.boe.es/buscar/doc.php?id=BOE-A-2018-15238 (accessed on 9 February 2023).

BOPA No. 169, of September 1. 2022. Decreto 59/2022, de 30 de agosto, por el que se regula la ordenación y se establece el Currículo de la Educación Secundaria Obligatoria en el Principado de Asturias. Available online: https://sede.asturias.es/bopa/2022/09/01/2022-06713.pdf (accessed on 9 February 2023).

Buzo, Isaac José. 2021. Aprendizaje inteligente y pensamiento espacial en Geografía. [Tesis doctoral, Universidad Complutense de Madrid]. Available online: https://eprints.ucm.es/id/eprint/70029/1/T43019.pdf (accessed on 9 February 2023).

Caballero, Manuela, Artemio Baigorri, and Mercedes Pardo. 2021. Educación y cambio climático: Una aproximación desde la ESO. *Athenea Digital: Revista de Pensamiento e Investigación Social* 21: e2293. [CrossRef]

Camarillo-Naranjo, Juan Mariano, José Ignacio Álvarez-Francoso, Natalia Limones-Rodríguez, María Fernanda Pita-López, and Mónica Aguilar-Alba. 2019. The Global Climate Monitor System: From Climate Data-Handling to Knowledge Dissemination. *International Journal of Digital Earth* 12: 394–414. [CrossRef]

Canarias por la Transición Ecológica. 2022. *Acción climática*. Available online: https://bit.ly/3FVsaHv (accessed on 9 February 2023).

Cities Footprint Project. 2014. Calcular tu huella de carboNo. Available online: https://huelladeciudades.com/AppHCCali/main.html (accessed on 9 February 2023).

Climate Central. 2023. Risk Zone Map. Available online: https://ss2.climatecentral.org (accessed on 9 February 2023).

CMMA [Comisión Mundial sobre el Medio Ambiente y el Desarrollo de las Naciones Unidas]. 1987. Informe de la Comisión Mundial sobre el Medio Ambiente y el Desarrollo "Nuestro futuro común". Available online: https://www.ecominga.uqam.ca/PDF/BIBLIOGRAPHIE/GUIDE_LECTURE_1/CMMAD-Informe-Comision-Brundtland-sobre-Medio-Ambiente-Desarrollo.pdf (accessed on 9 February 2023).

Consejería de Desarrollo Educativo y Formación Profesional de la Junta de Andalucía. 2022. Programa Aldea, Educación Ambiental para la Comunidad Educativa. Available online: https://www.juntadeandalucia.es/educacion/portals/web/aldea/programa-aldea/-/novedades/detalle/q8RTHRezWwbZ/catalogo-de-programas-para-la-innovacion-educativa-2022-1ay222l8ba6b1 (accessed on 9 February 2023).

Consejería de Educación, Universidades, Cultura y Deportes del Gobierno de Canarias. 2022. Programa de educación ambiental. Available online: https://www.gobiernodecanarias.org/educacion/web/programas-redes-educativas/programas-educativos/educa-ambiental/ (accessed on 9 February 2023).

Consejería de Educación y Cultura del Gobierno de La Rioja. 2022. *Centros Educativos Hacia la Sostenibilidad*. Available online: https://www.larioja.org/edu-innovacion-form/es/actividades-formacion-crie/proyectos-innovacion-educativa-pies/centros-educativos-hacia-sostenibilidad (accessed on 9 February 2023).

Consejería de Transición Ecológica, Lucha contra el Cambio Climático y Planificación Territorial del Gobierno de Canarias. 2021. Estudio de opinión en el proceso de elaboración de la Ley Canaria de Cambio Climático y Transición Energética y sus instrumentos. Resumen ejecutivo. Available online: http://www.gobiernodecanarias.org/medioambiente/piac/descargas/Cambio_climatico/Informacion-Publica/Iniciativas_tramitacion_Ley_de_Cambio_Climatico_de_Canarias/RESUMEN-EJECUTIVO-REV-CONCLUSIONES-ENCUESTA-ONLINE-PROPUESTAS-A-LA-LEY-DE-CAMBIO-CLIMATICO-DE-CANARIAS-1.pdf (accessed on 9 February 2023).

Corrales, Mario. 2021. Educar en sostenibilidad medioambiental desde la asignatura de geografía: Una experiencia práctica. In *Planificación Regional: Paisaje y patrimonio*. Edited by Julian Mora, José Castro and Jacinto (Coord) Garrido. Navarra: Aranzadi, pp. 185–96.

De Lázaro, María Luisa, and Miguel Ángel (Coord) Puertas. 2022. *La geoinformación en la educación para la sostenibilidad integrando los principios de Rosenshine. Situaciones de aprendizaje*. Madrid: UNED.

De Miguel, Mar. 2020. Los ecologistas demandan al Gobierno por incumplir sus compromisos para combatir el cambio climático. *El Mundo*. Available online: https://www.elmundo.es/ciencia-y-salud/ciencia/2020/09/15/5f6088ddfc6c838c6e8b4615.html (accessed on 9 February 2023).

DOCV No. 9486, of December 9. 2022. Ley 6/2022, de 5 de diciembre, de la Generalitat, del cambio climático y la transición ecológica de la Comunitat Valenciana. Available online: https://dogv.gva.es/datos/2022/12/09/pdf/2022_11735.pdf (accessed on 9 February 2023).

DOE No. 164, of August 25. 2022. Decreto 110/2022, de 22 de agosto, por el que se establecen la ordenación y el currículo de la Educación Secundaria Obligatoria para la Comunidad Autónoma de Extremadura. Available online: http://doe.juntaex.es/pdfs/doe/2022/1640o/22040165.pdf (accessed on 9 February 2023).

DOGC No. 7426, of August 3. 2017. Ley 16/2017, de 1 de agosto, del cambio climático. Available online: https://www.boe.es/buscar/act.php?id=BOE-A-2017-11001 (accessed on 9 February 2023).

Erviti, MaríaCarmen. 2020. Del "cambio climático" a la "emergencia climática": Análisis de El País y El Mundo. *Revista Prisma Social*, 65–81. Available online: https://revistaprismasocial.es/article/view/3866 (accessed on 9 February 2023).

European Environment Agency. 2022. *¿Qué diferencia hay entre adaptación y mitigación?* Available online: https://www.eea.europa.eu/es/help/preguntas-frecuentes/que-diferencia-hay-entre-adaptacion#:~:text=En%20resumen,%20la%20adaptaci%C3%B3n%20puede,(GEI)%20a%20la%20atm%C3%B3sfera (accessed on 9 February 2023).

European Investment Bank. 2022. The EIB Climate Survey 2022–23 (Edition V). Dataset of Full Results. Available online: https://www.eib.org/en/surveys/climate-survey/all-resources.htm (accessed on 9 February 2023).

FAO. 2019. Resiliencia climática rural en América Latina. Available online: https://www.fao.org/3/ca4632es/ca4632es.pdf (accessed on 9 February 2023).

Fernández-Mellizo, María. 2019. ¿Es posible el consenso educativo? La opinión pública sobre política educativa en España. *Panorama Social*, 107–24. Available online: https://www.funcas.es/wp-content/uploads/Migracion/Articulos/FUNCAS_PS/030art09.pdf (accessed on 9 February 2023).

Fortang, Amelia. 2021. *Countries and Territories Most Affected by Climate Change Also More Likely to Believe it to Be Personally Harmful*. Yale Programe on Climate Change Communication. Available online: https://climatecommunication.yale.edu/news-events/countries-and-territories-most-affected-by-climate-change-also-more-likely-to-believe-it-to-be-personally-harmful/ (accessed on 9 February 2023).

García-Pérez, Francisco. 1987. Qué Geografía enseñar en la adolescencia. Reflexiones a partir de una propuesta experimental. *Investigación en la Escuela*, 29–40. Available online: https://revistascientificas.us.es/index.php/IE/article/view/9327/8118 (accessed on 9 February 2023).

Germanwatch. 2022. Experts Demand a Phase out of Fossil Fuel Subsidies. Climate Change Performance Index. Available online: https://ccpi.org/country/esp/ (accessed on 9 February 2023).

GoConqr. 2023. *GoConqr*. Available online: https://www.goconqr.com/es (accessed on 9 February 2023).

González-Gaudiano, Edgar. 2007. Educación ambiental: Trayectorias, rasgos y escenarios. *Trayectorias* 25: 124–25. Available online: https://www.redalyc.org/articulo.oa?id=60715120013 (accessed on 9 February 2023).

GRAFCAN. 2022a. Visor del Plan de Impulso al Medioambiente (PIMA) Adapta Costas Canarias. Available online: https://grafcan1.maps.arcgis.com/apps/webappviewer/index.html?id=a1bc45dd09994ac1979479fcff4db989 (accessed on 9 February 2023).

GRAFCAN. 2022b. Sistema de información territorial de Canarias—IDECanarias. Visor web. Available online: https://visor.grafcan.es/visorweb/ (accessed on 9 February 2023).

Granados, Jesús, and Rosa María Medir, eds. 2021. *Enseñar y aprender geografía para un mundo sostenible*. Ulzama: España.

Gutiérrez, José Manuel, Richard Jones, Gemma Narisma, Lincoln Alves, Muhammad Amjad, Irina Gorodetskaya, Michael Grose, Nana Ama Klutse, Svitlana Krakovska, Jian Li, and et al. 2021. Atlas. In *Climate Change 2021: The Physical Science Basis. Contribution of Working Group I to the Sixth Assessment Report of the Intergovernmental Panel on Climate Change*. Edited by Valerie Masson-Delmotte, Panmao Zhai, Anna Pirani, Sarah Péan Connors, Berger Clotilde, Caud Sophie, Chen Nada, Goldfarb Yang, Gomis Melissa Leah and Ozge Yeleckçi. Cambridge: Cambridge University Press. Available online: http://interactive-atlas.ipcc.ch/ (accessed on 9 February 2023).

IGN. 2022a. Longitud de la línea de costa española por provincias. Available online: https://www.ign.es/web/ane-datos-geograficos/-/datos-geograficos/datosGenerales?tipoBusqueda=longCosta (accessed on 9 February 2023).

IGN. 2022b. Riesgos ambientales. Available online: https://atlasnacional.ign.es/wane/Riesgos_ambientales (accessed on 9 February 2023).

IGN. 2022c. Calidad e impacto ambiental. Available online: https://atlasnacional.ign.es/wane/Calidad_e_impacto_ambiental (accessed on 9 February 2023).

INE. 2018. Disponibilidad y origen del agua por comunidad autónoma y periodo. Available online: https://www.ine.es/jaxi/Tabla.htm?path=/t26/p067/p03/serie/l0/&file=02001.px&L=0 (accessed on 9 February 2023).

INE. 2022. Cifras oficiales de población resultantes de la revisión del Padrón municipal a 1 de enero. Resumen por comunidades autónomas. Available online: https://www.ine.es/jaxiT3/Datos.htm?t=2853 (accessed on 9 February 2023).

Inter-Agency Network for Education in Emergencies. 2022. Glosario de términos. Exposición. Available online: https://inee.org/es/eie-glossary/exposicion (accessed on 9 February 2023).

International Commission on the Futures of Education. 2021. Avances recientes de la Comisión Internacional sobre los Futuros de la Educación. Available online: https://unesdoc.unesco.org/ark:/48223/pf0000375746_spa.locale=en (accessed on 9 February 2023).

IPCC. 2013. Glosario [Planton, S. (ed.)]. In *Cambio Climático 2013. Bases físicas. Contribución del Grupo de trabajo I al Quinto Informe de Evaluación del Grupo Intergubernamental de Expertos sobre el Cambio Climático*. Edited by Valerie Masson-Delmotte, Panmao Zhai, Anna Pirani, Sarah Péan Connors, Berger Clotilde, Caud Sophie, Chen Nada, Goldfarb Yang, Gomis Melissa Leah and Ozge Yeleckçi. Cambridge: Cambridge University Press.

IPCC. 2018. Glosario [Matthews J.B.R. (ed.)]. In *Calentamiento global de 1,5 °C, Informe especial del IPCC sobre los impactos del calentamiento global de 1,5 oC con respecto a los niveles preindustriales y las trayectorias correspondientes que deberían seguir las emisiones mundiales de gases de efecto invernadero, en el contexto del reforzamiento de la respuesta mundial a la amenaza del cambio climático, el desarrollo sostenible y los esfuerzos por erradicar la pobreza*. Edited by Valérie Masson-Delmotte, Panmaom Zhai, Hans-Otto Pörtner, Debra Roberts, Jim Skea, Priyadarshi Shukla, Anna Pirani, Wilfran Moufouma-Okia, Clotilde Péan, Roz Pidcock and et al. Available online: https://www.ipcc.ch/site/assets/uploads/sites/2/2019/10/SR15_Glossary_spanish.pdf (accessed on 9 February 2023).

Lemberg, David, and Joseph Stoltman. 1999. Geography Teaching and the New Technologies: Opportunities and Challenges. *Journal of Education* 181: 63–76. [CrossRef]

Linés, Aalberto. 2010. Clima y cambio climático. *RAM (Revista del Aficionado a la Meteorología)*. Available online: https://www.divulgameteo.es/fotos/lecturas/Clima-CC-Lin%C3%A9s.pdf (accessed on 9 February 2023).

Lynas, Mark, Benjamin Houlton, and Simon Perry. 2021. Greater than 99% consensus on human caused climate change in the peer-reviewed scientific literature. *Environmental Research* 16: 114005. [CrossRef]

Martínez, Carlos, Antonio Iborra, Juan Pérez, and Carmen Figueras. 2016. El uso de SIG de software libre en una práctica de Biología y Geología de 4º de ESO: Los ecosistemas. *Didáctica de las Ciencias Experimentales y Sociales*, 103–16. Available online: http://hdl.handle.net/10550/55803 (accessed on 9 February 2023).

Maxwell, Janie, and Grant Blashki. 2016. Teaching about climate change in medical education: An opportunity. *Journal of Public Health Research* 5: 673. [CrossRef]

Ming-Lee, Tien, Ezra Markowitz, Peter Howe, Chia-Ying Ko, and Anthony Leiserowitz. 2015. Predictors of public climate change awareness and risk perception around the world. *Nature Climate Change* 5: 1014–20. [CrossRef]

Ministry of Education and Vocational Training. 2022a. Currículos en las Comunidades Autónomas. Available online: https://educagob.educacionyfp.gob.es/curriculo/curriculo-lomloe/curriculos-ccaa.html (accessed on 9 February 2023).

Ministry of Education and Vocational Training. 2022b. Calendario de implantación de la LOMLOE. Available online: https://educagob.educacionyfp.gob.es/lomloe/calendario-implantacion.html (accessed on 9 February 2023).

Ministry of Universities. 2022. Estadística de estudiantes. Available online: https://www.universidades.gob.es/estadistica-de-estudiantes/ (accessed on 9 February 2023).

MITECO [Ministry for Ecological Transition and the Demographic Challenge]. 2019a. Emisiones de GEI por Comunidades Autónomas a partir del Inventario Español—Serie 1990–2019. Available online: https://www.miteco.gob.es/es/calidad-y-evaluacion-ambiental/temas/sistema-espanol-de-inventario-sei-/emisionesgeiporccaaserie1990-2019_tcm30-524644.pdf (accessed on 9 February 2023).

MITECO [Ministry for Ecological Transition and the Demographic Challenge]. 2019b. Anuario de Estadística Forestal. Available online: https://www.miteco.gob.es/es/biodiversidad/estadisticas/aef2019_completo_estandar_tcm30-534526.pdf (accessed on 9 February 2023).

MITECO [Ministry for Ecological Transition and the Demographic Challenge]. 2022. Normativa relacionada con Cambio Climático. Available online: https://www.miteco.gob.es/es/cambio-climatico/legislacion/documentacion/normativa-y-textos-legales/default.aspx#:~:text=Real%20Decreto%201089/2020,%20de,en%20el%20periodo%202021-2030 (accessed on 9 February 2023).

MITECO [Ministry for Ecological Transition and the Demographic Challenge], ed. 2020. Plan Nacional de Adaptación al Cambio Climático 2021–30. Available online: https://www.miteco.gob.es/es/cambio-climatico/temas/impactos-vulnerabilidad-y-adaptacion/pnacc-2021-2030_tcm30-512163.pdf (accessed on 9 February 2023).

Monge, Cristina. 2019. Hacer de las aulas campos de cultivo de los ODS. Forum Aragón: Revista digital de FEAE-Aragón sobre organización y gestión educative. 6-8. Available online: https://dialnet.unirioja.es/descarga/articulo/8077542.pdf (accessed on 9 February 2023).

Morote, Álvaro-Francisco, and Jorge Olcina. 2021. La importancia de la enseñanza del cambio climático. Propuestas didácticas para la Geografía escolar. *Estudios Geográficos* 82: e078. [CrossRef]

Murga-Menoyo, María Ángeles. 2015. Competencias para el desarrollo sostenible: Las capacidades, actitudes y valores meta de la educación en el marco de la Agenda global post-2015. *Foro de Educación* 13: 55–83. [CrossRef]

Murga-Menoyo, María Ángeles Á. 2021. La educación en el Antropoce No. Posibilismo versus utopía. *Teoría de la Educación. Revista Interuniversitaria* 33: 107–28. [CrossRef]

Murga-Menoyo, María Ángeles, and Maria José Bautista-Cerro, eds. 2019. *Guía PRADO. Sostenibilizar el currículo de la Educación Secundaria.* Madrid: UNED. [CrossRef]

NASA. 2021. Sea Level Projection Tool. Available online: https://sealevel.nasa.gov/ipcc-ar6-sea-level-projection-tool (accessed on 9 February 2023).

National Association of Safety Directors. 2020. Riesgo: Peligrosidad, exposición y vulnerabilidad. Available online: https://directoresdeseguridad.es/2020/05/08/riesgo-peligrosidad-exposicion-y-vulnerabilidad/ (accessed on 9 February 2023).

Novella, Carlos, and Alexis Cloquell. 2021. Falta de consenso e inestabilidad educativa en España. *Revista Complutense de Educación* 33: 521–29. [CrossRef]

OJEU No. 243 of June 27. 2022. Recomendación del Consejo de 16 de junio de 2022 relativa al aprendizaje para la transición ecológica y el desarrollo sostenible. Available online: https://www.boe.es/buscar/doc.php?id=DOUE-Z-2022-70040 (accessed on 9 February 2023).

Our World in Data. 2020. GDP per Capita, 2020. Available online: https://ourworldindata.org/grapher/gdp-per-capita-worldbank?time=latest (accessed on 9 February 2023).

Peña-Gallardo, María, Mario Ferreras-Listán, and Olga Moreno-Fernández. 2020. Experiências e reflexões dos alunos do mestrado em ciências sociais sobre o ensino da Geografia nas salas de aula do ensino médio. Signos Geográficos—Boletim NEPEG de Ensino de Geografía, 2. Available online: https://revistas.ufg.br/signos/article/view/64276 (accessed on 9 February 2023).

RAE. 2022a. Diccionario Panhispánico del Español Jurídico. Desarrollo sostenible. Available online: https://dpej.rae.es/lema/desarrollo-sostenible (accessed on 9 February 2023).

RAE. 2022b. Diccionario Panhispánico del Español Jurídico. Prevención. Available online: https://dle.rae.es/prevenci%C3%B3n (accessed on 9 February 2023).

Rosenshine, Barak. 2010. Principles of instruction. International Academy of Education and International Bureau of Education. Available online: http://www.ibe.unesco.org/fileadmin/user_upload/Publications/Educational_Practices/EdPractices_21.pdf (accessed on 9 February 2023).

Sachs, Jeffrey, Guillaume Lafortune, Christian Kroll, Grayson Fuller, and Finn Woelm. 2022. *From Crisis to Sustainable Development: The SDGs as Roadmap to 2030 and Beyond. Sustainable Development Report 2022*; Cambridge: Cambridge University Press. Available online: https://dashboards.sdgindex.org/map (accessed on 9 February 2023).

Sánchez-Almodóvar, Esther, Jorge Olcina, Javier Martí, Antonio Oliva, and Samuel Biener Camacho. 2022. La enseñanza del clima, el cambio climático y los extremos atmosféricos: Propuestas para la ciudadanía desde la universidad. In *Ensino de Geografia e a Redução do Risco de Desastres em espaços urbanos e rurais*. Edited by Lourenço Magnoni, María Da Graça, Mayleen Cabral, Mardilson Machado, Claudio Eduardo De Castro, Luciana De Resende, Roberto Serrano-Notivoli, Wellington Dos Santos, Humberto Alves and José Ignacio Prieto. São Paulo: CPS.

Sherrington, Tom. 2019. *Rosenshine's Principles in Action*. Woodbridge: John Catt Educational.

Siperstein, Stephen, Shane Hall, and Stephanie LeMenager, eds. 2017. *Teaching Climate Change in the Humanities*. New York: Routledge, p. 112.

Torices, Alfonso. 2020. Un Congreso fracturado aprueba la 'ley Celaá'. El Correo. Available online: https://www.elcorreo.com/sociedad/educacion/congreso-aprueba-lomloe-ley-celaa-20201119202757-ntrc.html (accessed on 9 February 2023).

Toro, Francisco Javier. 2007. El desarrollo sostenible: Un concepto de interés para la Geografía. *Cuadernos Geográficos*, 149–81. Available online: https://www.ugr.es/~cuadgeo/docs/articulos/040/040-008.pdf (accessed on 9 February 2023).

Torres, Victor Manuel. 2019. Tiempo, clima y los fenómenos atmosféricos: Desde torbellinos hasta cambio climático. *Revista Digital Universitaria* 20. [CrossRef]

UNDP. 2022. Los ODS en acción. Available online: https://www.undp.org/es/sustainable-development-goals (accessed on 9 February 2023).

UNESCO. 2016. Educación 2030: Declaración de Incheon y Marco de Acción para la realización del Objetivo de Desarrollo Sostenible 4: Garantizar una educación inclusiva y equitativa de calidad y promover oportunidades de aprendizaje permanente para todos. Available online: https://unesdoc.unesco.org/ark:/48223/pf0000245656_spa (accessed on 9 February 2023).

United Nations. 1992. Convención Marco de las Naciones Unidas sobre el cambio climático. Available online: https://unfccc.int/resource/docs/convkp/convsp.pdf (accessed on 9 February 2023).

United Nations. 2015. Objetivos y metas de desarrollo sostenible. 17 objetivos para transformar nuestro mundo. Available online: https://www.un.org/sustainabledevelopment/es/sustainable-development-goals/ (accessed on 9 February 2023).

United Nations. 2022. Objetivos de Desarrollo Sostenible. Available online: https://www.un.org/sustainabledevelopment/es/ (accessed on 9 February 2023).

United Nations. 2021. Cambio climático. Available online: https://www.un.org/sustainabledevelopment/es/climate-action/ (accessed on 9 February 2023).

WMO. 2017. Directrices de la Organización Meteorológica Mundial sobre la generación de un conjunto definido de productos nacionales de vigilancia del clima. Available online: https://library.wmo.int/doc_num.php?explnum_id=4215#:~:text=La%20anomal%C3%ADa%20de%20la%20temperatura,en%20los%20informes%20de%20vigilancia (accessed on 9 February 2023).

WMO. 2022. Preguntas Frecuentes—Clima. Available online: https://public.wmo.int/es/preguntas-frecuentes-clima (accessed on 9 February 2023).

Wolf, Martina, John Emerson, Daniel Esty, Alex De Sherbinin, Zachary Wendling, Quinn Nortonsmith, Sam Brigham, Slaveya Zaharieva, Iris Li, Mathew Jacob, and et al. 2022. *Environmental Performance Index*. New Haven: Yale Center for Environmental Law and Policy.

World Bank Group. 2021. Climate Change Knowledge Portal. Available online: https://climateknowledgeportal.worldbank.org/ (accessed on 9 February 2023).

World Bank Group. 2023. Cambio Climático. Available online: https://datos.bancomundial.org/tema/cambio-climatico (accessed on 9 February 2023).

social sciences

MDPI

Article

Climate Change and Extreme Weather Events in the Education of the Citizens of the Twenty-First Century: The Perception of Secondary Education Students

Esther Sánchez-Almodóvar [1], Isabel María Gómez-Trigueros [2,*] and Jorge Olcina-Cantos [1,3]

[1] Laboratory of Climatology, Interuniversity Institute of Geography, University of Alicante, 03690 San Vicente del Raspeig, Spain

[2] Department of General Didactics and Specific Didactics, Faculty of Education, University of Alicante, 03690 San Vicente del Raspeig, Spain

[3] Department of Regional Geographical Analysis and Physical Geography, University of Alicante, 03690 San Vicente del Raspeig, Spain

* Correspondence: isabel.gomez@ua.es

Abstract: Within the current context of climate emergency, the topic of climate change has become more prominent in secondary education in Spain. However, in general, conceptual confusions arise which should be clarified due to the social importance of climate issues. The objectives of this study, focused on third and fourth year students of ESO (Obligatory Secondary Education) in state schools in a town in the province of Alicante (Region of Valencia, Spain), seek to reveal the perception of the students regarding climate change and extreme weather events, in accordance with the subject in which these contents are taught; and to analyse whether the students have acquired a basic knowledge of the topic at the end of their secondary education. In order to fulfil these objectives, a non-experimental, descriptive, cross-sectional and survey-based correlational study has been conducted. The sample was made up of 784 students, surveyed during the academic year 2021–2022. The results indicate that the principal subject in which climate change is taught is Geography and History. The students consider that climate change is a threat to human beings and believe that anthropogenic action is the principal cause. They also perceive an increase in extreme weather events, although it is necessary to qualify this aspect. Therefore, this study defends the need to address this topic in the third and fourth years of ESO, as it is one of the major challenges faced by society and one in which students should be educated within the framework of the 2030 Agenda and the Sustainable Development Goals (SDGs).

Keywords: climate change; natural risks; Obligatory Secondary Education; 2030 Agenda; SDGs

Citation: Sánchez-Almodóvar, Esther, Isabel María Gómez-Trigueros, and Jorge Olcina-Cantos. 2023. Climate Change and Extreme Weather Events in the Education of the Citizens of the Twenty-First Century: The Perception of Secondary Education Students. *Social Sciences* 12: 27. https://doi.org/10.3390/socsci12010027

Academic Editor: Nigel Parton

Received: 16 November 2022
Revised: 28 December 2022
Accepted: 28 December 2022
Published: 30 December 2022

1. Introduction

Geography is the most suitable scientific discipline for teaching climate change. It combines the knowledge and interpretation of natural and human factors that intervene in the territory, offering a causal explanation of the processes occurring within it and the effects on the space and society, while contributing mitigation and adaptation actions. Since the end of the nineteenth century, the Spanish education system has incorporated contents on weather and climate in different study plans (Tonda-Monllor and Sebastiá-Alcaraz 2003). It has constituted one of the branches of Geography of greatest interest in the academic world in recent years due to its growing social prominence as a result of its importance for understanding the current climate crisis. Climate experts indicate that teaching about climate change is a complex task (Olcina Cantos 2017), as it requires a broad knowledge of climate and the weather system, which is necessary for analysing climate variables and the influence of geographic factors and for adopting measures to address the new environmental realities (Martín Vide 2009).

Currently, climate change is one of the most important challenges faced by humankind. Therefore, it is important that the teaching-learning process is conducted with scientific rigour and based on the principal source of information on climate change, the Intergovernmental Panel on Climate Change (IPCC). Since 1990, the IPCC reports have shown the current state of climate and the effects of climate change, both on nature and society. Furthermore, this organisation has proposed a series of measures to limit global warming and to adapt and mitigate the effects of climate change. All of these measures include education as a fundamental piece for raising awareness in society, among all age groups, through the dissemination of data and rigorous scientific information, without recurring to extremist, alarmist and catastrophic messages. Title 8 of Law 7/2021 of 20 May on climate change and energy transition, "Education, Research and Innovation in the fight against climate change and energy transition", addresses the importance of education to ensure the involvement of Spanish society in the responses to climate change and in training for sustainable development and climate care (Jefatura del Estado 2021, p. 62041).

The greater presence of this topic in the Spanish Geography and History curriculum of Spanish teenagers (ESO and Baccalaureate) is particularly important due to the current climate change context, in which the increase in global warming (IPCC 2018) caused by human action is leading to an increase in extreme weather events (IPCC 2022). One of the greatest impacts in the Region of Valencia is the change in precipitation patterns, with an increase in torrential rains and the hourly intensity of the events (Moutahir et al. 2014; Olcina Cantos and Vera-Rebollo 2016; Serrano-Notivoli et al. 2018). It is vitally important to establish a new horizon in the teaching-learning process in order to fulfil Objective 13 of climate action within the 2030 agenda in order to meet the Sustainable Development Goals (hereafter, SDGs) (United Nations General Assembly 2015).

This topic has become increasingly relevant in recent years, with the ever-more visible effects of climate change. In view of the present scenario, the political arena has increased its efforts to improve the teaching of climate change and the Sustainable Development Goals (SDGs). This increased interest in studying this issue in the educational domain is due to several reasons. One is that this topic forms part of the geographic content included in the Social Sciences curriculum: Geography at Obligatory Secondary Education and Baccalaureate levels, included in Royal Decree 217/2022 of 29 March (Ministerio de Educación y Formación Profesional 2022). The new Organic Law 3/2020 of 29 December on Education (LOMLOE) incorporates substantial changes related to how to teach contents relating to the environment and natural catastrophes in secondary classrooms (Jefatura del Estado 2020). Among these changes, the law underlines the need to promote a competency-based, autonomous, significant and reflexive learning in all subjects. In this respect, the LOMLOE indicates as "basic knowledge" in the subject of Geography and History, included in Block A: Current global challenges for the 1st and 2nd years of ESO the study of: "The climate emergency: elements and factors that condition climate and the impact of human activities. [. . .] Climate risks and catastrophes in the present, past and future. Vulnerability, prevention and resilience of the population to natural catastrophes and the effects of climate change" (Ministerio de Educación y Formación Profesional 2022, pp. 41683–6484). Similarly, for the third and fourth years of ESO, the "basic knowledge" included in Block A refers to: Current global challenges are focused on: "Sustainable Development Goals. Climate emergency and sustainability. Relationship between natural and anthropogenic factors on Earth" (Ministerio de Educación y Formación Profesional 2022, p. 41688) and in Block C: Local and global civil commitment: "Involvement in the defence and protection of the environment. Action and position with respect to the climate emergency" (Ministerio de Educación y Formación Profesional 2022, p. 41689).

The area of study of this research, the Spanish Mediterranean coast bathed by the Mediterranean, has become a high-risk region, due to its weather conditions and the current context of climate change and also due to the increase in urbanisation and occupation of floodable areas (Pérez-Morales et al. 2022). For these two reasons, it has become a study area for citizens of the twenty-first century, where it is necessary to educate society to raise awareness with respect to climate change with tangible effects through more frequent and intense extreme weather events that are affecting the Spanish Mediterranean (Sánchez-Almodóvar et al. 2022b). The current process of global warming increases the risk that already exists in this territory due to the increase in extreme weather events predicted by the climate models (Figure 1). In this respect, education has become a corner stone for raising awareness about and training in the causes and actions to implement to combat these extreme events, which generate considerable economic damage and the loss of human lives.

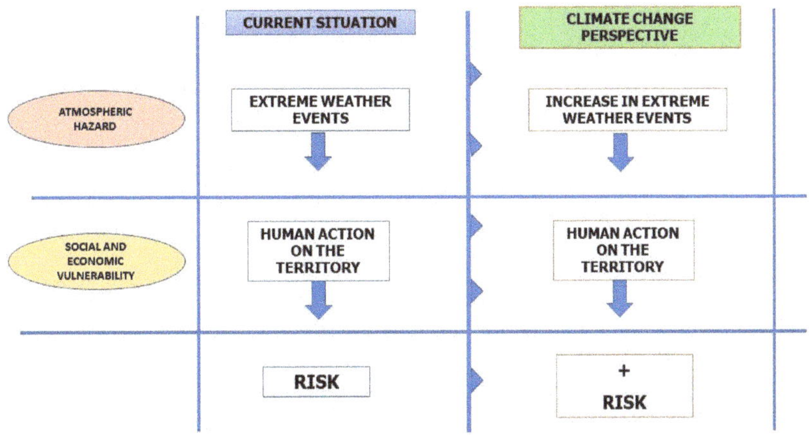

Figure 1. Societies at risk in the current context of global warming. Source: Own elaboration.

Within this context, education, sensitising and information and awareness-raising campaigns in society regarding the natural risks that exist in a territory constitute a fundamental pillar for the adaptation and mitigation of extreme natural events resulting from climate change and its effects. The science of geography seeks to educate critical citizens with a holistic view of the climate crisis which enables them to rigorously analyse its causes and consequences, an essential objective to create a more resilient society which is prepared to face this challenge.

1.1. Theoretical Framework

Before addressing the issue of climate change in the classroom, it is necessary to establish a conceptual corpus with which to facilitate the understanding and good practice in the most technical concepts which are usually used as synonyms, giving rise to errors. According to Martín Vide (2009), the concepts related to climate change (Figure 2), which are often treated as synonyms are: (1) global warming is the increase in the average temperature of the earth's surface, partly due to anthropogenic action. (2) Climate change includes human caused global warming, with the increase in the average temperature of the earth, the variation of precipitations and the retreat of the glaciers. Article 1.2 of the United Nations Framework Convention on Climate Change defines climate change as: *"a change of climate which is attributed directly or indirectly to human activity that alters the composition of the global atmosphere and which is in addition to natural climate variability observed over comparable time periods"* (Naciones Unidas 1992, p. 9). (3) Global change is a broader concept that covers all phenomena that involve changes on a global scale, such as climate change, the loss of biodiversity, changes in land use, etc.

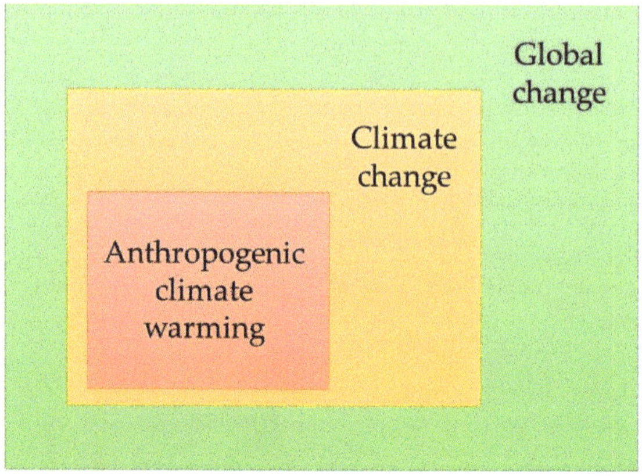

Figure 2. Outline of the concepts used in climate change. Source: modification of (Martín Vide 2009), 64.

1.2. The State of the Question

The geography discipline plays a highly important role in the teaching of climate change, as, in order to understand this concept, it is necessary to know the territory, its physical and human characteristics and the interaction between the two. As indicated by Morote Seguido and Moltó Mantero (2017), geographic science should seek to ensure that through their training, students can acquire the competencies to allow them to understand, analyse and interpret the space in which they live and the most relevant events taking place within it.

On an international level, recent studies have been published that address the perception of climate change and natural risks. In China, the perception of flood risk among primary school children has been analysed (Zhong et al. 2021), and in Canada, the individual knowledge of secondary education students has been assessed regarding their actions to mitigate climate change (Pickering et al. 2020). In the United Kingdom, Kurup et al. (2021) analyse how the conception about global warming and climate change alters among secondary education students, before and after a class on the topic. Jamelske et al. (2013) conduct a study comparing the perceptions of climate change among American and Chinese university students. For Taiwan, the opinions of higher education students with respect to the challenge of climate change are examined (Li and Liu 2021). Furthermore, the perception of climate change among secondary and primary teachers has been analysed for the case of Brazil (Marchezini and Londe 2020) and in South Africa's Western Cape the knowledge of climate change of secondary teachers has been studied (Anyanwu and Grange 2017).

In Spain, the teaching of climate and weather has attracted academic attention and social interest in recent years in the Spanish education system, as found in the scientific production of prestigious authors who address the teaching of Climatology, such as Martín Vide (2009), Olcina Cantos (2017), Tonda-Monllor and Sebastiá-Alcaraz (2003) and Martínez Fernández and Olcina Cantos (2019). In this respect, particularly noteworthy is the teaching initiative carried out by Morote Seguido (2016), Morote Seguido and Moltó Mantero (2017) and Sánchez Almodóvar et al. (2022) who propose visits to the Climatology Laboratory of the University of Alicante, weather museums and educational exhibitions as an innovative resource and methodology for teaching climatology.

The teaching-learning process of climate change, together with its causes and consequences, is one of the principal challenges of the current Spanish education system, from primary (Morote Seguido 2019a; Morote and Olcina 2021), and secondary education (Olcina Cantos 2017; Morote et al. 2022) to university training (Morote Seguido 2020; Morote Seguido and Hernández Hernández 2020; Sureda-Negre et al. 2014). The non-existence of a solid line of research on the teaching of climate change from the subject of Geography and Social Sciences in Spain indicates the need for scientific production on the teaching of climate change.

Recent studies reveal the situation of Geography teaching with respect to climate change and natural risks, analysing the school Social Science textbooks in Primary Education (Morote and Olcina 2021; Morote Seguido and Olcina Cantos 2020; Morote Seguido 2019a, 2021; Morote and Olcina 2022) and in those of Obligatory Secondary Education (Serantes-Pazos 2015; Navarro Díaz et al. 2020; Hernández Carretero et al. 2018), which are mostly characterised by a lack of scientific rigour.

From the experimental sciences, there are more proposals and didactic experiences that address the climate crisis from the didactics of Biology and Geology or Physics and Chemistry. There are various proposals with practical laboratory experiences for the different levels of Secondary Education, such as those presented by Sóñora et al. (2009), which include: the simple representation of convention currents, testing the regular effect of water temperature, simulating the greenhouse effect and the melting of the poles, and calculating the school's carbon footprint. The latter can cover a whole school year and be developed as a research project for ESO students (Martín Díaz 2009). Specific laboratory experiences can also be addressed, such as the development of a CO_2 sink (Boronat Gil et al. 2018). In addition, the dissemination work carried out by the sciences is essential (Escrivà i Garcia 2021).

The scientific community has difficulty in finding a simple explanation for the evolution of the climate features on a planetary scale and its current status due to the amalgam of factors that intervene (Özdem et al. 2014). To these factors, we should add the stereotypes and reports in the media (Morote et al. 2021a), which can be confusing, as shown by the results of studies that analyse the effect of *fake news* on climate change (Lutzke et al. 2019) and its exposure to the general public depending on the cultural constructs (Hong 2020).

In Spain, two decades ago, Souto González (1998) indicated the lack of studies on the perception of students regarding the teaching of Geography and the lack of training of teachers in the Social Science field. In this respect, in recent years, research has been carried out to determine the social representations and perceptions of the future teachers of Primary Education who are in training (Morote Seguido 2019a, 2019b; Morote Seguido and Hernández Hernández 2020; Morote et al. 2021a) and the knowledge they have on the topic (Morote Seguido and Souto González 2020), as this will have an impact on the perception of primary (Morote and Hernández 2022) and secondary (Sánchez-Almodóvar et al. 2022a) students. Furthermore, the interpretations of postgraduate students have also been analysed (Morote Seguido 2020; Morote et al. 2021b) and those of the whole community of the University of Alicante (Ramos Ribeiro et al. 2014).

1.3. Hypothesis and Objectives

The working hypothesis on which this research is based is that, in light of the existing evidence of climate change, this environmental and social problem is one of the principal social challenges of the twenty-first century, but the current paradigm in secondary education reveals that the students do not adequately receive this message through the teaching-learning process and, therefore, are not aware and do not become involved in this global problem. The principal objectives of this study are: (a) to reveal the perception of the students of climate change and natural risks in accordance with the subject in which these contents are taught to them; (b) to analyse whether the students of the third and fourth years of ESO acquire the basic knowledge related to climate change at the end of this cycle, taking into account that these contents are not address in later stages.

2. Materials and Methods

2.1. Design of the Research

This research is based on a descriptive, non-experimental and cross-sectional study as the information analysed has been gathered at a specific moment (academic year 2021–2022) and refers to a case study, in which four public secondary education centres of the province of Alicante in the Region of Valencia are analysed. To do this, the frequency and percentage descriptive of the items studied have been extracted. Similarly, correlational analyses have been conducted in order to predict or learn the degree of association (relationship) between two or more variables through the Pearson correlation coefficient of the items.

2.2. Context and Survey Participants

With respect to the context and survey participants, the selection procedure was carried out through a non-probability sampling (availability or convenience sampling). The participants in this study were students of Obligatory Secondary Education (third and fourth years; 13–16 years old or more). The total number of students enrolled in the participating centres for these years was 1248. With respect to the representativeness of the sample and taking into account the total number of students enrolled, in the whole of the area analysed, of the third and fourth years of ESO (n = 1248) a minimum of 295 students was required in order to obtain a representative sample so as to achieve a confidence interval of 99% and a margin of error of 5% (Acuña et al. 2020). Finally, given that the total number of survey participants was 784, a representative number was obtained in order to draw general and not partial conclusions of the phenomenon studied (Table 1).

Table 1. Students participating in the research.

Gender	Age				Total
	13 Years	14 Years	15 Years	16 Years or More	
Female	4	136	173	44	357
Male	2	142	175	52	371
Prefer not to say	0	21	22	13	56
Total	6	299	370	109	784

Source: Results of the questionnaire Own elaboration.

With respect to the socio-demographic characteristics (gender and age), the figures are similar from a gender point of view: man (47.3%; n = 371); woman (45.5%; n = 357); prefer not to say (7.1%; n = 56).

In order to ensure the reliability and validity of the instrument of analysis, different tests were conducted. On the one hand, the reliability of the model or instrument constructed has been confirmed through the Cronbach's alpha coefficient. This coefficient consists of the mean of the correlations between the variables that form part of the scale and can be calculated in two ways: based on the variances (Cronbach's alpha) or on the correlations of the items (standardised Cronbach's alpha) (Hair et al. 2011). In this case, the Cronbach's alpha was analysed between the items that made up the instrument (Table 2). A result of 0.789 was obtained, indicating an internal consistency of the instrument, with a value close to 1 (Martínez Arias et al. 2014).

Table 2. Cronbach's alpha reliability statistic.

Reliability Statistics	
Cronbach's alpha	0.789
No. of items	13

Source: Own elaboration.

In the same way, Pearson's Chi-squared index has been found with results of p-value < 1 = Sig. 0.001 (Cohen et al. 2017), indicating the high correlation of the questions asked and illustrative of the validity of the items and structure of the instrument used in the research.

2.3. Questionnaire

The instrument designed to carry out the research was based on a questionnaire to obtain the necessary data and fulfil the proposed objectives. The questionnaire was elaborated expressly for this research, following the model of other studies on social perception (Morote et al. 2021b; Morote Seguido 2019b; López-Fernández and Oller Freixa 2019). It was adapted to the non-university school stage and was made up of different items: on the one hand, there was a Likert scale (items 5, 6, 7, 10, 13, 14 and 15); and on the other, questions with diverse response options (items 4, 8, 9, 11 and 12) and socio-demographic questions in order to characterise the participating sample (items 1, 2 and 3) and one open question (item 16) (see Appendix A).

The questionnaire was validated by Secondary Education teachers and researchers from the Experimental and Social Sciences Department of the Social Sciences Teaching division of the University of Valencia; the Social Sciences Teaching Department of Geography of the University of Salamanca; the Specific Teaching Department of the Social Sciences Teaching Division of the University of Burgos; the Mathematical and Social Science Education Department of the University of Murcia; and the Specific Teaching Department of the Social Sciences Teaching Division of the University of Zaragoza.

The validation was carried out through the two-phase "Expert Judgement" method (Galicia Alarcón et al. 2017). In the first phase, the constructed instrument was sent to the experts participating in the validation via email. In this email the objectives of the research were explained, together with the context of the implementation and a period of two weeks was proposed for its evaluation. Subsequently, after receiving the indications of the judges, the items of the instrument were modified. Then, the modified instrument was resent to the same judges. It should be noted that the comments of the judges referred to the wording of the questions and two initial items were merged together.

As a result of this validation, the instrument included 16 items.

2.4. Procedure

Before beginning the survey process, authorisation was requested from the Regional Department of Education, Culture and Sport of the Regional Government of Valencia to visit the Secondary Education centres in order to conduct the survey. The request was accepted on 23 December 2021 by the Regional Secretary of Education and Professional Training of the Regional Department of Education, Culture and Sport; under the terms of Decree 173/2020 of 30 October, of the Regional Department, approving the Organic and functional regulations of the Regional Department of Education, Culture and Sport.

The survey process was conducted between December 2021 and June 2022. Contact had been previously made with the head teachers of the educational centres and Social Sciences departments: Geography and History. The questionnaire was administered in collaboration with the teachers of this discipline in one session, with a response time of ten minutes. The students were not previously informed, although the objectives of the research were explained to them once they had been presented with the instrument. Finally, the participants remained anonymous throughout the whole procedure and confidentiality was guaranteed, as indicated in point 5 of the resolution authorising this activity (Resolution 23/12/2021 91K7VMAE: TVL2D461: D85M5MPM).

2.5. Data Analysis

The data analysis procedure was conducted using version 27.0 of the statistics software SPSS. Different statistical-inferential analyses were carried out (non-parametric tests) of the frequencies and percentages with which the sample was characterised and to determine the perception of the participants of the key concepts of the study. The Pearson's chi-squared test was performed in order to determine the correlation of the validated instrument, which generated a result of p-value < 1 = Sig. 0.001, indicating the high correlation of the questions asked, illustrating the validity of the items and the structure of the instrument. Furthermore, the response to the items was coded so as to conduct a cross tabulation analysis and bivariate correlations in order to characterise and establish differences and correlations with respect to the concept of climate change and its relationship with the greenhouse effect, in accordance with the subject in which these contents were taught to the students.

3. Results

First, a frequency (*f*) and percentages (%) analysis was carried out of the responses to item 4: "Subject in which you have been taught about climate change", in order to determine the subject perceived by the participating sample of third- and fourth-year ESO students (item 2: "Gender"; item 3: "Which year are you in"), that teaches the contents relating to the concept of study (climate change). The global data (a total of 784 responses) reveal that the principal subject in which this topic is taught is Geography and History (Table 3).

Table 3. Frequency and percentages of the perception of the students by year and gender of the subject in which climate change is addressed.

Subject in Which You Have Been Taught about Climate Change	Group												TOTAL	
	Third Year of ESO						Fourth Year of ESO							
	Gender						Gender							
	w		m		*pns*		w		m		*pns*			
	f	%	*f*	%	*f*	%	*f*	%	*f*	%	*f*	%	*f*	%
Geography and History	111	14.1	111	14.1	12	1.6	62	8	55	7	3	0.4	354	45.3
Physics or Chemistry	15	1.9	8	1	6	0.7	0	0	2	0.3	1	0.1	32	4
Biology or Geology	30	3.8	24	3	5	0.6	34	4.3	47	6	2	0.25	142	18.1
I have not learnt about climate change	63	8	80	12.2	25	3.2	42	5.3	44	5.5	2	0.25	256	32.6
TOTAL	219	28	223	28.4	48	6.1	138	17.6	148	18.8	8	1	784	100

Notes: Legend: *f* = frequency; *w* = woman; *m* = man; *pns* = prefer not to say. Source: Results of the questionnaire Own elaboration.

Therefore, it is observed that the students perceive the subject of Geography and History as the one in which the concept of climate change is addressed (*f* = 354), with a representation of 45.3% of the responses of the participating sample. In second place, the students consider that the subject of Biology and Geology addresses these contents (*f* = 142) with a representation of this answer option of 18.1%. Finally, the participants gave a lower response percentage to the subject of Physics and Chemistry (*f* = 32), with a significance with respect to the total percentage of responses of 4%.

It is worth pointing out the option "I have not learnt about climate change" with a high representation among the responses given by the sample of 32.6% of the total (*f* = 256). By gender, the above-mentioned responses are the same, with the subject of Geography and History being the most indicated equally by men and women than those who prefer not to define themselves as either a man or woman. Similarly, the response "I have not learnt about climate change" is in second place from a gender point of view.

If we observe the responses by the academic year of the students, the same trend is observed. In first place, Geography and History is perceived as the subject in which the contents relating to climate change are addressed (third-year ESO *f* = 234; fourth-year ESO *f* = 120). In addition, the response "I have not learnt about climate change" is the second most popular option selected by the students of third-year ESO (*f* = 168) and fourth-year ESO (*f* = 88) students.

In order to analyse whether the students perceive climate change as a threat and whether they relate it to extreme weather events, item 7 "Is climate change a threat to human beings" has been interrelated with item 10 "Do you think that more extreme weather events are occurring now? (extreme weather events can be hurricanes, torrential rains, heatwaves, etc.)." (Table 4). In this respect, it should be noted there is considerable quorum among the students (f = 586) who believe that climate change is a threat ant that extreme phenomena are increasing, representing 74.8%. However, some students (f = 86), while believing that climate change is a threat, are not sure that more extreme weather phenomenon are occurring now (10.9%): A small number of the sample (f = 20), representing 2.5%, express uncertainty for both of these questions. With respect to the total data, it should be pointed out that 5.7% believe that climate change is not a threat (f = 45).

Table 4. Frequency and percentages of the perception of students in relation to climate change and the development of associated extreme weather phenomena.

Item 7. Is Climate Change a Threat for Human Beings?	Item 10. Do You Think That Now There Are More Extreme Weather Events? (Extreme Weather Events Can Be Hurricanes, Torrential Rains, Heatwaves, etc.).						Total	
	Yes		No		Don't Know/No Answer			
	f	%	f	%	f	%	f	%
Yes	586	74.8	16	2	86	10.9	688	87.8
No	26	3.3	8	1	11	1.5	45	5.7
Don't know/ no answer	28	3.5	3	0.4	20	2.5	51	6.5
Total	640	81.6	27	3.4	117	14.9	784	100

Legend: f = frequency. Source: Results of the survey. Own elaboration.

In order to determine the linear dependency between the responses given by the sample for item 7 and item 10, the Pearson's correlation was found (Table 5). The results show that there is a strong correlation between the two questions referring to climate change. Furthermore, the responses to both items show a statistically significant linear relationship (r = 0.683, $p < 0.001$). The direction of the relationship is positive, which means that these variables tend to increase together in relation to climate change. As we can observe in Table 5, the magnitude or strength of the association between the perception of risk for human beings of climate change and the ever-greater proliferation of extreme weather events as a consequence of this climate change (between 0.5 and 1) is strong.

Table 5. Pearson's correlations for the responses related to climate change (items 7 and 10).

		Item 7. Is Climate Change a Threat for Human Beings?	Item 10. Do You Think That Now There Are More Extreme Weather Events? (Extreme Weather Events Can Be Hurricanes, Torrential Rains, Heatwaves, etc.).
Item 7. Is climate change a threat for human beings?	Pearson's correlation	1	0.683 **
	Sig. (bilateral)		0.000
	N	784	784
Item 10. Do you think that now there are more extreme weather events? (extreme weather events can be hurricanes, torrential rains, heatwaves, etc.).	Pearson's correlation	0.683 **	1
	Sig. (bilateral)	0.000	
	N	784	784

Note: ** means the correlation is significant at the 0.01 level (bilateral). Source: Results of the questionnaire Own elaboration.

With respect to item 8 of the research instrument (Appendix A), four responses are proposed (natural factors; factors derived from human action; there is no climate change; do not know, no answer) in relation to the principal causes of the current process of climate change. The majority of the participating sample responded to this question with the option "factors derived from human action" with 91.3% of the answers (*f* = 716). The second most chosen answer was "natural factors" with 5.3% of answers (*f* = 42) and the third was the option "dk/na" with 1.7% of answers (*f* = 14). It is striking that the response option "there is no climate change", which, while being the least chosen response had a percentage very close to "dk/na" with 1.5% of the responses obtained (*f* = 12).

When we observe the responses by groups (third year of ESO and fourth year of ESO), we can see that, in both years the majority option selected is "factors derived from human action" representing 56% of responses from third-year ESO students (*f* = 439) and 35.5% of responses of fourth-year ESO students (*f* = 277) (Table 6).

Table 6. Response options regarding factors that influence climate change by school year analysed.

Item 8. If the Following Factors That Contribute to Climate Change, Which Do You Consider to Be the Principal Cause of the Current Climate Change Process?										
Group	Natural Factors		Factors Derived from the Action of Human Beings		Climate Change Does Not Exist		Dk/Na		Total	
	f	%	*f*	%	*f*	%	*f*	%	*f*	%
Third-year ESO	32	4	439	56	7	0.8	12	1.5	490	62.5
Fourth-year ESO	10	1.3	277	35.3	5	0.7	2	0.2	294	37.5
Total	42	5.3	716	91.3	12	1.5	14	1.7	784	100

Legend: Dk/na—don't know/no answer Source: Results of the questionnaire Own elaboration.

Item 9 refers to the consequences derived from climate change. With the objective of determining the consequences that the students perceive as being most important, the item is presented as a multiple response question where the participants must select the three most important options in accordance with their criteria. Figure 3 shows the absolute frequencies of the responses selected as first, second and third options. In the first response option, the most frequent answer was "Increase in sea level" (*f* = 489), representing 62.4%, followed by "Changes in the ecosystems" (*f* = 205) representing 26.1%, and finally, the option with the lowest frequency in this group was "Melting of the mountain glaciers" (*f* = 54) with a representation of 6.9%. As a second response option, the most selected by the participants was "Increase in the average temperature of the planet (*f* = 319), representing 40.7%, in this group the option "Change in the systems" was repeated (*f* = 162), representing 27.8%, while "Melting of the mountain glaciers" (*f* = 162) was third with 20.7% of the answers. In the third response option, the most selected by the students was "Increase in hurricanes, droughts, fires, etc." (*f* = 378), with 48.2%. This result is related to the 81.6% (*f* = 640) (Table 4) who consider that now extreme weather events occur more frequently. With very little difference, the option "Increase in the average temperature of the planet" (*f* = 307) represented 39.2% of responses. The least frequently selected option was "Melting of the mountain glaciers" (*f* = 34) with just 4.3%.

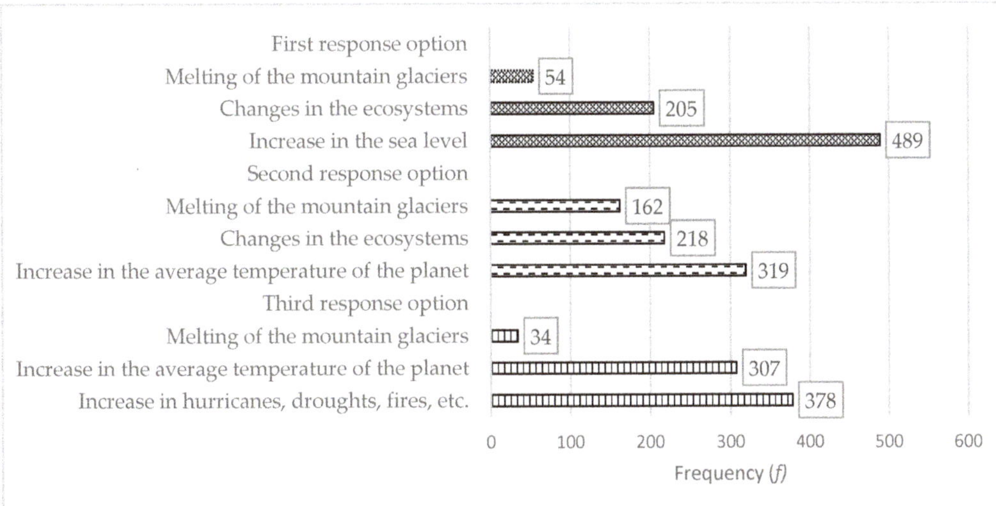

Figure 3. Response options regarding the consequences of climate change. Source: Results of the questionnaire Own elaboration.

4. Discussion

The teaching and dissemination of climate change and natural risks is essential for developing social awareness, which help to enhance our understanding of the causes and consequences of these processes so as to sensitise the population about prevention and adaptation measures.

The results obtained in this research reveal the importance that geographic science has in the teaching of climate change and natural risks, given that 45.3% of the students surveyed have worked on this topic in the subject of Geography and History in their secondary education. In spite of this, 32.6% of the participants claimed that they had not seen contents related to climate change. Similar results were obtained in the study conducted by Özdem et al. (2014) with primary education students in Turkey, where 24% of the sample had not heard of climate change and 23% were not sure, as opposed to 52% who responded positively. In the case of young Canadian people, the level of confidence was not high with respect to the question about how sure they were of the education they had received at school about climate change (Pickering et al. 2020). A study on the perception of Brazilian primary and secondary teachers revealed that 60.3% of the total participants had not received training in climate change when they were undergraduate students (Marchezini and Londe 2020).

With respect to the question about whether climate change represents a threat to human beings, in this study the majority of the sample (87.8%) responded affirmatively. This result coincides with that obtained in the case of the Brazilian teachers, with 83.5% considering that climate change is affecting the municipality (Marchezini and Londe 2020). With respect to the statement that climate change is a threat to life, the results of this study coincide with those obtained by Gómez Trigueros (2020) where the university students participating in the classroom intervention considered that climate change is produced and expressed in our daily lives. However, only 41% of the primary students (Turkey) perceive climate change as a threat, while 40% are not sure and 19% consider that they are not affected or will not be affected (Özdem et al. 2014).

The findings obtained in this research show that 81.6% of secondary students perceive an increase in extreme weather events as a result of climate change. These are similar to those obtained in other studies analysing perception, in this case students of the Primary Education Degree of the University of Valencia, who were asked whether climate change is influencing the increase in flood risk, to which 63.52% of the participants responded affirmatively (Morote Seguido 2019b, p. 78). In the same area of study but with a different context, 70% perceive that rainfall episodes are more intense (Morote Seguido and Hernández Hernández 2020).

According to the IPCC (2021, p. 425), the hypothesis of human influence as the cause of the current climate change process has been reinforced with the evaluations conducted from the second report of the IPCC in 1995 to the fifth in 2014 which reveal the clear human influence on the climate system, due to the increase in greenhouse gas emissions into the atmosphere. This idea has gained force and has been demonstrated in the last report. Therefore, in this research, it is interesting to determine the perception of the students regarding this fact, with the majority (91.3%) considering that human factors influence climate change. A recent study conducted for the Region of Valencia (Spain) analyses the perception of primary, secondary and baccalaureate students, revealing that the students consider that the principal cause of this phenomenon is attributed to pollution (70.1%) (Morote and Hernández 2022). Similarly, Jamelske et al. (2013) compare the opinion of Chinese and American university students regarding whether the principal causes of climate change are human or natural. The Chinese students principally believe (86.4%) that climate change has anthropogenic causes, while barely 59% of the American students also believe this to be the case. In the same way, 82% of Brazilian teachers consider that the principal cause is due to the increase in carbon emissions (Marchezini and Londe 2020). In contrast, only 38% of university students in Taiwan identify anthropogenic activity as a cause of global warming (Li and Liu 2021).

When young people hear about climate change their first thought is principally related to the consequences, as shown by González-Gaudiano and Maldonado-González (2014) in their study on the social representations of university students in Mexico. Similarly, Bello Bello Benavides et al. (2021) examine the social representations of baccalaureate students in Mexico, where the increase in temperature, the greater presence of extreme weather events and the increase in sea level are the most recognised consequences of climate change. With respect to this study of the consequences, the results of the survey carried out among secondary students show that the responses most frequently selected were "Increase in sea level" (62.4%), "Increase in the average temperature" (40.7%) and "Increase in hurricanes, droughts, fires, etc." (48.2%). Similar results were obtained by Morote and Hernández (2022), where the participants indicated that the principal effect of climate change is the "Increase and changes in temperature" (61.7%), in second and third place they indicated "Melting" (8.7%) and "Increase in natural threats" (5.6%).

Taking as a base the new Organic Law 3/2020, of 29 December on Education (LOMLOE), which bestows greater importance to the teaching of the contents related to the environment, natural disasters and climate change, this research reveals the need to address climate change and avoid catastrophism and present the evidence realistically, based on reports and scientific data. It shares the proposal made by Martínez Fernández and Olcina Cantos (2019), with the introduction of new contents on the risks associated with climate and climate change for primary and secondary education and the proposal made for the level of third year of ESO is particularly interesting for this study. In the same way, the results obtained from the in-depth reviews of the existing contents on climate change in primary education textbooks (Morote and Olcina 2021; Morote Seguido and Olcina Cantos 2020; Morote Seguido 2019a, 2021; Morote and Olcina 2022) and, principally secondary education textbooks (Serantes-Pazos 2015; Navarro Díaz et al. 2020; Hernández Carretero et al. 2018), highlight the need to address this topic with more scientific rigour. Meanwhile, in a study on the subject of climate change at university level, Gómez Trigueros (2020), conducts an intervention in a classroom of future primary education teachers, giving a prominent role to GIS as a didactic innovation in teaching climate change, stressing again the importance of addressing this problem on a local scale (Gómez-Trigueros et al. 2019) and taking into account in the teaching-learning process the interrelationship between human beings and the natural or artificial environment (Martínez

Castillo 2010). With a more informative nature, the proposal made by Nelles and Serrer (2020) is a clear example of how to address the communication and teaching of climate change simply and directly, but with a high level of scientific rigour through illustrations that clearly transmit the causes and consequences of the phenomenon, its effects and even response actions to the climate emergency, which are useful as a teaching resource in the classroom.

5. Conclusions

This study analyses the perceptions of the students of third and fourth year of ESO with respect to the topic of climate change and natural risks. It explores the subjects in which the students consider that these contents are addressed more precisely. The results show that Geography is the predominant subject, which ratifies its suitability to address issues related to climate change.

The knowledge acquired by the students in their obligatory education on climate change, the associated extreme weather events and consequences has been assessed. The study shows the need to study current climate change in greater depth in the classroom, the influence of human beings on the development of this climate change, the territorial and socio-economic effects of global warming and the future evolution of this process. Although in recent years there has been an increasing mobilisation of students in aspects related to the current climate crisis, it is necessary improve the contents taught in the classroom on climate change, which should not focus on extreme or catastrophic messages but should be based on rigorous scientific data and official reports on the topic.

All of this has the objective of generating a society better adapted to the future climate change scenarios. The development of materials on climate change and the associated extreme weather events contemplated in the new Education Law (LOMLOE; Law 3/2020 of 29 December) can constitute an ideal framework for learning about this topic at pre-university levels. Furthermore, it is vitally important to establish a new horizon in teaching, within the framework of the 2030 Agenda and the development of the SDGs (Naciones Unidas Asamblea General 2015), and the Sendai Framework for Disaster Risk Reduction 2015–2030 (United Nations General Assembly 2015), as a tool to raise awareness, change attitudes, focus on the importance of reducing existing risk and on the strengthening of a sustainable and resilient society.

However, we should be aware that the implementation of these proposals in the classroom could be hindered by: (1) a lack of training received by Geography and History teachers (it should be noted that in Spain many of the teachers of this subject have a History background); (2) the scientific rigour of the explanations in the textbooks; a matter widely discussed by Morote Seguido and Olcina Cantos (2020) or (3) the influence of the media and social networks. Therefore, when the topic of climate change is brought to the classroom it should be based on a series of premises which avoid the recurrence to catastrophism and the formulation of sensationalist theories. To do this, it is necessary to educate not only the students, as a basic part of the general population, avoiding the propagation of *fake news* and misinformation by the different media. The emergence of the youth movement "*Friday For Future*" represents a step forward for this population group to become involved and participate in the defence of the planet and the fight against climate change, as they will inherit the initiatives that are currently being implemented.

Therefore, the role played by the academic institutions (universities, research centres) is important for divulging climate science, with the development, using clear and accessible language, of materials, talks or guided tours that bring students of basic levels of education into contact with the climate reality of our planet and can already be observed on scales close to them (regional and local). The new information tools (ICTs, viewers, apps) and social networks can play an important complementary role to the methods traditionally used in the classroom, due to their educational capacity and degree of social penetration in the youngest layers of society. These actions should always pursue the improvement of environmental culture and the respect for the environment.

Among the future development in this line of research is the analysis of other concepts related to climate change, such as the greenhouse effect, the analysis of the perception of flood risk and the decision-making processes involved in them both. Another line is to determine the opinion of students regarding the governance of climate change and their personal action proposals.

Author Contributions: Conceptualization, E.S.-A.; I.M.G.-T. and J.O.-C.; methodology, E.S.-A. and I.M.G.-T.; software, I.M.G.-T. and E.S.-A.; validation, I.M.G.-T. and J.O.-C.; formal analysis, E.S.-A. and I.M.G.-T.; investigation, E.S.-A.; I.M.G.-T. and J.O.-C.; resources, E.S.-A.; I.M.G.-T. and J.O.-C.; data curation, I.M.G.-T. and E.S.-A.; writing—original draft preparation, E.S.-A. and I.M.G.-T.; writing—review and editing, J.O.-C.; supervision, I.M.G.-T. and J.O.-C.; funding acquisition, E.S.-A.; I.M.G.-T. and J.O.-C. All authors have read and agreed to the published version of the manuscript.

Funding: This research was funded by the Vice-Rectorate for Research of the University of Alicante (predoctoral fellowship to E.S.-A., reference FPUA2019-54); This work is the result of the ongoing investigation of the Emerging Project GV/2021/077 "La brecha digital de género y el modelo TPACK en la formación del profesorado: análisis de la capacitación digital docente" led by Isabel María Gómez-Trigueros.: Emerging Project GV/2021/077.

Institutional Review Board Statement: Not applicable.

Informed Consent Statement: Not applicable.

Data Availability Statement: Data are contained within the article.

Conflicts of Interest: The authors declare no conflict of interest.

Appendix A

Item	Type of Response/Variable
Item 1. Gender	Closed question: Female/Male/Prefer not to say
Item 2. Age	Closed question: 13/14/15/16 or more
Item 3. Year	Closed question: 3rd year ESO/4th year ESO
Item 4. Subject in which you have been taught about climate change.	Closed question: Geography and History/Physics or Chemistry/Biology or Geology/I have not learnt about climate change
Item 5. Is the greenhouse effect a natural process?	Likert scale: (1) Yes/(2) No/(3)Don't know/no answer
Item 6. Is the greenhouse effect positive for life on Earth?	Likert scale: (1) Yes/(2) No/(3)Don't know/no answer
Item 7. Is climate change a threat for human beings?	Likert scale: (1) Yes/(2) No/(3)Don't know/no answer
Item 8. If the following factors that contribute to climate change, which do you consider to be the principal cause of the current climate change process?	Closed question: Natural factors/factors derived from the action of human beings/Climate change does not exist/Dk/Na
Item 9. Indicate the consequences of climate change (Mark the 3 most important options in accordance with your criteria)	Multiple question: Increase in the sea level/Changes in the ecosystems/Increase in volcanic activity/eruptions/Melting of the mountain glaciers/Increase in the average temperature of the planet/Increase in hurricanes, droughts, fires, etc./Climate change does not have any consequences/Increase in earthquakes
Item 10. Do you think that now there are more extreme weather events? (extreme weather events can be hurricanes, torrential rains, heatwaves, etc.).	Likert scale: (1) Yes/(2) No/(3)Don't know/no answer
Item 11. Which natural phenomenon do you consider most important in your town and the province of Alicante?	Closed question: Floods/Droughts/Earthquakes/Fires
Item 12. With respect to floods in towns, what do you think is the principal cause?	Closed question: The urban growth of the towns has occupied floodable areas/The rivers overflow/The amount of rainfall is higher/There are no floods in urban environments

Item	Type of Response/Variable
Item 13. Would you know how to act in a situation of flood alert?	Likert scale: (1) Yes/(2) No/(3)Don't know/no answer
Item 14. Do you consider that the government authorities are implementing solutions to the problem of climate change?	Likert scale: (1) Yes/(2) No/(3)Don't know/no answer
Item 15. Do you think that you could do something to fight against climate change?	Likert scale: (1) Yes/(2) No/(3)Don't know/no answer
Item 16. If yes, indicate what you would do to fight against climate change. If no, why cannot you fight against climate change?	Open question

References

Acuña, Guillermo, Cristián Echeverría, and Cristian Pinto-Gutiérrez. 2020. Consumer Confidence and Consumption: Empirical Evidence from Chile. *International Review of Applied Economics* 34: 75–93. [CrossRef]

Anyanwu, Raymond, and Lesley Le Grange. 2017. The Influence of Teacher Variables on Climate Change Science Literacy of Geography Teachers in the Western Cape, South Africa. *International Research in Geographical and Environmental Education* 26: 193–206. [CrossRef]

Bello Benavides, Laura O., Gloria E. Cruz Sánchez, Pablo Á. Meira Cartea, and Edgar J. González Gaudiano. 2021. El Cambio Climático En El Bachillerato. Aportes Pedagógicos Para Su Abordaje. *Enseñanza de Las Ciencias* 39: 137–56. [CrossRef]

Boronat Gil, Raquel, Margarita Gómez Tena, and y José Pedro López Pérez. 2018. Diseño experimental de un sumidero de CO_2 y sus implicaciones en el cambio climático. Una experiencia de trabajo con alumnos en el laboratorio de Educación Secundaria. *Revista Eureka sobre enseñanza y divulgación de las ciencias* 15: 1202. Available online: https://dialnet.unirioja.es/servlet/articulo?codigo=6236804&orden=0&info=link (accessed on 18 October 2022). [CrossRef]

Cohen, Louis, Lawrence Manion, and Keith Morrison. 2017. *Research Methods in Education*, 8th ed. London: Routledge. [CrossRef]

Escrivà i Garcia, Andreu. 2021. Encara no és tard? *Mètode: Revista de difusió de la investigació* 3: 112. Available online: https://dialnet.unirioja.es/servlet/articulo?codigo=8043987&orden=0&info=link (accessed on 18 October 2022).

Galicia Alarcón, Liliana Aidé, Jorge Arturo Balderrama Trápaga, and Rubén Edel Navarro. 2017. Validez de Contenido Por Juicio de Expertos: Propuesta de Una Herramienta Virtual. *Apertura* 9: 42–53. [CrossRef]

Gómez Trigueros, Isabel María. 2020. El Cambio Climático y La Enseñanza Interdisciplinar: Una Propuesta Didáctica Al Reto Educativo Del Siglo XXI. In *The Climate Crisis in Mediterranean Europe: Cross-Border and Multidisciplinary Issues on Climate Change*. Edited by Jonathan Gómez Cantero, Carolina Morán Martínez, Justino Losada Gómez and Fabio Carnelli. Lago: Il Sileno Edizioni, vol. 3, pp. 163–81.

Gómez-Trigueros, Isabel María, Mónica Ruiz-Bañuls, and Delfín Ortega-Sánchez. 2019. Digital Literacy of Teachers in Training: Moving from Icts (Information and Communication Technologies) to Lkts (Learning and Knowledge Technologies). *Education Sciences* 9: 274. [CrossRef]

González-Gaudiano, Edgar J., and Ana Lucía Maldonado-González. 2014. ¿Qué Piensan, Dicen y Hacen Los Jóvenes Universitarios Sobre El Cambio Climático?: Un Estudio de Representaciones Sociales. *Educar Em Revista* 3: 35–55. [CrossRef]

Hair, Joe F., Christian M. Ringle, and Marko Sarstedt. 2011. PLS-SEM: Indeed a Silver Bullet. *Journal of Marketing Theory and Practice* 19: 139–52. [CrossRef]

Hernández Carretero, Ana María, Mario Burgui Burgui, Federico Velázquez de Castro, and José María Corrales Vázquez. 2018. ¿Responden Los Libros de Texto a Las Demandas de La Educación Ambiental? Un Análisis Para La Educación Secundaria. *Boletín de La Asociación de Geógrafos Españoles* 77: 80–110. [CrossRef]

Hong, Seong Choul. 2020. Presumed Effects of "Fake News" on the Global Warming Discussion in a Cross-Cultural Context. *Sustainability* 12: 2123. [CrossRef]

IPCC. 2018. Global Warming of 1.5 °C. An IPCC Special Report on the Impacts of Global Warming of 1.5 °C above Pre-Industrial Levels and Related Global Greenhouse Gas Emission Pathways, in the Context of Strengthening the Global Response to the Threat of Climate Change. Available online: https://www.cambridge.org/core/books/global-warming-of-15c/summary-for-policymakers/31C38E590392F74C7341928B681FF668 (accessed on 18 October 2022).

IPCC. 2021. Climate Change 2021: The Physical Science Basis. Contribution of Working Group I to the Sixth Assessment Report of the Intergovernmental Panel on Climate Change. Available online: https://www.ipcc.ch/report/ar6/wg1/downloads/report/IPCC_AR6_WGI_FrontMatter.pdf (accessed on 18 October 2022).

IPCC. 2022. Climate Change 2022: Impacts, Adaptation and Vulnerability. Contribution of Working Group II to the Sixth Assessment Report of the Intergovernmental Panel on Climate Change. Available online: https://www.ipcc.ch/report/ar6/wg2/downloads/report/IPCC_AR6_WGII_Citation.pdf (accessed on 18 October 2022).

Jamelske, Eric, James Barrett, and James Boulter. 2013. Comparing Climate Change Awareness, Perceptions, and Beliefs of College Students in the United States and China. *Journal of Environmental Studies and Sciences* 3: 269–78. [CrossRef]

Jefatura del Estado. 2020. Ley Orgánica 3/2020, de 29 de Diciembre, Por La Que Se Modifica La Ley Orgánica 2/2006, de 3 de Mayo, de Educación. *Boletín Oficial Del Estado* 340: 122868–22953. Available online: https://www.boe.es/eli/es/lo/2020/12/29/3 (accessed on 18 October 2022).

Jefatura del Estado. 2021. Ley 7/2021, de 20 de Mayo, de Cambio Climático y Transición Energética. *Boletín Oficial Del Estado* 121: 62009–52. Available online: https://www.boe.es/eli/es/l/2021/05/20/7 (accessed on 18 October 2022).

Kurup, Premnadh M., Ralph Levinson, and Xia Li. 2021. Informed-Decision Regarding Global Warming and Climate Change Among High School Students in the United Kingdom. *Canadian Journal of Science, Mathematics and Technology Education* 21: 166–85. [CrossRef]

Li, Yuh Yuh, and Shu Chiu Liu. 2021. Examining Taiwanese Students' Views on Climate Change and the Teaching of Climate Change in the Context of Higher Education. *Research in Science & Technological Education* 40: 515–28. [CrossRef]

López-Fernández, José Antonio, and Montserrat Oller Freixa. 2019. Los Problemas Medioambientales En La Formación Del Profesorado de Educación Primaria. *REIDICS. Revista de Investigación En Didáctica de Las Ciencias Sociales* 4: 93–109. [CrossRef]

Lutzke, Lauren, Caitlin Drummond, Paul Slovic, and Joseph Árvai. 2019. Priming Critical Thinking: Simple Interventions Limit the Influence of Fake News about Climate Change on Facebook. *Global Environmental Change* 58: 101964. [CrossRef]

Marchezini, Victor, and Luciana R. Londe. 2020. Looking to Future Perceptions about Climate Change in Brazil: What Children's Teachers Think, Learn and Teach About? *Natural Hazards* 104: 2325–37. [CrossRef]

Martín Díaz, M. 2009. Cambio climático y consumo energético en un instituto de educación secundaria. *Enseñanza de las ciencias: Revista de investigación y experiencias didácticas*: 2125–28. Available online: https://dialnet.unirioja.es/servlet/articulo?codigo=6905776&orden=0&info=link (accessed on 18 October 2022).

Martín Vide, Javier. 2009. Conceptos Previos y Conceptos Nuevos En El Estudio Del Cambio Climático Reciente. *Investigaciones Geográficas* 49: 51–63. [CrossRef]

Martínez Arias, María del Rosario, María Victoria Hernández, and María José Hernández Lloreda. 2014. *Psicometría*. Edited by Alianza. Madrid: Alianza.

Martínez Castillo, Roger. 2010. La Importancia de La Educacion Ambiental Ante Los Retos Actuales. *Revista Electrónica Educare* XIV: 97–111. Available online: https://www.redalyc.org/pdf/1941/194114419010.pdf (accessed on 18 October 2022).

Martínez Fernández, Luis Carlos, and Jorge Olcina Cantos. 2019. La Enseñanza Escolar Del Tiempo Atmosférico y Del Clima En España: Currículo Educativo y Propuestas Didácticas. *Anales de Geografía de La Universidad Complutense* 39: 125–48. [CrossRef]

Ministerio de Educación y Formación Profesional. 2022. Real Decreto 217/2022, de 29 de Marzo, Por El Que Se Establece La Ordenación y Las Enseñanzas Mínimas de La Educación Secundaria Obligatoria. *Boletín Oficial Del Estado* 76: 41571–789. Available online: https://www.boe.es/eli/es/rd/2022/03/29/217/con (accessed on 18 October 2022).

Morote, Álvaro-Francisco, and Jorge Olcina. 2021. Cambio Climático y Sostenibilidad En La Educación Primaria. Problemática y Soluciones Que Proponen Los Manuales Escolares de Ciencias Sociales. *Sostenibilidad: Económica, Social y Ambiental* 3: 25–43. [CrossRef]

Morote, Álvaro-Francisco, Jorge Olcina, and María Hernández. 2022. Teaching Atmospheric Hazards in the Climate ChangeContext—Environmental Didactic Proposals in the Mediterranean Region for Secondary Schools. *Environments* 9: 29. [CrossRef]

Morote, Álvaro Francisco, Benito Campo, and Juan Carlos Colomer. 2021a. Percepción Del Cambio Climático En Alumnado de 4º Del Grado En Educación Primaria (Universidad de Valencia, España) a Partir de La Información de Los Medios de Comunicación. *Revista Electrónica Interuniversitaria de Formación Del Profesorado* 24: 131–44. [CrossRef]

Morote, Álvaro Francisco, and María Hernández. 2022. What Do School Children Know about Climate Change? A Social Sciences Approach. *Social Sciences* 11: 179. [CrossRef]

Morote, Alvaro Francisco, and Jorge Olcina. 2022. El Tratamiento de Los Riesgos Naturales En Los Libros de Texto de Ciencias Sociales (Educación Primaria). Una Aproximación a Las Definiciones y Problematización de Las Actividades. *Cuadernos Geográficos* 61: 223–46. [CrossRef]

Morote Seguido, Álvaro-Francisco, and María Hernández Hernández. 2020. La Formación y Percepción Sobre El Riesgo de Inundación. Una Exploración a Partir de Las Representaciones Sociales Del Futuro Profesorado de Educación Primaria. In *Riesgo de Inundación En España: Análisis y Soluciones Para La Generación de Territorios Resilientes*. Edited by López Ortiz, María Inmaculada and Joaquín Mergarejo Moreno. Orihuela: Universidad de Alicante, pp. 1143–152.

Morote Seguido, Álvaro Francisco. 2016. El Laboratorio de Climatología de La Universidad de Alicante. Enseñanza, Divulgación e Investiación de La Geografía. In *La Investigación e Innovación En La Enseñanza de La Geogafía*. Edited by Rafael Sebastiá Alcaraz and Emilia María Tonda Monllor. San Vicente del Raspeig: Publicaciones de la Universidad de Alicante, pp. 275–89.

Morote Seguido, Álvaro Francisco. 2019a. La Enseñanza Del Cambio Climático En La Educación Primaria. Exploración a Partir de Las Representaciones Sociales Del Futuro Profesorado y Los Manuales Escolares de Ciencias Sociales. *ENSAYOS, Revista de La Facultad de Educación de Albacete* 34: 213–28. Available online: http://www.revista.uclm.es/index.php/ensayos (accessed on 18 October 2022).

Morote, Álvaro Francisco, María Hernández, and Jorge Olcina. 2021b. Are Future School Teachers Qualified to Teach Flood Risk? An Approach from the Geography Discipline in the Context of Climate Change. *Sustainability* 13: 8560. [CrossRef]

Morote Seguido, Álvaro Francisco. 2019b. Percepción de Los Futuros Maestros de Primaria Sobre El Riesgo de Inundación. La Geografía Como Herramienta Para Lograr Una Sociedad Más Resiliente Al Cambio Climático. *Papeles de Geografía* 65: 67–88. [CrossRef]

Morote Seguido, Álvaro Francisco. 2020. La Interpretación de Los Riesgos Naturales y El Cambio Climático En Postgrado. Ejemplos de Casos de Estudios En La Provincia de Alicante (España). *Cuadernos de Geografía de La Universitat de València* 103: 47. [CrossRef]

Morote Seguido, Álvaro Francisco. 2021. La Explicación Del Riesgo de Sequía En La Geografía Escolar: Una Exploración Desde Los Manuales Escolares de Ciencias Sociales (Educación Primaria). *Boletin de La Asociacion de Geografos Espanoles* 88. [CrossRef]

Morote Seguido, Álvaro Francisco, and Enrique Moltó Mantero. 2017. El Museo Del Clima de Beniarrés (Alicante). Propuesta de Un Recurso Didáctico Para La Enseñanza de La Climatología. *Didáctica de Las Ciencias Experimentales y Sociales* 4379: 109. [CrossRef]

Morote Seguido, Álvaro Francisco, and Jorge Olcina Cantos. 2020. El Estudio Del Cambio Climático En La Educación Primaria: Una Exploración a Partir de Los Manuales Escolares de Ciencias Sociales de La Comunidad Valenciana. *Cuadernos Geograficos* 59: 158–77. [CrossRef]

Morote Seguido, Álvaro Francisco, and Xosé Manuel Souto González. 2020. Educar Para Convivir Con El Riesgo de Inundación. *Estudios Geográficos* 81: 1–14. [CrossRef]

Moutahir, Hassane, Martín de Luis, Roberto Serrano-Notivoli, Issam Touhami, and Juan Bellot. 2014. Analisis De Los Eventos Climáticos Extremos En La Provincia De Alicante, Sureste De España. *Cambio Climático y Cambio Global* 9: 457–66. [CrossRef]

Naciones Unidas. 1992. Convención Marco de Las Naciones Unidas Sobre El Cambio Climático' 62301. Available online: https://unfccc.int/resource/docs/convkp/convsp.pdf (accessed on 18 October 2022).

Naciones Unidas Asamblea General. 2015. Resolución aprobada por la Asamblea General el 25 de septiembre de 2015. A/RES/70/1 Transformar Nuestro Mundo: La Agenda 2030 Para El Desarrollo Sostenible. p. 40. Available online: https://undocs.org/es/A/RES/70/1 (accessed on 18 October 2022).

Navarro Díaz, Miriam, Olga Moreno Fernández, and Ana Rivero García. 2020. El Cambio Climático En Los Libros de Texto de Educación Secundaria Obligatoria. *Revista Mexicana de Investigación Educativa* 25: 957–85. Available online: https://dialnet.unirioja.es/servlet/extart?codigo=7604721 (accessed on 18 October 2022).

Nelles, David, and Christian Serrer. 2020. *El Pequeño Manual Del Cambio Climático*. Barcelona: Grijalbo.

Olcina Cantos, Jorge. 2017. La Enseñanza Del Tiempo Atmosférico y Del Clima En Los Niveles Educativos No Universitarios. Propuestas Didácticas. In *Enseñanza y Aprendizaje de La Geografía Para El Siglo XXI*. Edited by Rafael Sebastiá and Emilia Tonda. Alicante: Universidad de Alicante, pp. 119–48.

Olcina Cantos, Jorge, and José Fernando Vera-Rebollo. 2016. Adaptación Del Sector Turístico Al Cambio Climático En España. La Importancia de Las Acciones a Escala Local y En Empresas Turísticas. *Anales de Geografía de La Universidad Complutense* 36: 321–52. [CrossRef]

Özdem, Yasemin, Burçkin Dal, Nilay Öztürk, Duygu Sönmez, and Umut Alper. 2014. What Is That Thing Called Climate Change? An Investigation into the Understanding of Climate Change by Seventh-Grade Students. *International Research in Geographical and Environmental Education* 23: 294–313. [CrossRef]

Pérez-Morales, Alfredo, Salvador Gil-Guirado, and Jorge Olcina Cantos. 2022. La Geografía de Los Riesgos En España (1992–2022): Cambios y Oportunidades Enuna Temática de Trabajo Consolidada y En Alza. In *La Geografía Española Actual. Estado de La Cuestión: Aportación Española Al Congreso de La UGI. París 2022. El Tiempo de Los Geógrafos*. Edited by Asociación Española de Geografía. Madrid: Comité Español de la Unión Geográfica Internacional, pp. 183–202.

Pickering, Gary J., Kaylee Schoen, Marta Botta, and Xavier Fazio. 2020. Exploration of Youth Knowledge and Perceptions of Individual-Level Climate Mitigation Action. *Environmental Research Letters* 15: 104080. [CrossRef]

Ramos Ribeiro, Rodrigo Rudge, Jorge Olcina Cantos, and Sergio Molina Palacios. 2014. Análisis de La Percepción de Los Riesgos Naturales En La Universidad de Alicante. *Investigaciones Geográficas* 61: 147. [CrossRef]

Sánchez-Almodóvar, Esther, Isabel María Gómez-Trigueros, Jorge Olcina-Cantos, and Javier Martí-Talavera. 2022a. Percepción Del Cambio Climático y de Los Extremos Atmosféricos Asociados En El Alumnado de ESO y Bachillerato. In *XVII Coloquio Ibérico de Geografía. Nuevas Fronteras y Nuevos Horizontes En La Geografía Ibérica: Políticas y Transformaciones Territoriales. Libro de Actas*. Edited by Asociación Española de Geografía. Salamanca: Asociación Española de Geografía, pp. 1087–99.

Sánchez-Almodóvar, Esther, Javier Martin-Vide, Jorge Olcina-Cantos, and Marc Lemus-Canovas. 2022b. Are Atmospheric Situations Now More Favourable for Heavy Rainfall in the Spanish Mediterranean? Analysis of Episodes in the Alicante Province (1981–2020). *Atmosphere* 13: 1410. [CrossRef]

Sánchez Almodóvar, Esther, Jorge Olcina Cantos, Javier Martí Talavera, Antonio Oliva Cañizares, and Samuel Biener Camacho. 2022. La Enseñanza Del Clima, El Cambio Climático y Los Extremos Atmosféricos: Propuestas Para La Ciudadanía Desde La Universidad. In *Ensino de Geografia e e a Redução Do Risco de Desastres Em Espaços Urbanos e Rurais*. São Paulo: CPS, pp. 59–96. [CrossRef]

Serantes-Pazos, Araceli. 2015. Como Abordan o Cambio Climático Os Libros de Texto Da Ensinanza Secundaria Obligatoria Na España. *AmbientalMENTEsustentable: Revista Científica Galego-Lusófona de Educación Ambiental* 20: 249–62. [CrossRef]

Serrano-Notivoli, Roberto, Santiago Beguería, Miguel Ángel Saz, and Martín De Luis. 2018. Recent Trends Reveal Decreasing Intensity of Daily Precipitation in Spain. *International Journal of Climatology* 38: 4211–24. [CrossRef]

Sóñora, Francisco, Ma Mercedes Rodríguez-Ruibal, and y Raquel Troitiño. 2009. Un modelo activo de educación ambiental: Prácticas sobre cambio climático. *Enseñanza de las Ciencias de la Tierra* 17: 196–206.

Souto González, Xosé Manuel. 1998. *Didáctica de La Geografía*. Edited by Ediciones del Serbal. Barcelona: Ediciones del Serbal.

Sureda-Negre, Jaume, Miquel Oliver-Trobat, Albert Catalan-Fernändez, and Rubén Comas-Forgas. 2014. Environmental Education for Sustainability in the Curriculum of Primary Teacher Training in Spain. *International Research in Geographical and Environmental Education* 23: 281–93. [CrossRef]

Tonda-Monllor, Emilia-María, and Rafael Sebastiá-Alcaraz. 2003. Las Dificultades En El Aprendizaje de Los Conceptos de Tiempo Atmosférico y Clima: La Elaboración e Interpretación de Climogramas. *Revista de Educación de La Universidad de Granada* 16: 47–69.

United Nations General Assembly. 2015. Resolution 69/283. Sendai Framework for Disaster Risk Reduction 2015–2030. June 3, p. 24. Available online: www.unisdr.org/we/inform/terminology (accessed on 18 October 2022).

Zhong, Shuang, Qiu Cheng, Shuwei Zhang, Cunrui Huang, and Zhe Wang. 2021. An Impact Assessment of Disaster Education on Children's Flood Risk Perceptions in China: Policy Implications for Adaptation to Climate Extremes. *Science of The Total Environment* 757: 143761. [CrossRef]

social sciences

MDPI

Article

Conflicting Knowledge Paradigms: Competence Discourse and Disciplinary Reality in Social Sciences Teaching

Diego Luna * and José Antonio Pineda-Alfonso

Department of Experimental and Social Sciences Didactics, Faculty of Education Sciences, University of Seville, 41013 Seville, Spain
* Correspondence: dluna@us.es

Abstract: The purpose of this study was to offer an inside look at the coexistence of different ways of conceiving and organising the knowledge of social sciences in current schools. To this end, the methods of autoethnography and critical discourse analysis were used in a case study focused on the teaching experience of one of the authors. The system of categories constructed allowed us to empirically verify the existence of a conflict between two major epistemological paradigms, competency and disciplinary, in clear correspondence with a gap between innovative educational discourses and traditional school practices. In the case analysed, this conflict led to a curricular over-dimensioning, which aimed to expand the elements to be worked on by teachers and their students qualitatively and quantitatively. The causes of this phenomenon are related to neo-liberal pedagogical parameters and allowed us to conclude that we should not conceive of competency-based learning either as an educational change by itself or as a strategy that guarantees change.

Keywords: social sciences; competency-based learning; knowledge; case study

Citation: Luna, Diego, and José Antonio Pineda-Alfonso. 2022. Conflicting Knowledge Paradigms: Competence Discourse and Disciplinary Reality in Social Sciences Teaching. *Social Sciences* 11: 553. https://doi.org/10.3390/socsci11120553

Academic Editor: Alvaro Morote

Received: 31 October 2022
Accepted: 25 November 2022
Published: 28 November 2022

1. Introduction

> A methodological approach based on key competences and learning outcomes entails important changes in the conception of the teaching-learning process, changes in school organisation and culture; it requires close collaboration between teachers in curriculum development and in the transmission of information on student learning, as well as changes in working practices and teaching methods. (Orden ECD/65/2015, de 21 de enero 2015, de 21 de enero 2015, p. 6988)

Competency-based learning is an important pedagogical pillar of innovative schools (Guzmán Marín 2017; Hou 2022; Marcotte and Gruppen 2022), a model to which all schools, both private and public, seem to aspire today. This can be seen, for example, in the strong commitment to the preparation or competency training of educational communities, which are understood as the "human capital" of schools (Valencia Álvarez and González 2017, p. 131). The prominence of competences in the desirable school model is also perceived in the argument for a globalising approach to learning content, with the supposed aim of developing an interpretative perspective of reality in pupils that is much more elaborate and complex than the traditional one.

According to Spanish educational laws, the curricular integration of competency-based learning would be perfectly compatible with the classic objectives of areas such as social sciences: using different strategies (observation, data collection, measurement, interpretation, synthesis, drawing conclusions, and so on) to analyse the past in order to understand the present, applying social concepts to resolve significant problems of current reality, developing democratic habits through discussion and teamwork tools, incorporating new values and social attitudes to improve coexistence with others, acquiring intellectual maturity and developing a well-argued critical conscience, and so on.

Basically, competence is know-how or applied knowledge, around which old teaching and learning strategies need to be reconsidered. This approach should start from the elements of the curriculum, establishing meaningful relationships between different knowledge or subjects precisely by working on the so-called "key competences". Currently, the Ministerio de Cultura y Formación Profesional (formerly the Ministerio de Educación, Cultura y Deporte) of the Spanish Government officially considers the following eight "key competences", which have been integrated in the related curricular developments, from pre-school to baccalaureate: competence in linguistic communication; digital competence; entrepreneurial competence; multilingual competence; personal, social, and learning to learn competence; competence in cultural awareness and expression; competence in mathematics, science, technology, and engineering; and citizenship competence.

The competency-based learning model, very close to vocational training, has actually been trying to penetrate schools for more than a decade through the latest educational laws. According to Cañadell (2018, p. 106), its origins date back to the end of 2007, when the European Union began to address the need for "new skills for new jobs". Some of these "soft" skills were communication, analysis, problem solving, and entrepreneurship skills. In fact, the economicist perspective embedded in the reformist philosophy of laws such as the Ley Orgánica para la Mejora de la Calidad Educativa (de España 2013) (Organic Law for the Improvement of the Quality in Education) is especially linked to the argument for useful skills to compete in the labour market.

> A more open, global and participatory society demands new, more sophisticated and diversified profiles of citizens and workers, as well as alternative ways of organisation and management that prioritise collaboration and teamwork, and proposals that assume that true strength lies in the mix of different skills and knowledge. (de España 2013, p. 5)

Together with the LOMCE, within the Spanish regulatory framework, we can highlight the importance of texts such as Orden ECD/65/2015, de 21 de enero, por la que se describen las relaciones entre las competencias, los contenidos y los criterios de evaluación de la educación primaria, la educación secundaria obligatoria y el bachillerato (Order ECD/65/2015, of 21 January, which describes the relationships among the competences, the contents, and the evaluation criteria of primary education, compulsory secondary education, and baccalaureate). Addressed primarily to teachers, this document attempts to justify, clarify, and specify various aspects, both theoretical and operational, relating to the work and assessment of competences. At first glance, the objectives appear to be quite ambitious, since, among other issues, it states that: "competences should be developed in the fields of formal, non-formal and informal education throughout primary education, compulsory secondary education and baccalaureate, and in lifelong learning throughout life" (Orden ECD/65/2015, de 21 de enero 2015, de 21 de enero 2015, p. 6989). A close correspondence has also been established between competences and student motivation: "One of the key elements in teaching by competences is to awaken and maintain motivation for learning in students, which implies a new approach to the role of the student, active and autonomous, aware of being responsible for their learning" (Orden ECD/65/2015, de 21 de enero 2015, de 21 de enero 2015, p. 7002).

These kinds of arguments, together with the numerous allusions to the international context with which the Order begins, seem to evoke a new pedagogical meta-narrative that is impossible to avoid. The most remarkable part of the text is undoubtedly its Appendix I, which includes rich descriptions of each of the key competences that are "essential for the well-being of European societies, economic growth and innovation, and describes the essential knowledge, skills and attitudes linked to each of them" (Orden ECD/65/2015, de 21 de enero 2015, de 21 de enero 2015, p. 6986). In this sense, authors such as Kozma (2012) have devoted their efforts to identifying the so-called "21st century skills": responding flexibly to complex problems, communicating effectively, managing information, working in teams, using technology, and producing knowledge.

In any case, the sub-discourse of competency-based knowledge does not only affect students and their learning, but also the functions of agents such as school principals or teachers. The former must now possess new competences that go beyond the "insufficient" skills of traditional principals (Jolonch 2019, p. 217). In fact, it is no longer enough to be "a mere manager, and even less so to hold the status of government" (Jiménez Abad 2015, p. 172). On the contrary, the new leaders must "be optimistic, resilient, outward-looking, curious, collaborative and committed to social justice, as well as being pragmatic," while being "efficient and effective in leading the school to create an orderly environment" (Greany 2019, p. 64).

Regarding teachers, today's desirable teacher must have a vast range of new resources and skills, which complement or directly rethink some of the traditional teaching strategies. Their digital competence (Martínez-Abad et al. 2017; Redecker 2020; Rossi Cordero and Barajas Frutos 2018) depends not only on the design of new formats for teaching and learning but also the possibility of instilling in students a set of profoundly valuable knowledge in a highly technologised socio-cultural context. Hence, the need to reduce the digital divide that still exists between a large part of the teaching profession and digital natives is urgent (Liesa et al. 2018; Mohamed et al. 2017; Pérez-Escoda et al. 2016; Rubio and Tejada 2017). Along with the rest of the competences, digital or information literacy undoubtedly requires an in-depth review of traditional assessment formulas (Gulikers et al. 2018; Robinson and Dervin 2019).

However, the argument for competency-based teaching and learning has not been without controversy in recent years (Cachinho and Reis 2007; Espinoza Aros 2014; Solé Blanch 2020; Soto and Pérez-Milans 2018), as the alleged drawbacks have not been few. Some of the voices against the competence approach have been the following.

- Hirtt (2010) criticised the supposed displacement of humanistic learning in favour of the prominence of a set of skills clearly designed to be productive in professional settings.
- From a pessimistic point of view, Barrio (2013) warned that: "If education were reduced to the provision of 'skills', it would be no different from the training of an irrational animal" (p. 111).
- Jiménez Abad (2015) called for what he called "knowing how to be", "that is, values and inclusive personal development" (p. 167). As an example, he referred to the case of digital competence, which "cannot be reduced to knowing how to operate a computer. It includes among other things a critical and reflective attitude to the information available and a responsible use of interactive media. It is not just a question of keys, it is above all a question of ideas, criteria and ethical values" (p. 165).
- Tricot (2019) regretted the fact that "nothing should be taught that cannot be described in the form of competences" (p. 95). The same author argued that "this concept corresponds much more to an awareness than to an innovation" (p. 95), suggesting that competency learning would already have been implicit, in one form or another, in the traditional learning model.

Assuming the existence of these critical perspectives on the chosen topic, the specific objective of this work, within the framework of a much broader doctoral research project, was to develop an internal investigation into the true place of competency-based learning in classroom practice, which is related to the traditional way of conceiving and working with school knowledge.

2. Methodology

2.1. A Case Study

To carry out this project, we opted for a case study focused on all the converging discourses in the Geography and History classes in Spanish secondary education of one of the authors. This case was chosen for several reasons: (1) it was set in a discursive and practical context, a private school in the city of Seville, whose predominant school culture required an in-depth analysis; (2) it was undergoing a major epistemological crisis as a result of the didactic transformations developed in recent years; and (3) it was a professional environment full of difficulties and contradictions, which had to be understood before it could be exposed to any kind of improvement proposal.

The corpus compiled during the fieldwork combines a total of 12 texts from four different discursive fields: the legislative framework, the school, the Social Sciences Department, and the classroom itself (see Table 1). The research participants, the teacher–researcher and 44 students from the four years of secondary school, belong to the last of these contexts.

Table 1. Composition and organisation of the research corpus.

Sub-Corpora	Text
Spanish and Andalusian education policies	(1) Ley Orgánica 8/2013, de 9 de diciembre (de España 2013)
	(2) Real Decreto 1105/2014, de 26 de diciembre
	(3) Orden ECD/65/2015, de 21 de enero
	(4) Decreto 111/2016, de 14 de junio
School	(5) Orden de 14 de julio de 2016
	(6) School Educational Project
	(7) School Regulation on Organisation and Operation
Department of Social Sciences	(8) Compilation of materials on the school's methodology
	(9) Geography and history subject guide for students
	(10) Annual reports for the academic years 2017–2018 and 2018–2019
Geography and History class	(11) Interviews with 44 students (11–15 years old)
	(12) Teacher–researcher's diary

2.2. Methods, Techniques, and Instruments

In this research, the analysis methods chosen were autoethnography (Adams et al. 2015; Ellis 2016; Mitra 2010; Starr 2010; Wamsted 2012) and critical discourse analysis (Fairclough 2003; Pardo Abril 2013; Pini 2009; Rogers 2011, 2018; Wodak and Meyer 2003). In this way, the descriptive potential of narrative methodologies was combined with a set of varied procedures to compare the underlying intentions of the texts with their actual consequences. In this sense, it is worth highlighting the use of techniques that are applicable to both quantitative and qualitative data and that are both deductive and inductive in nature, always taking the demands of the object of study and its particular discursive–verbal approach into account (see Table 2). This methodological proposal was directly inspired by the assumptions established in the theoretical framework by previous research and teaching experiences as well as by the parameters of Grounded Theory (Strauss and Corbin 2008).

Table 2. Relationships among the methods, techniques, and instruments used in the research.

Methods	Data Collection Techniques	Analysis Techniques	Instruments
Autoethnography	Participant observations		Teacher–researcher's diary
	Structured interviews		Interview script
Critical discourse analysis		Pre-analysis	
		Lexicometric analysis	MAXQDA and Sketch Engine
		Content analysis	Mixed category system
		Linguistic analysis	

All the methodological elements of our case study were based on these parameters. In this sense, a constant work of observation, comparison, reflection and reading was developed throughout the research process, assuming the attributes of all qualitative researchers; for example, "the ability to live with ambiguity" or "an acceptance of the self as a research instrument" (Strauss and Corbin 2008, p. 13).

2.3. Strategies of Analysis

The analysis, inspired by the proposals of several authors (Pardo Abril 2013; Wodak and Meyer 2003), although specifically designed for this research, involved four complementary steps. Each of these should be understood as a different, albeit complementary, approach to the research corpus.

1. Pre-analysis, where the texts of the corpus were characterised in order to discover their heuristic potential in relation to the research objectives. In this sense, we observed how the regulatory texts, with the LOMCE as the protagonist, found their raison d'être in a concept of business quality, which is in line with the ideology of the central government at the time. The sub-discourse on competences was perfectly aligned with this. On the other hand, the school's texts responded precisely to the aims of a business project whose product was the educational model defended by its promoters, while the texts of the Department of Social Sciences reflected the existence of a certain discursive gap in terms of the Geography and History subject approach. Finally, we were able to note a strong impact of the previous discourses in the Geography and History class, according to the testimonies of its participants, but also an important margin for questioning. Although intuitive and preliminary, this corpus characterisation was the first step in laying the foundations for the subsequent quantitative and qualitative analyses.
2. Lexicometric analysis using techniques based on textual statistics (word count, identification of key segments, recognition of co-texts, and so on) made it possible to extract previously unnoticed information on frequencies and associations. For example, we were able to confirm the high quantitative presence of students in the four sub-corpora. With regard to the concept of innovation, the central focus of the global research, the identification of its corresponding associations allowed us to characterise this concept in the following way: (a) its meaning was taken for granted and worked as an argument to justify the changes of the LOMCE educational reform; (b) in its main collocations, it coincided with particularly ethereal words; (c) its co-occurrences and concordances only suggested positive connotations; (d) it was closely linked to business, entrepreneurship and economics; and (e) it appeared closely linked to new technologies and little to new methodologies. All these results encouraged us to go beyond the textual materiality to capture the meanings, visible and invisible, deductive and inductive, thematic and non-thematic, with which the discursive units were used in our corpus.
3. Content analysis, which materialised in a system that combined both deductive categories (identified in the theoretical framework) and inductive categories (from the analysed data). Our system of categories fulfilled the objective of deconstructing the corpus from a thematic perspective, allowing us to know what the main themes and topics present in our texts were, thus approaching the way in which such contents are interpreted according to each discursive field. In this sense, we managed to locate the central problem of our study, the impact of educational innovation from the teaching experience, in a broad, complex and contradictory context, where this phenomenon conflicted with different discourses and practices related to the educational model defended both in the laws and in the organisational documents of the school.
4. Linguistic analysis, which focused on identifying the grammatical resources and strategies used. The analysis of the coherence and consistency of the discourses present in the corpus revealed diverse conceptions of educational innovation, highlighting the dissonance represented by the teaching voice regarding the uncritical, distanced and

indifferent perspective of the rest of the discourses. The analysis of naming strategies allowed us to know how the discursive actors are recognised in the texts, according to the perspective of each discursive field. Finally, the analysis of legitimation processes showed us a wide range of persuasive strategies, which contributed to convince of the appropriateness and urgency of the proposals for change coming from the highest spheres. Particularly interesting was the game of recognition between students and their teacher. Each in their own way, with their own processes and linguistic resources, tried to construct their own discourse in response to the domination that was being exercised over them.

Although these four phases were totally conducted, this paper focuses specifically on the third of these phases, an intermediate point between lexicometric exploration and the linguistic deconstruction of the different levels of meaning present in the discourses under analysis. In this phase, we can find the most appropriate response to the challenge of managing and organising a large and diverse corpus, providing a completer and more elaborate vision of the discursive topics than that offered in the pre-analysis phase.

The category of analysis this study focused on directly linked us to the didactic model observed in the Geography and History classes that constitute our case study. More specifically, this category allowed us to understand the typology of the contents that the students and the teacher–researcher involved had to work on (see Table 3). The name of this category, *confronting knowledge paradigms*, refers, primarily, to the fact that the elements that make up this typology, which are all related to school knowledge identified in practical reality, are far from being homogeneous but, as we will see in the following section, are diverse and even contrary to each other.

Table 3. Sections of the general category system on which this work is focused.

Category	Code	Subcategory	Code
Confronting knowledge paradigms	COKPA	Competence discourse	CODI
		Disciplinary reality	DIRE

3. Results

3.1. Competence Discourse

Our first subcategory reflects the prominence of competency-based learning in the educational discourse contained in the normative texts. This element is presented as the central axis of the curriculum of the different educational levels and as a definitive impetus for the educational changes that must take place in the classroom. This is expressed in the following fragment: "This Real Decreto is based on the promotion of learning by competences, integrated into the curricular elements in order to promote a renewal in teaching practice and in the teaching and learning process" (CODI-T2).

First of all, it is worth noting the variety of definitions of the term "competence" in texts which, in theory, should be consonant with each other. On the one hand, competence is defined as "a combination of practical skills, knowledge, motivation, ethical values, attitudes, emotions, and other social and behavioural components that are mobilised together to achieve effective action" (CODI-T2). Moreover, it is stated that competences "are conceptualised as 'know-how' that applies to a variety of academic, social and professional contexts" (CODI-T2). Particular emphasis is also placed on the importance of a proper understanding of the typology of knowledge present in competences.

> Competence knowledge integrates a conceptual knowledge base: concepts, principles, theories, data and facts (declarative knowledge-knowing how to say); knowledge related to skills, referring to both observable physical action and mental action (procedural knowledge-knowing how to do); and a third component that has a strong social and cultural influence, and which involves a set of attitudes and values (knowing how to be). (CODI-T3)

Secondly, several benefits are associated with the interrelationships of the different components that characterise competence learning. One of these is the fact that this type of learning "favours the learning processes themselves and the motivation to learn" (CODI-T3). In this way, a direct and unquestionable correlation is established between work based on competences and the supposed stimulation of students. Another of the great virtues of competency-based learning is expressed in Orden ECD/65/2015, de 21 de enero, as "an integral training of people", which enables them to "reorganise their thinking and acquire new knowledge, improve their actions and discover new forms of action and new skills that enable them to perform tasks efficiently, favouring lifelong learning" (CODI-T3). From this perspective, competency-based learning is presented as a definitive and revolutionary approach to renew and improve the quality of any educational process.

As a complement to this apologetic approach, the same text indicates some of the keys to developing an effective process of competency-based learning at school.

> Since competency-based learning is characterised by its transversality, its dynamism and its comprehensive nature, the competency-based teaching-learning process must be approached from all areas of knowledge and by the various entities that make up the educational community, in both formal and non-formal and informal spheres. Its dynamism is reflected in the fact that competences are not acquired at a given moment and remain unalterable, but involve a process of development through which individuals gradually acquire higher levels of performance in the use of these competences. (CODI-T3)

Focusing now on the process of integrating the competency-based learning model into the official Spanish curriculum, we can point out the international references identified in the normative texts. On the one hand, it was stated that UNESCO was the organisation that established the precursor principles for the application of competency-based teaching "by identifying the basic pillars of lifelong learning for the 21st century, consisting of 'learning to know', 'learning to do', 'learning to be' and 'learning to live together'" (CODI-T3). On the other hand, in line with what was noted in the theoretical framework, the direct reference for Spain is the European Union. According to the text, the guidelines of this institution raise the relevance of competences by defending the acquisition of these "as an indispensable condition for individuals to achieve full personal, social and professional development that meets the demands of a globalised world and makes economic development possible, linked to knowledge" (CODI-T3).

In fact, Spanish legislation adopts the concept of *key competences* as defined by the European Union: "those that all people need for their personal fulfilment and development, as well as for active citizenship, social inclusion and employment" (CODI-T2), adding, at another point, that these are "essential for the well-being of European societies, economic growth and innovation" (CODI-T2). The precedent, in the Spanish case, would be the Ley Orgánica de Educación (LOE) (2006), which included the term "basic competences" for the first time. In this sense, the LOMCE and the different regulatory texts that developed it claimed to go "further" in terms of the implementation of the curriculum of competences by including the term "competences" within the definition of the "basic curriculum" (CODI-T3).

Specifically, the two main sources with which the Orden ECD/65/2015 de 21 de enero claimed to align itself are "the results of educational research" and "the European trends set out in Recommendation 2006/962/EC of the European Parliament and of the Council of 18 December 2006 on key competences for lifelong learning" (CODI-T3). On this basis, Article 2 of the Orden established that the key competences in the Spanish education system are the following: (a) linguistic communication, (b) mathematical competence and basic competences in science and technology, (c) digital competence, (d) learning to learn, (e) social and civic competences, (f) a sense of initiative and entrepreneurship, and (g) cultural awareness and expressions (CODI-T3). Article 4 ("Key competences and objectives of the stages") contains certain requirements for the integration of competences in the different educational stages. Among them, due to the originality it represents with respect to the ideas already discussed, the following should be highlighted: "the design of integrated learning activities that enable progress to be made towards the learning outcomes of more than one competence at the same time" (CODI-T3).

At this point, it is necessary to know more specifically what the guidelines are for the integration of key competences in all curricular areas or subjects, which can be found in Article 5 of the same Orden we have just mentioned. These guidelines, which are summarised in the following table, constitute certain basic requirements to be taken into account by teachers when drawing up or redefining teaching programmes in terms of key competences (see Table 4).

Table 4. Selection of excerpts from the Orden ECD/65/2015, de 21 de enero, on the topic *guidelines for the integration of key competences in the curriculum.*

Excerpt	Code
The learning outcomes to be achieved by pupils should be "defined, made explicit and sufficiently developed in the subject areas".	CODI-T3
"(. . .) Since the evaluable learning standards are linked to competences, this profile will identify those competences that are developed through that area or subject".	CODI-T3
"All areas and subjects must contribute to the development of competences. The set of evaluable learning standards of the different areas or subjects that relate to the same competence gives rise to the profile of that competence (competence profile). The elaboration of this profile will facilitate the competency assessment of students".	CODI-T3
"The assessment of the competency development must be integrated with the assessment of contents, insofar as being competent means mobilising knowledge, skills, attitudes and values to respond to the situations posed, providing learning with functionality and applying what is learnt from an integrated approach".	CODI-T3

Considering all these requirements, this is the right moment to find out how the competency approach is included in the subject of Geography and History, according to the regulatory discourses. In this sense, one of the objectives of the subject contemplated in the Real Decreto 1105/2014, de 26 de diciembre can be used as an introduction to this new topic: "to continue acquiring the competences necessary to understand the reality of the world in which they live, past and present collective experiences, their orientation in the future, as well as the space in which life in society develops" (CODI-T2).

In the Orden de 14 de julio de 2016, which, among other aspects, develops the curriculum of secondary education in the region of Andalusia, we found a characterisation of the subject which specified that our subject will train students in certain "skills" for, on the one hand, "the understanding of the complexity of current societies" and, on the other, "the analysis and implementation of the strategies required to exercise responsible, participatory citizenship, aware of their identity, rights and obligations, in a plural and globalised environment" (CODI-T5). This last idea is mentioned again in the supposed transversal "contributions" of Geography and History to the overall ESO curriculum, referring to "the development of personal competences and social skills for the exercise of participation through the assessment of the various channels of citizen action and empowerment" (CODI-T5). At first glance, in the context of our subject, the concept of competence acquires a new significance that moves away from what has been said about key competences to focus on knowledge that is fundamentally attitudinal and related to the development of civic education.

In order to understand the supposed contribution of the subject of Geography and History to the development of competences, we must move forward in examining the same Orden. Below (see Table 5), we list some excerpts that allowed us to explore this issue in greater depth.

Table 5. Selection of excerpts on the topic *contribution of the subject of Geography and History to the learning of competences, according to the Orden de 14 de julio de 2016.*

Competence	Excerpt	Code
Linguistic communication	"By working on oral and written comprehension and expression through source analysis, preparation of papers and participation in debates".	CODI-T5
Mathematics and basic competences in science and technology	"Through the handling and analysis of numerical information as well as the assessment of scientific and technological advances for social progress".	CODI-T5
Digital competence	"Use of applications and software that allow the collection, organisation, presentation and editing of information and conclusions of content and projects related to this subject".	CODI-T5
Cultural awareness and expressions	"Thanks to the understanding of the cultural fact, its relationship with personal and social identity, its most outstanding manifestations and the importance and benefits of its conservation, dissemination and enhancement".	CODI-T5
Learning to learn	"Through case studies, research papers, projects and the use of cognitive skills involving comparison, organisation and analysis".	CODI-T5
Sense of initiative and entrepreneurship	"Thanks to the knowledge of the world of economics, business and the functioning of society and politics, as well as the development of personal and social skills in group work".	CODI-T5
Social and civic	"With the knowledge and skills that promote personal and collective well-being through the assumption of the codes of conduct, rules of operation and rights and obligations of citizenship that govern social and democratic states".	CODI-T5

All these discursive propositions have an impact on the way in which the "Objectives" (the title of the section) or "competences" (the term by which the list of objectives is finally referred to) are linked to how the subject of Geography and History is formulated. In this respect, it is interesting to note the use of the latter word as a synonym for "specific competences" of the subject, which may cause some confusion for teachers when interpreting the official curriculum. In fact, it would seem that the development of these "competences" would imply the implementation of several competences at the same time. As an example of this last idea, one of the 16 objectives or capacities established in the Andalusian curriculum is reproduced below: "To know and analyse the ways in which human society transforms the environment, and how the territory influences the organisation and identity of this society, reflecting on the dangers that human intervention generates in the environment, with special emphasis on the case of Andalusia" (CODI-T5).

3.2. Disciplinary Reality

> In the area of Social Sciences, in particular, the contents are almost strictly conceptual, to the extent that there is a certain obsession with facts, dates, characters, data, features or geographical places on the map. Bearing in mind the importance we give to memorised exams, both in History and Geography, it is clear that these contents are understood as closed, almost timeless and, of course, classified in disciplinary compartments. Proof of this is that, from time to time, the Department asks us to include a series of notes and materials dating back fifty years. (DIRE-T12)

Our second subcategory represents the predominance of disciplinary contents in school practice as opposed to a competency-based approach to teaching and learning. Paradoxically, this situation arises in the very regulatory texts responsible for regulating the official curriculum, through the existence of certain contradictions. On the one hand, it is stated, as we saw a moment ago, that competences must be integrated into all curricular elements; on the other, we find that the way of stating certain elements may be indicative of a purely conceptual dimension rather than a competence dimension. In other words, the suspension of the discourse on competences, in some parts of the normative texts, implies the appearance of a discourse that is not very different from the traditional disciplinary organisation of curricular content.

As an example of this, we can reproduce some of the terms used in the Orden de 1 de julio de 2016 to characterise the most appropriate way of working on the elements considered to be "relevant" in the subject of Geography and History: "range of scientific references", "study", "conceptualising society", "global vision", "knowledge of historical facts", "understanding the position and relevance of Andalusia in the rest of Spain", "knowing and handling the vocabulary, and research and analysis techniques specific to the social sciences", "causes of war and conflict", "relevant characteristics of the natural environment both in Andalusia and the rest of the world", "gathering information", and so on (DIRE-T5).

The indications about the possible competences of the "objectives" of Geography and History are few and diffuse. In fact, among the consequences of working in this subject, the text cited does not highlight the development of any specific key competences but highlights the "appreciation" of the following aspects: "the connections between past and present and human beings and nature; the importance of the notions of change and continuity in social structure and dynamics and the value of comparative methodology together with diachronic and synchronic analysis" (DIRE-T5). The apparent abandonment of the competency discourse can also be verified by looking at the main actions that are mentioned in the objectives of our subject (not including those that make up subordinate clauses), which are grouped in Table 6 below according to their frequency of appearance.

Table 6. Main actions contemplated in the objectives of the subject of Geography and History, according to the Orden de 14 de julio de 2015.

Frequency of Appearance	Actions
More than five times	Analyse (10), know (6), evaluate (6)
Two to five times	Understand (4), value (4)
One time only	Conceptualise, situate, classify, reflect, acquire, manifest, manifest, compare, contextualise, appreciate, explain, expose, master, argue, debate, handle, lend, perform, participate, employ, respect, express

The following excerpt, focused on the task of knowing, is a good example of the type of actions that articulate the objectives, which do not seem to have undergone any kind of transformation under the competence approach that is supposed to be present in all curriculum subjects: "Knowledge of society, its organisation and functioning over time is essential in order to understand today's world. Knowing the space where societies develop, the natural resources and the use that has been made of them, provides us with data about the past (...)" (DIRE-T2). Nor in the following words do we discover any clue as to the place that competences should occupy in these changes that the secondary stage represents with respect to primary education: "pupils will delve, in a more systematic, organised and profound way than in Primary Education, into the foundations of the identity and functioning mechanisms of human society, and the forms of relationship between them and with the environment, as well as the spatial dimension in which these arise and develop" (DIRE-T5).

Even when speaking of "transversal elements", no reference to competences is made, as such, but instead to what appear to be themes to be included in the conventional content blocks. Some examples are: "the situation of women throughout history and the struggle for the recognition of their rights … ; the identity, projection and space of Andalusian culture in the rest of Spain and the world … ; the economic growth and development of Andalusia in history and the present" (DIRE-T5). In general, in the regulatory texts, there is a predominant use of terms such as "disciplines" and a recognition of the usefulness of both Geography and History as well as all those disciplines linked to the area of Social Sciences, "for a better understanding of social reality" (DIRE-T2).

Despite the complexity and interdisciplinarity advocated in the normative texts, the organisation of curricular content is, at least in appearance, rather conventional: "Geography is organised, in the first cycle, in the blocks 'The physical environment' and 'Human space', and in the fourth year it focuses on globalisation. History studies societies over time, following a chronological criterion throughout the two cycles of ESO" (DIRE-T2).

Regarding the discourses produced by the school's Department of Social Sciences, the first thing to note is that the question of competency-based learning appears, as in the normative texts, to be diluted and/or directly displaced. This can be seen in the discourse represented in the Geography and History subject guide for students in their first year of secondary education, where expressions such as "ask ourselves questions and draw conclusions", "write detailed answers and essays", "give a voice to historical figures", "broaden both our knowledge of our environment and our critical thinking", and so on (DIRE-T9) are used. One of the fragments where the survival of a habitual way of approaching teaching and learning processes can be best appreciated is the following one:

> Our History of Art contents will run parallel to the History topics, and will also be worked on in relation to the general characteristics of each artistic style, from some well-known examples, as well as from the specific vocabulary that allows us to formally describe an artistic work. (DIRE-T9)

Leaving aside the didactic criteria of the Department, the particular way of conceiving and working on competences at the School (in this case, in relation to the assessment instruments used) was interpreted from the teaching perspective in a critical way:

> I remember how stunned I was last year when I asked about the assessment of competences in our subjects. One of my senior colleagues said to me—completely convinced—: "Sure, here they are next to each question", pointing out to me the acronyms of the key competences, which he put next to questions that were 100% conceptual, the same ones he had been asking for thirty years. This is practically all the questions in our exams, with the exception of those on historical maps or some Geography graphs. (DIRE-T12)

The following diary note, again a critical one, provides further insight into the Department's priorities in teaching plans:

> When I arrived at the school, I was struck by the fact that in 3rd Secondary level there were also History subjects. My Department argues that it is important that pupils should not spend a year without seeing historical content, so the Middle Ages are left for 2nd and the Modern Age for 3rd. The truth is that I think this is a good approach, although it follows the same chronological criteria as always and the weight of Geography is somewhat reduced in order to go deeper into data and references which, in my opinion, completely escape the purpose of the subject at these levels. (DIRE-T12)

To conclude this inside look at what was happening in our case study, the best way to confirm what type of content was being worked on in the subject of Geography and History is through the pupils' answers to the first question in the interview, specifically the second part: "Do you consider that the content of Social Sciences (History, Geography, Art, Music …) requires a lot of effort? What do you find most difficult?". The next graph shows which aspects are the most difficult for students. The number of times each of these aspects was mentioned was taken into account, as well as the total number of aspects mentioned, in order to calculate the respective percentages. In Figure 1, the aspects are ordered from the highest to the lowest number of mentions received.

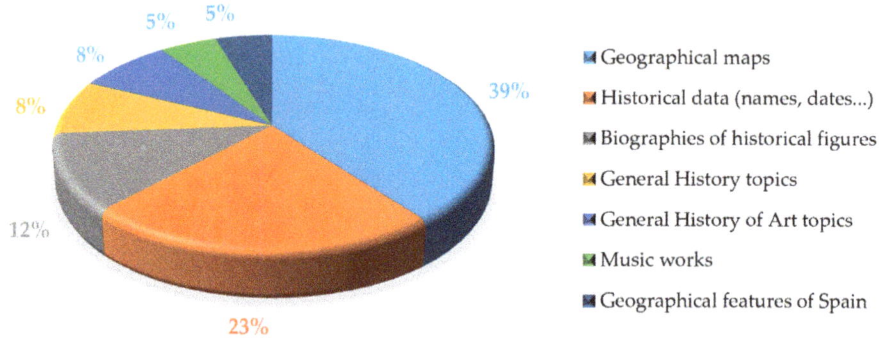

Figure 1. Typology of Social Sciences contents that generate greater difficulty for students.

Table 7 provides a representative selection of the responses from which the data shown in the graph above were obtained.

Table 7. Selection of student responses to the first interview question.

Excerpt	Code
"For my part, I find maps and history subjects (Greece, Rome) the most difficult contents. For me, it does require effort".	DIRE-T11
"In my opinion, it does. I think the things that cost the most are: maps, historical characters and important dates".	DIRE-T11
"What I find most difficult is to learn the theory, but it is not very difficult, as I learn everything like a movie of life. It is true that it is very difficult for me to learn the names and dates of events. I don't like art very much, I find it difficult because I don't like it and I'm not capable of learning it, as it doesn't attract my attention. The same thing happens to me with music".	DIRE-T11
"Well, partly yes and partly no. Yes because there is a lot of theory, especially in History, and because when you have to examine maps you have to dedicate a lot of time to it. And no because if you do the answers and you work a bit on the project you more or less manage it, even if there is a lot of theory, then well . . . Generally yes. What I find most difficult are the maps".	DIRE-T11
"Yes, and I think it's the subject we do more exams in than any other. I find it very difficult to study the physical maps, although I get them right later on, but I find it hard to remember the names".	DIRE-T11
"For me it doesn't require much effort, but it is true that there are some subjects in History, Art and Music that I find more difficult. What I find most difficult is Art and some History subjects. Maps are a problem for me, especially the political maps of the world".	DIRE-T11

As we can see, the tasks that demand the most effort and time on the part of pupils have nothing to do with competences. Almost half of the pupils point to the memorised study of maps as the main challenge they face in the subject of Geography and History. This practice, in our case study, did not involve the use of any kind of digital technology, as the only resource needed to tackle it were paper maps of Spain and the world, both physical and political. As far as the History content is concerned, the development of the topics, paradoxical as it may seem, did not represent a more important challenge than learning the abundant "names" and "dates" asked about in the written exams.

4. Discussion and Conclusions

This work has allowed us to approximate the true place of the competence discourse in many educational schools, revealing that in the reality of the classroom, it can go completely unnoticed beneath the great influence that the disciplinary paradigm still exerts when it comes to organising and working with educational content. Complementing some of the existing discussions in this regard (Guzmán Marín 2017; Liesa et al. 2018), we have been able to confirm the existence of a close link between the sub-discourse in favour of competency-based learning, the educational use of new technologies, and the promotion of emotional competences from what Solé Blanch (2020) characterised as the influence of positive psychology and the "happiness industry".

Despite their prominence in innovative discourses, in which the regulatory texts themselves participate, competences are not properly integrated into the curriculum of subjects such as Geography and History, nor are there significant relationships between different strands of knowledge or subjects, nor is there any specific work on the supposed "key competences". Proof of this is the absence of any mention of pupils' digital competence in the documents of the Department of Social Sciences, or in the testimonies of the research subjects. According with other authors (Pérez-Escoda et al. 2016), working on digital competence is fundamental for promoting the correct use of new technological resources.

Surprisingly, in the analysed regulations, we have found several slightly different definitions of the term "competence". As far as our subject is concerned, this concept has acquired a new meaning in these texts that moves away from the discourse on key competences by international institutions to focus on knowledge, mainly attitudinal knowledge, related to the development of civic education. This relative suspension or abandonment of the discourse on competences, the consequence of which is none other than the maintenance of the traditional disciplinary logic, is also perfectly palpable in the way in which the general objectives of the subject are formulated and how the curricular content is organised. These and other circumstances reflect certain tensions with regard to the supposed requirements of competency-based educational practice which teachers must take into account.

Similarly, competency-based learning did not appear in the guide of the Geography and History subject, either. This fact was confirmed by two other sources. On the one hand, the interviews conducted with the pupils revealed that the six most demanding tasks had nothing to do with working on key competences but with the memorisation of conceptual content (facts, dates, names, places, and so on); on the other hand, numerous observations recorded in the diary pointed to the excessively disciplinary nature of the knowledge worked on in the classroom as the major pedagogical and didactic problem in that school context. Some of the most representative didactic strategies of the traditional didactic model were the "respondones", which is a battery of 50 questions and short answers about each topic to memorise, or the numerous exams on geographical maps.

These dynamics based on the repetition of positive and decontextualised data fit perfectly with the so-called "teaching to the test" culture (Hursh 2013; Robinson and Dervin 2019). This is a pedagogical philosophy that, on the other hand, also responds to the technocratic concept of quality, which is characteristic of neo-liberal education and its interest in the acquisition of specific and perfectly measurable skills (Slater 2015; Sondel 2015). One of the best arguments to justify this neo-disciplinary model is, of course, the curricular overload that forces teachers to give priority to what is easiest to measure. The contradiction that this model presents with respect to many of the slogans of current educational innovation, including those related to competency-based, active, and personalised learning, is quite remarkable. The deviation towards traditional didactic formulas, despite pro-innovative intentions, has been identified by researchers such as Sigurðardóttir and Hjartarson (2015).

There is no doubt that the conscious maintenance of certain traditional educational objectives, content, and processes in the 21st century, despite the supposed discursive embrace of new educational slogans and approaches, has its own ideological connotations. In this respect, authors such as Cachinho and Reis (2007) have openly expressed their misgivings about "the need to change the paradigm of knowledge transmission for the development of competences" (p. 187) in a political context conditioned by "the most rancid conservatism" (p. 201). This is a debate that can only be approached from the premise that, as Soto and Pérez-Milans (2018) stated in their ethnographic work, the commodification of any type of pedagogical element is taking place on a global scale today.

In any case, the incorporation of the competency-based teaching model into the area of Social Sciences should not eclipse or suppress the work of the specific objectives and contents of this field but enrich them by rethinking the logic that organises the way they are worked on. In the words of Palma Valenzuela (2020), the disciplines "must facilitate attitudes, skills and abilities; together with basic conceptual schemes that allow for a deeper understanding of the individual and social reality and to intervene in a transformative way in both" (p. 91).

As our results have shown, we should not conceive of competency-based learning either as an educational change (since, at least in our case study, this approach had not yet been integrated) or as a strategy that guarantees change (since the factors that determine the possibility of undertaking change, e.g., the guidelines of a department, go far beyond the mere reproduction of a novel discourse). This contradicts, for example, the studies by Gulikers et al. (2018) or Poth and Searle (2021), which focused on how to assess competency-based learning, regarding the assumption that, in educational contexts, we can find the necessary conditions to work from this approach.

As the case study presented here shows, the fact of talking about competences does not imply either automatic motivation of the students or, even less, the development of a comprehensive education of the students. As an explanation for the propagandistic use of innovation by schools such as the one in this research, Pascual (2019) alluded to a need not for them to be competent "but only to demonstrate that they are", adding that "unfortunately, this mimicry can even reach the point where teachers cheat—intentionally or not—to ensure their permanence in the system" (p. 185).

In any case, an analysis focused on the new competences that teachers and principals should possess today (Greany 2019; Mohamed et al. 2017) is still pending, which will broaden the view of the professional knowledge currently required to teach. Furthermore, we must acknowledge that this work has not allowed us to observe the supposed drawbacks of the competency approach pointed out by several authors (Barrio 2013; Espinoza Aros 2014; Hirtt 2010; Jiménez Abad 2015; Tricot 2019), since this constitutes a perspective reduced to the purely discursive dimension of our object of study. In this sense, we could ask to what extent the competency-based approach is nothing more than a new criterion for classifying pupils using the new categories of "competent" and "incompetent" instead of "clever" and "clumsy", respectively. In any case, this new vocabulary allows us to keep talking about a desirable student in the context of the neo-liberal school.

All these questions encourage us to continue to broaden the perspective on what we have identified here as a conflict between the particularities of a discourse which attempts to impose itself from the top down, such as that of competency-based learning, and a practice, the classroom practice, which is still completely dominated by the traditional disciplinary culture.

Author Contributions: Conceptualization, D.L. and J.A.P.-A.; methodology, D.L. and J.A.P.-A.; software, D.L.; formal analysis, D.L.; investigation, D.L.; resources, D.L.; writing—original draft preparation, D.L.; writing—review and editing, J.A.P.-A.; visualization, D.L. and J.A.P.-A.; supervision, J.A.P.-A. All authors have read and agreed to the published version of the manuscript.

Funding: This research received no external funding.

Data Availability Statement: Not applicable.

Conflicts of Interest: The authors declare no conflict of interest.

References

Adams, Tony E., Stacy Holman Jones, and Carolyn Ellis. 2015. *Autoethnography*. New York: Oxford University Press.
Barrio, José María. 2013. *La innovación educativa pendiente: Formar personas*. Barcelona: Erasmus.
Cachinho, Herculano, and João Reis. 2007. Educación geográfica y ciudadanía en Portugal: De los discursos a la práctica en los centros escolares. *Didáctica geográfica* 9: 185–204.
Cañadell, Rosa. 2018. El asalto neoliberal a la educación. *Con-Ciencia Social (Segunda Época)* 1: 103–17.
de España, Gobierno. 2013. Ley Orgánica 8/2013, de 9 de diciembre, para la Mejora de la Calidad Educativa (LOMCE). *Boletín Oficial del Estado (BOE)* 295: 27548–62.
Ellis, Carolyn. 2016. *Revision. Autoethnographic Reflections on Life and Work*. New York: Routledge.
Espinoza Aros, Olga. 2014. Análisis crítico del discurso de las competencias en la formación inicial docente en Chile. *Estudios Pedagógicos* 40: 147–59. [CrossRef]
Fairclough, Norman. 2003. El análisis crítico del discurso como método para la investigación en ciencias sociales. In *Métodos de análisis crítico del discurso*. Compilated by Ruth Wodak and Michael Meyer. Barcelona: Gedisa, pp. 179–204.
Greany, Toby. 2019. Liderar la mejora y la innovación en las escuelas y en los sistemas educativos: Evidencias e implicaciones para las políticas y las prácticas. In *Las paradojas de la innovación educativa*. Coordinated by Miquel Martínez Martín and Anna Jolonch i Anglada. Barcelona: Horsori, pp. 53–82.
Gulikers, Judith T. M., Runhaar Piety, and Martin Mulder. 2018. An assessment innovation as flywheel for changing teaching and learning. *Journal of Vocational Education y Training* 70: 212–31. [CrossRef]
Guzmán Marín, Francisco. 2017. Problemática general de la educación por competencias. *Revista Iberoamericana de Educación* 74: 107–20. [CrossRef]
Hirtt, Nico. 2010. La educación en la era de las competencias. *Revista Electrónica Interuniversitaria de Formación del Profesorado* 13: 108–14.
Hou, Su-I. 2022. Advancing a Competency-Based Mixed Methods Tool to Assess a Course-Based Service-Learning Model Integrating Real-World Experience. *Canadian Journal of Program Evaluation* 36: 391–407. [CrossRef]
Hursh, David. 2013. Raising the stakes: High-stakes testing and the attack on public education in New York. *Journal of Education Policy* 28: 574–88. [CrossRef]
Jiménez Abad, Andrés. 2015. Motivación docente y factores impulsores de la innovación. In *Innovación educativa y tradición*. Coordinated by María Teresa Cid Vázquez and Edited by Lydia Jiménez. Madrid: Fundación Universitaria Española, pp. 155–80.
Jolonch, Anna. 2019. Las comunidades profesionales de aprendizaje: Liderazgo e innovación educativa. In *Las paradojas de la innovación educativa*. Coordinated by Miquel Martínez Martín and Anna Jolonch i Anglada. Barcelona: Horsori, pp. 215–38.
Kozma, Robert. 2012. *Les TIC i la transformació de l'educació en l'economia del coneixement*. Barcelona: Fundació Jaume Bofill.
Liesa, Eva, Monserrat Castelló, and Lorena Becerril. 2018. Nueva escuela, ¿nuevos aprendizajes? *Revista de Estudios y Experiencias en Educación* 2: 15–29. [CrossRef]
Marcotte, Kayla M., and Larry D. Gruppen. 2022. Competency-Based Education as Curriculum and Assessment for Integrative Learning. *Education Sciences* 12: 267. [CrossRef]
Martínez-Abad, Fernando, Marcos Bielba-Calvo, and María Esperanza Herrera-García. 2017. Evaluación, formación e innovación en competencias informacionales para profesores y estudiantes de Educación Secundaria. *Revista de educación* 376: 110–34. [CrossRef]
Mitra, Rhul. 2010. Doing ethnography, being an ethnographer: The autoethnographic research process and I. *Journal of Research Practice* 6: M4.
Mohamed, Zulaikha, Martin Valcke, and Bram De Wever. 2017. Are they ready to teach? Student teachers' readiness for the job with reference to teacher competence frameworks. *Journal of Education for Teaching* 43: 151–70. [CrossRef]
Orden ECD/65/2015, de 21 de enero, por la que se describen las relaciones entre las competencias, los contenidos y los criterios de evaluación de la educación primaria, la educación secundaria obligatoria y el bachillerato. 2015. *Boletín Oficial del Estado (BOE)* 25: 6986–7003.
Palma Valenzuela, Andrés. 2020. Enseñar Ciencias Sociales entre riesgos e incertidumbres. *Didácticas Específicas* 22: 88–103. [CrossRef]
Pardo Abril, Neyla Graciela. 2013. *Cómo hacer análisis crítico del discurso. Una perspectiva latinoamericana*. Bogotá: Universidad Nacional de Colombia.
Pascual, Javier. 2019. Relaciones de poder y Empoderamiento Docente para la Innovación Educativa. Ph.D. dissertation, Universitat Autònoma de Barcelona, Barcelona, Spain.
Pérez-Escoda, Ana, Ignacio Aguaded, and María José Rodríguez-Conde. 2016. Digital Generation vs. Analogic School. Digital Skills in the Compulsory Education Curriculum. *Digital Education Review* 30: 165–83. [CrossRef]
Pini, Mónica, comp. 2009. *Discurso y educación. Herramientas para el análisis crítico*. Buenos Aires: UNSAM EDITA.
Poth, Cheryl, and Michelle Searle. 2021. Competency-based evaluation education: Four essential things to know and do. *Canadian Journal of Program Evaluation* 35: 296–309. [CrossRef]
Redecker, Christine. 2020. *Marco Europeo para la Competencia Digital de los Educadores: DigCompEdu*. Madrid: Secretaría General Técnica, Centro de Publicaciones, Ministerio de Educación y Formación Profesional.
Robinson, Bonnie K., and Alicia Dervin. 2019. Teach to the student, not the test. *Urban Education Policy and Research Annuals* 6: 34–44.
Rogers, Rebecca, ed. 2011. *An Introduction to Critical Discourse Analysis in Education*, 2nd ed. New York and Abingdon: Routledge.

Rogers, Rebecca. 2018. Critical discourse analysis and educational discourses. In *The Routledge Handbook of Critical Discourse Studies*. Edited by John Flowerdew and John E. Richardson. Abingdon and New York: Routledge, pp. 465–79.

Rossi Cordero, Andrea S., and Mario Barajas Frutos. 2018. Competencia digital e innovación pedagógica: Desafíos y Oportunidades. *Profesorado. Revista de Currículum y Formación de Profesorado* 22: 317–39. [CrossRef]

Rubio, Vanesa, and José Tejada. 2017. Las competencias informacionales de los docentes y alumnos de Educación Secundaria Obligatoria. *Journal for Educators, Teachers and Trainers* 8: 127–40.

Sigurðardóttir, Anna K., and Torfi Hjartarson. 2015. The idea and reality of an innovative school: From inventive design to established practice in a new school building. *Improving Schools* 19: 62–79. [CrossRef]

Slater, Graham B. 2015. Education as recovery: Neoliberalism, school reform, and the politics of crisis. *Journal of Education Policy* 30: 1–13. [CrossRef]

Solé Blanch, Jordi. 2020. El cambio educativo ante la innovación tecnológica, la pedagogía de las competencias y el discurso de la educación emocional. Una mirada crítica. *Teoría de la Educación* 32: 101–21. [CrossRef]

Sondel, Beth. 2015. Raising Citizens or Raising Test Scores? Teach For America, No Excuses Charters, and the Development of the Neoliberal Citizen. *Theory & Research in Social Education* 43: 289–313. [CrossRef]

Soto, Carlos, and Miguel Pérez-Milans. 2018. Language, Neoliberalism, and the Commodification of Pedagogy. *Language and Intercultural Communication* 18: 490–506. [CrossRef]

Starr, Lisa J. 2010. The Use of Autoethnography in Educational Research: Locating Who We Are in What We Do. *Canadian Journal for New Scholars in Education/Revue Canadienne des Jeunes Chercheures et Chercheurs en Éducation* 3: 1–9.

Strauss, Anselm, and Juliet Corbin. 2008. *Basics of Qualitative Research: Techniques and Procedures for Developing Grounded Theory*, 3rd ed. Thousand Oaks, London, New Delhi and Singapore: Sage Publications.

Tricot, André. 2019. *Innovar en educación. Sí, pero ¿cómo? Mitos y realidades*. Madrid: Narcea.

Valencia Álvarez, Adriana B., and Jaime R. Valenzuela González. 2017. Innovación disruptiva, innovación sistemática y procesos de mejora continua … ¿Implican distintas competencias por desarrollar? In *Innovación educativa: Investigación, formación, vinculación y visibilidad*. Edited by María S. Ramírez Montoya and Jaime R. Valenzuela González. Madrid: Síntesis, pp. 109–34.

Wamsted, John O. 2012. Borges & Bikes Riders: Toward an Understanding of Autoethnography. *Qualitative Research in Education* 1: 179–201. [CrossRef]

Wodak, Ruth, and Michael Meyer, comps. 2003. *Métodos de Análisis Crítico del Discurso*. Barcelona: Gedisa.

social sciences

MDPI

Article

Trainee Teachers' Perceptions of Socio-Environmental Problems for Curriculum Development

Roberto García-Morís [1] and Ramón Martínez-Medina [2,*]

[1] Department of Specific Teaching Training and Research and Diagnosis Methods in Education, University of A Coruña, 15008 A Coruña, Spain
[2] Department of Specific Didactics, University of Córdoba, 14071 Córdoba, Spain
* Correspondence: rmartinez@uco.es

Abstract: Socio-environmental problems are some of the major concerns of today's society, and education is an essential area to raise awareness and mould future citizens who will be committed to sustainability. The purpose of this paper is to ascertain how future primary and secondary education teachers in the area of social sciences perceive the socio-environmental problems that affect today's society, for their subsequent inclusion in the curriculum. To this end, a study was carried out by means of a questionnaire, which showed that trainee primary and secondary teachers have a high consideration of these environmental problems, with hardly any differences noted according to sex or type of degree studied. This perception is positive in terms of the subsequent inclusion of socio-environmental problems in the development of the curriculum by the participating trainee teachers, who largely prefer an autonomous model for curricular implementation.

Keywords: teacher training; socio-environmental problems; environment; sustainability; social sciences

Citation: García-Morís, Roberto, and Ramón Martínez-Medina. 2022. Trainee Teachers' Perceptions of Socio-Environmental Problems for Curriculum Development. *Social Sciences* 11: 445. https://doi.org/10.3390/socsci11100445

Academic Editor: Alvaro Morote

Received: 22 July 2022
Accepted: 21 September 2022
Published: 28 September 2022

1. Introduction

In 2020, Spain approved a new educational law (Ministerio de Educación 2022) that proposes an important renewal of the curriculum towards a more competence-based model in line with European recommendations and the Sustainable Development Goals (Moya and Zubillaga del Río 2020). However, the success of the new curriculum framework will be determined by the issues raised during teacher training, since initial training is essential to generate competent teachers for curricular development that includes environmental problems also from a social perspective (Moya and Zubillaga del Río 2020; López 2021; López-Fernández and Oller 2019), as is proposed for the training of students from other fields of knowledge (Boca and Saraçlı 2019). Therefore, it is important to start from the perceptions of trainee primary and secondary school teachers regarding the SDGs in a broad sense, and particularly with regard to the environment and climate change that we are experiencing on a global scale, as well as the role of environmental education (Tahsin 2003; Boon 2010; Fuster et al. 2021; Ministerio de Educación 2022; Massip et al. 2021; Medina and Castro 2021; Mosothwane 2002).

The environment has a strong presence in Spain's new curricular framework, given that it is included as an interdisciplinary and cross-cutting component of the competence-based approach and school contents (Ministerio de Educación 2022). There is no doubt that for the environmental perspective to be successful in the development of the curriculum in schools, we must take action with regard to pre-service and in-service teacher training. Therefore, it is essential to understand the perceptions of trainee teachers regarding environmental problems in order to implement successful educational approaches (Morote et al. 2022; López 2021; Cucurachi et al. 2017). This study uses the concept of socio-environmental problems rooted in a holistic understanding of the concept: environmental problems are social, and most social problems also have an environmental perspective.

Subjects included within the area of Social Sciences—Social Sciences at primary school and Geography and History at secondary school—are paramount when addressing socio-environmental issues and promoting attitudes favourable to social change with an environmental perspective (Fuster et al. 2021). This global conception of the social and environmental components that characterise the environment cannot be alien to the curricular approach taken to social subjects, which is why this research was proposed, as it seeks to understand the perceptions of social and environmental problems held by future social science teachers, and to see whether there any differences between the views of future primary compared to future secondary school teachers.

The analysis of perceptions regarding the environment within the field of education has sparked interest for decades in different countries, analysing misconceptions held (Tahsin 2003), with recommendations on educational approaches in teacher training (Van Petegem et al. 2005), perceptions about Creative Thinking for environmental education (Daskolia et al. 2012) and environmental education it-self (Mosothwane 2002). Studies of the perceptions held by inservice teachers are also a constant thread running through the literature on the subject (Flogaitis and Agelidou 2003; Benavides-Lahnstein and Ryder 2020; Galvis-Riaño et al. 2020; Munoz et al. 2009).

With regard to Spain, the contributions of López-Fernández and Oller (2019) and López (2021) are particularly important, used as a reference here when selecting the social and environ-mental problems put forward in our research. These authors identified a good predisposition among trainee primary school teachers towards the inclusion of socio-environmental problems in school, although the proposals they design show little theoretical knowledge. Álvarez-García et al. (2015) presented a review of the existing literature in which they observed that research on Environmental Education has increased in recent years, highlighting the weak knowledge of trainee teachers on the subject (Tal 2010; Puk and Stibbards 2010; Kyridis et al. 2005) and suggesting its inclusion in learning and development plans for competence-based training as environmental educators (Boca and Saraçlı 2019; Yavetz et al. 2014).

Environmental Education is usually given a high level of consideration by future teachers, especially in the areas of Natural Sciences and Social Sciences, although knowledge of the subject is limited (Yavetz et al. 2014). In addition, it is developed more fully when addressing local environmental problems (Campo and García-Monteagudo 2020) and those influenced by the media (Tal 2010). Field trips are presented as favourable experiences (Tal 2010) for Environmental Education, whose introduction in teacher training plans is very useful for future curricular inclusion (Yates et al. 2019; Alcántara and Medina 2019).

The presence of Environmental Education in Spain's curricular framework has been gaining ground over the years. If we focus on the Primary curriculum, we see that issues related to Environmental Education were only broached very timidly until the enactment of the General Organic Law on the Education System (LOGSE) in 1990 (Ministerio de Educación 1990). The curricular development set out in this Law does not explicitly include any content related to the environment, although one of the objectives of the Law refers to the manifestations of human interventions, their scope, possible solutions, ecological balance, and the conservation of natural resources. In 2006, with the enactment of the new Organic Law on Education (LOE) (Ministerio de Educación 2006), other concepts related to the environment, such as sustainability and climate change, appear explicitly. In the recently repealed Law for the Improvement of Education Quality (LOMCE) (Ministerio de Educacíon 2013), environmental issues focus on the field of Social Sciences teaching. Sustainable development appears as a concept to be learned, and its teaching focuses on human intervention in the environment, environmental problems, and the sustainable use of resources (Martínez-Medina and Arrebola 2019).

The Organic Law amending Organic Law 2/2006, of 3 May, on Education (LOMLOE) of 2020 (Ministerio de Educación 2020) makes great strides towards facing the challenges of society in the 21st Century, focusing the whole process of teaching subjects on the achievement of the Sustainable Development Goals (SDGs) (Moya and Zubillaga del

Río 2020). In the new curriculum, the teaching of Natural and Social Sciences is unified once more in a single integrated area of knowledge in Primary Education. The subject is structured into specific competencies, evaluation criteria linked to these competencies, and core knowledge (Ministerio de Educación 2022).

Specific competency number 6 addresses all environmental issues. It aims to identify the causes and consequences of human intervention in the local and global environment, from various points of view (environmental, economic, social, etc.), seeking solutions and changes in the way we act to implement sustainable and respectful lifestyles to care for and protect people and planet.

Core knowledge is structured into three main blocks of contents: A. Scientific culture, B. Technology and digitisation and C. Societies and territories. The teaching of Social Sciences is integrated into this last block, which in turn is subdivided into four main sections (1. Challenges of today's world, 2. Societies in time, 3. Civic literacy and 4. Ecosocial awareness).

The Ecosocial awareness block proposes a gradation of learning according to the stages into which Primary Education is divided. Issues ranging from knowledge of the environment, landscapes, and human action on the environment to conservation actions, the responsible use of resources, and climate change are all tackled. Pupils progressively explore these issues in greater depth and breadth right through to the third stage, when they explore sustainable development, the SDGs, sustainable urban development, and the green economy.

As we can see, the curriculum change in Spain aims to build a society presided over by social justice and environmental sustainability, and education plays an important role in this, in terms of moulding citizens that are competent in this area. In other words, in line with the SDGs, the idea is that students gain knowledge and skills to promote sustainable development, SDG 4 Goal 7 (Moya and Zubillaga del Río 2020; Ministerio de Educación 2022). To achieve this goal, the new curriculum framework is not enough. Teachers must be trained and conscientiously in favour of Environmental Education. The purpose of this study is to ascertain the perceptions of future primary and secondary school teachers about socio-environmental problems and their curricular development.

2. Materials and Methods

2.1. Objectives

The general objective of this paper is to ascertain the perceptions of future Social Sciences teachers in Spain regarding socio-environmental problems. To this end, an analysis questionnaire has been prepared based on the results obtained previously in other reference works of research (López-Fernández and Oller 2019; López 2021). This overarching aim is developed through the following specific objectives:

- To analyse the perception of trainee primary school teachers regarding environmental and social problems.
- To analyse the perception of trainee secondary school geography and history teachers regarding environmental and social issues.
- To identify how future social sciences teachers will address socio-environmental issues within the classroom.

2.2. Participants

The study was carried out with students enrolled on the bachelors' degree in Primary Education and the master's degree in Secondary Education Teaching at the universities of Cordoba, A Coruña, and Murcia (Spain). Sampling was based on accessibility and convenience, as individuals were easily accessible by the authors.

The final study sample consisted of 168 students, of whom 36.9% (n = 62) were men and 61.3% (n = 106) were women, aged between 20 and 59 years with a mean age of 23.6 (SD = 5.2).

Students were also divided by type of degree studied. In the case of Spain, this classification is relevant given that the training of primary school teachers is via the bachelor's degree in Primary Education (4 years) and the training of secondary teachers is via a postgraduate degree (master's degree in Compulsory Secondary Education Teaching) lasting one academic year. This postgraduate course is completed after the students receive their degree, which, in the case of Social Science teachers, is usually a degree in Geography, History or Art History (4 years). 50.6% of the participants were enrolled on the master's degree in Compulsory Secondary Education Teaching and 49.4% on the bachelor's degree in Primary Education (Table 1).

Table 1. Descriptive education variables.

	n	%
Type of degree studied		
Bachelor's degree in Primary Education	83	49.4
Master's degree in Secondary Education Teaching (Social Sciences, Geography and History)	85	50.6
Previous degree studied by students on the master's course		
Geography and History	21	24.71
Geography, Geography and Spatial Planning, Geography and the Environment	2	2.35
History	35	41.18
History of Art	14	16.47
Humanities	10	11.76
Other	3	3.53

2.3. Instrument

For the purposes of this research, a questionnaire was designed, the Survey to Evaluate Socio-Environmental Problems in the Initial Training of Social Sciences Teachers, which is based on the results of the study by López-Fernández and Oller (2019) and López (2021), on environmental problems in Primary Education Teacher Training. Beginning with the problems identified in this previous research, the study was extended to students enrolled on a master's degree in Secondary Education Teaching, with the intention of identifying differences both in the perception of socio-environmental problems and in their curricular inclusion.

The questionnaire contains a total of 31 questions organised into four analytical dimensions. Of the total number of questions, one is open ended, and the others are closed. In dimensions 1, 2 and 3, the closed questions were designed using a Likert scale ranging from 1 to 10, with 10 being the highest degree of importance and 1 being the lowest degree of importance.

The first dimension, which only contains three questions, seeks to ascertain whether future teachers attach greater importance to environmental or social problems, and their assessment of each one of them. The second dimension of the questionnaire tackles the participants' evaluation of environmental problems (climate change, pollution, greenhouse effect, etc.). The third dimension is structured in a similar way to the previous one and tackles social problems, their importance and assessment. In the last dimension, through an open-ended question, student teachers are asked about how they will address socio-environmental issues in the classroom. In order to establish comparisons in the assessment of the different problems, the social and environmental problems were presented separately, also asking about the greater importance of some over others. The separate analysis of socio-environmental problems is based on the results obtained in the research carried out by López-Fernández and Oller (2019). Although these problems are conceived in an integrated way as socio-environmental for their curricular development, it was decided to present them in this way for the comparative analysis of the perceptions of future teachers.

2.4. Data Analysis

For the descriptive statistical analysis of the sample, basic descriptive methods were used. For qualitative variables, absolute and relative frequencies were calculated; and for quantitative variables, minimum, maximum, mean and standard deviation values were calculated.

Groups of quantitative variables between two groups were compared using Student's *t*-test for independent samples, having first verified the assumptions of normality (Kolmogorov–Smirnov test) and homogeneity of variances (Levene test). The effect size was studied by means of Cohen's d, considering 0.2 to be a small effect, 0.5 to be medium, and 0.8 to be large.

Statistical analysis was carried out using SPSS 26.0 for Windows. Differences were considered statistically significant with $p < 0.05$.

In the open-ended question of the last dimension, responses were categorised according to the models of curricular development (Canals and González-Monfort 2011), making a descriptive analysis of that categorisation.

3. Results

3.1. Perception of Social and Environmental Problems

With regard to the participants' perception of the most important problems, when asking about social and environmental problems in a differentiated way, 75.6% (n = 127) of the future teachers participating in the study believe that the main problems are social, whereas 24.4% (n = 41) believe they are related to the environment. In addition, looking at the importance attached to each type of problem, environmental problems have an average importance of 8.87 points out of 10 (Min–Max: 5–10, SD = 1.12), while social problems reach more relevant values, with a mean of 9.45 points (Min–Max: 7–10, SD = 0.85).

The questionnaire asked students to assess, on a scale of 0 to 10, a number of environmental issues. The results show that there is strong concern for all environmental problems, with air pollution and climate change being the top priority, while problems related to the use of fossil fuels and noise pollution are given a lower score, below 9, with the latter receiving the lower of the two scores, although still above 7 (Table 2). In other words, all the problems were rated highly on the scale.

Table 2. Relevance of environmental problems.

Environmental Problems	Min–Max	Mean (SD)
Climate change	3–10	9.36 (0.99)
Noise pollution	2–10	7.87 (1.65)
Water pollution	5–10	9.23 (1.11)
Air pollution	5–10	9.37 (0.98)
Deforestation	6–10	9.20 (1.01)
Greenhouse effect	4–10	9.12 (1.15)
Water shortages/drought	3–10	9.27 (1.21)
Overexploitation of resources	5–10	9.27 (1.08)
Use of fossil fuels	2–10	8.70 (1.34)

With regard to perceived social problems, the participants once again rate them very high on the scale, above 7 (Table 3). Interestingly, gender-based violence is perceived as the most important social problem. The fact that gender-based violence receives the highest score as a social problem by future teachers is very positive, since education is one of the most important fields for acting against this blight.

The participants also have a strong perception of poverty and racism, but interestingly illegal immigration receives one of the lowest scores. Among the social problems included in the questionnaire, there were some that significantly affect the school population, both primary and secondary, and these were rated from highest to lowest as follows:

bullying, unemployment, eating disorders, drug addiction, youth emigration, alcoholism, consumerism, and overweight.

Table 3. Importance of social problems.

Social Problems	Min–Max	Mean (SD)
Access to universal health care	5–10	9.44 (1.10)
Bullying	2–10	9.45 (1.04)
Alcoholism	1–10	8.51 (1.62)
Consumerism	1–10	8.37 (1.70)
Social inequality	2–10	9.43 (1.07)
Drug addiction	2–10	8.66 (1.48)
Youth emigration	2–10	8.62 (1.53)
Lack of values	2–10	9.18 (1.23)
Illegal immigration	1–10	8.09 (2.01)
Gender violence	5–10	9.72 (0.66)
Marginality	2–10	9.01 (1.25)
Unemployment	4–10	9.09 (1.09)
Poverty	7–10	9.55 (0.77)
Racism	2–10	9.52 (0.99)
Overweight	2–10	8.14 (1.63)
Unclean streets	2–10	8.18 (1.62)
Overpopulation (unchecked population growth)	1–10	7.96 (1.74)
Eating disorders	2–10	8.73 (1.47)

It is likely that the main reason bullying is classed as one of the top four social problems is because the respondents are studying to become teachers. However, this is a social problem, not just an educational one, and it is perceived and rated as such by the students participating in the study. It is also striking that overpopulation, that is to say, the unchecked growth of the world's population, is the least valued social problem of all, since it scores lower than an 8 in quantitative terms, when it is a problem that is closely linked to other social and also environmental problems.

When analysing perceived social and environmental problems by sex (Table 4), with regard to environmental problems, the only difference observed is in relation to noise pollution, a problem that women consider to be more important than men, but in both cases this problem receives the lowest score.

With regard to social problems there are differences in relation to alcoholism, social inequality, drug addiction, lack of values, illegal immigration, marginality, unemployment, poverty, overpopulation and eating disorders, which women consider to be more important than men.

Gender-based violence is given a higher score by women, but one particularly positive finding is that the perception of men in the study is very close to that of women, with little difference between the two, and in both cases, it is the highest rated problem.

The lack of values is a problem that ties in very well with the teaching of Social Sciences, since one of the aims of this field is to provide an education in values. This lack of values as a social problem receives an average score of 9.38 among women and 8.85 among men, a more significant difference.

When analysing the data according to the degree course studied (Table 5), environmental problems only yield differences in their assessment in the case of deforestation, which students on the master's degree in secondary education teaching consider to be more important than students on the primary education degree. The former rank deforestation as the main environmental problem whereas the latter rank air pollution as the highest. Grouping responses by degree course studied, noise pollution is once again placed last.

In terms of social problems, there are no differences between undergraduate and master's degree students with regard to any of the problems. Gender-based violence is once again at the top of the leader board and receives a higher score among future primary teachers than secondary teachers, as does bullying and a lack of values, although with unsignificant differences. While overpopulation continues to appear in both cases as one of the least relevant problems.

Table 4. Descriptive and comparative statistics of the importance given by men and women to environmental and social problems.

	Sex, Mean (SD)		Difference of Means	Student's *t*-Test		d
	Male	Female		t(166)	*p*-Value	
Environmental problems	8.66 (1.16)	8.99 (1.09)	−0.33	−1.85	0.067	−0.29
Climate change	9.26 (0.92)	9.42 (1.03)	−0.16	−0.99	0.324	−0.16
Noise pollution	7.53 (1.60)	8.07 (1.65)	−0.54	−2.05	0.042	−0.33
Water pollution	9.08 (1.23)	9.31 (1.03)	−0.23	−1.30	0.194	−0.21
Air pollution	9.26 (0.97)	9.43 (0.98)	−0.17	−1.13	0.261	−0.18
Deforestation	9.06 (1.13)	9.27 (0.92)	−0.21	−1.30	0.194	−0.21
Greenhouse effect	9.03 (1.13)	9.17 (1.16)	−0.14	−0.75	0.455	−0.12
Water shortages/drought	9.15 (1.21)	9.35 (1.21)	−0.2	−1.05	0.294	−0.17
Overexploitation of resources	9.11 (1.20)	9.36 (1.00)	−0.25	−1.43	0.156	−0.23
Use of fossil fuels	8.45 (1.29)	8.85 (1.36)	−0.4	−1.87	0.064	−0.30
Social problems	9.34 (0.94)	9.51 (0.80)	−0.17	−1.25	0.212	−0.20
Access to universal health care	9.29 (1.29)	9.53 (0.97)	−0.24	−1.26	0.176	−0.22
Bullying	9.27 (0.96)	9.55 (1.07)	−0.28	−1.66	0.1	−0.26
Alcoholism	8.06 (1.74)	8.76 (1.50)	−0.7	−2.75	0.007	−0.44
Consumerism	8.13 (1.59)	8.51 (1.75)	−0.38	−1.40	0.163	−0.22
Social inequality	9.19 (1.30)	9.58 (0.87)	−0.39	−2.27	0.024	−0.36
Drug addiction	8.34 (1.50)	8.85 (1.45)	−0.51	−2.18	0.031	−0.35
Youth emigration	8.47 (1.30)	8.71 (1.65)	−0.24	−0.98	0.329	−0.16
Lack of values	8.85 (1.48)	9.38 (1.01)	−0.53	−2.46	0.007	−0.43
Illegal immigration	7.35 (2.21)	8.52 (1.76)	−1.17	−3.75	<0.001	−0.60
Gender violence	9.63 (0.68)	9.77 (0.64)	−0.14	−1.36	0.169	−0.22
Marginality	8.60 (1.48)	9.25 (1.03)	−0.65	−3.05	0.001	−0.53
Unemployment	8.76 (1.25)	9.28 (0.93)	−0.52	−2.87	0.002	−0.49
Poverty	9.34 (0.92)	9.67 (0.63)	−0.33	−2.51	0.006	−0.44
Racism	9.35 (1.27)	9.61 (0.78)	−0.26	−1.45	0.103	−0.26
Overweight	7.89 (1.49)	8.29 (1.69)	−0.4	−1.56	0.12	−0.25
Unclean streets	7.92 (1.66)	8.34 (1.58)	−0.42	−1.63	0.104	−0.26
Overpopulation	7.58 (1.85)	8.18 (1.64)	−0.6	−2.18	0.031	−0.35
Eating disorders	8.37 (1.54)	8.94 (1.40)	−0.57	−2.47	0.015	−0.39

SD: Standard deviation. d: Cohen's d (effect size).

Table 5. Descriptive and comparative statistics regarding the importance given by students on the undergraduate and master's degree courses to environmental and social problems.

	Degree Course Studied, Mean (SD)		Difference of Means	Student's *t*-Test		d
	Bachelor's Degree	Master's Degree		t(166)	*p*-Value	
Environmental problems	8.84 (1.11)	8.89 (1.15)	−0.05	−0.29	0.771	−0.05
Climate change	9.37 (0.87)	9.34 (1.11)	0.03	0.21	0.834	0.03
Noise pollution	7.98 (1.41)	7.76 (1.85)	0.22	0.83	0.408	0.13
Water pollution	9.31 (0.88)	9.14 (1.29)	0.17	1.01	0.316	0.16
Air pollution	9.41 (0.90)	9.33 (1.05)	0.08	0.53	0.596	0.08
Deforestation	9.01 (1.05)	9.38 (0.93)	−0.37	−2.38	0.018	−0.37
Greenhouse effect	9.13 (0.99)	9.11 (1.29)	0.02	0.15	0.881	0.02
Water shortages/drought	9.29 (1.19)	9.26 (1.25)	0.03	0.16	0.872	0.02
Overexploitation of resources	9.20 (1.08)	9.33 (1.08)	−0.13	−0.75	0.457	−0.12
Use of fossil fuels	8.73 (1.20)	8.67 (1.48)	0.06	0.31	0.757	0.05

Table 5. *Cont.*

	Degree Course Studied, Mean (SD)		Difference of Means	Student's *t*-Test		d
	Bachelor's Degree	Master's Degree		t(166)	*p*-Value	
Social problems	9.48 (0.82)	9.41 (0.89)	0.07	0.53	0.596	0.08
Access to universal health care	9.48 (1.09)	9.40 (1.12)	0.08	0.48	0.63	0.07
Bullying	9.55 (0.82)	9.34 (1.21)	0.21	1.34	0.184	0.21
Alcoholism	8.70 (1.44)	8.32 (1.77)	0.38	1.53	0.128	0.24
Consumerism	8.55 (1.56)	8.19 (1.82)	0.36	1.40	0.164	0.22
Social inequality	9.46 (0.85)	9.41 (1.25)	0.05	0.28	0.78	0.04
Drug addiction	8.81 (1.35)	8.52 (1.60)	0.29	1.27	0.207	0.20
Youth emigration	8.67 (1.53)	8.56 (1.54)	0.11	0.46	0.643	0.07
Lack of values	9.28 (1.09)	9.09 (1.35)	0.19	0.97	0.335	0.15
Illegal immigration	8.28 (1.72)	7.91 (2.26)	0.37	1.20	0.233	0.18
Gender violence	9.81 (0.45)	9.64 (0.80)	0.17	1.72	0.089	0.26
Marginality	9.12 (1.02)	8.89 (1.44)	0.23	1.18	0.242	0.18
Unemployment	8.98 (1.18)	9.20 (0.99)	−0.22	−1.34	0.183	−0.21
Poverty	9.49 (0.80)	9.60 (0.73)	−0.11	−0.90	0.37	−0.14
Racism	9.60 (0.80)	9.44 (1.15)	0.16	1.09	0.276	0.17
Overweight	8.25 (1.48)	8.04 (1.76)	0.21	0.87	0.388	0.13
Unclean streets	8.34 (1.34)	8.04 (1.85)	0.3	1.21	0.227	0.19
Overpopulation	7.94 (1.56)	7.98 (1.90)	−0.04	−0.14	0.892	−0.02
Eating disorders	8.89 (1.41)	8.58 (1.52)	0.31	1.39	0.167	0.21

SD: Standard deviation. d: Cohen's d (effect size).

3.2. *Curricular Approach to Social and Environmental Problems for Curricular Development*

Having analysed the perceptions of socio-environmental problems held by future primary and secondary social science teachers, the fourth and final dimension of the questionnaire dealt with how they would address these problems in their future teaching practice.

In the previous section, we found that, despite the differences in the degree courses studied by teachers for Primary and Secondary education, their perceptions of socio-environmental problems did not differ significantly. This is significant because the course required to become a primary school teacher is a 4-year honours degree that spans a very diverse curriculum of subjects, but all related to education. On the other hand, future secondary teachers take a teaching specialisation master's degree, following on from a four-year honours degree in History and Geography mainly devoted exclusively to these disciplines. Therefore, a priori, trainee primary school teachers would have less disciplinary training related to the environment than trainee secondary school teachers. Therefore, this last part of the investigation seeks to ascertain how the different problems will be addressed in the classroom, and whether there is a difference according to the degree studied by the participants.

Given that the question was open-ended, the responses were categorised by models of curricular development in Social Sciences, taking into account three models: traditional, autonomous, and critical (Canals and González-Monfort 2011). The traditional model is the positivistic model, which has been the prevailing model used in schools for the longest time, consisting of the reproduction of knowledge with pupils playing a fairly passive role. The autonomous or active teaching model gives greater prominence to pupils and their relationship with the environment, with procedural proposals predominating. Finally, the critical model introduces relevant social problems, thinking about the future and promoting social change (Canals and González-Monfort 2011; Cebrián et al. 2021; Massip et al. 2021) to foster "the formation of reflective, critical and creative thinking, the analysis of values and social practices, and learning about decision making and problem solving" (Canals and González-Monfort 2011, p. 53).

The critical model has an important presence (30.49%), the autonomous model is the dominant one (57.32%), and the traditional model is proposed in 12.19% of cases (Table 6).

This model has a token presence in the case of trainee primary school teachers (3.66%) and more important in the case of trainee secondary school teachers (20.73%). In part, it is understandable that students who have spent less time focusing on education-related studies would have a more traditional view of curricular approaches, probably due to their previous experiences, because of the disciplinary training they have received over the course of their studies.

Table 6. Educational proposals by curricular models and by type of degree studied.

Curricular Models	Degree Primary Ed.	%	Master's Degree Secondary	%	Total	%
Traditional model	3	3.66	17	20.7	20	12.19
Autonomous model	55	67.07	39	47.6	94	57.32
Critical Model	24	29.27	26	31.7	50	30.49
Total	82	100	82	100	164	100

Proposals to address these topics categorised under the traditional model link socio-environmental problems with the other contents in the curriculum for a better understanding of them. As we can see in the following narratives, socio-environmental problems are not at the heart of the didactic proposal itself but are linked to the conclusions of the topics or so as to highlight them throughout history, but always related to the subject contents themselves or to improve the assimilation thereof, never as the central elements of the didactic proposal, and always through a transmissive approach to socio-environmental problems.

> Trying to relate them to the subject contents covered in class, with activities that help them to understand the subject contents and socio-environmental problems. (Cod. 16, Secondary, Male)

> Correlating these problems with the Social Sciences content covered in class. Especially when we design activities for students to highlight the relationships between theoretical contents. (Cod. 35, Secondary, Male)

> Based on a thorough analysis of the causes and consequences of any historical or geographical issue, conclusions can be drawn that affect the present day at all levels. (Cod. 96, Secondary, female)

> Explaining in class the evolution of ecosystems throughout history, [...] up to the present day. [...] In Geography class, working in class on factors that pollute rivers, what a drainage basin is, etc. (Cod. 103, Secondary, female)

When considering how to approach these issues in school through the subjects of Social Sciences, we observe a tendency to follow the autonomous curricular model, that is, focusing interventions more on students and methodological aspects rather than from the critical approach, which focuses on decision-making and problem solving. This model is more present in primary (67.07%) than secondary school (47.56%) and mainly emphasises practical training and student motivation.

> I would develop them in a fun way, such as making models, school trips, etc. so that students learn about the importance of these problems in a didactic and visual way. (Cod. 55, Secondary, female)

> With a programme based on experiences and visual explanations, and a lot of support material, such as documentaries or films, experiences told by first-hand by those affected ... (Cod. 164, Primary, male)

In addition to the examples focused on methodological strategies and motivating resources, special importance is given to field trips around the local area, which is very characteristic of the autonomous model. Although some of the responses categorised in this model might approximate the critical model, inclusion in the autonomous model was chosen when the main argument of the proposal focused on methodological issues more than the problematisation of content, the search for solutions, and the promotion of social participation.

> Asking them questions to find out the children's point of view and knowledge and then taking a trip to a recycling centre. (Cod. 84, Primary, female)

> Through activities and field trips related to the local area where they can directly see how these issues affect their daily lives, to achieve greater awareness. (Cod. 32, Secondary, male)

The critical model is present in 30.49% of the proposals made by students, which is very positive. The representativeness of this model is found in a fairly balanced proportion between primary (29.27%) and secondary (31.71%) teachers. In other words, despite the greater presence of the traditional model among future secondary school teachers, the critical model also has an important representation at this stage.

> Through activities that will develop critical thinking in these areas so that students can begin to play an active role in their resolution or, at least, in more sustainable maintenance. (Cod. 93, Secondary, female)

> Through issues that affect their closest environment and foster real student involvement. Taking advantage of something that is happening in the city in order to be able to work from within the problem, going on trips, analysing, proposing solutions, putting them into practice ... (Cod. 107, Primary, female)

> Through environmental awareness activities (recycling, sustainability) as well as social activities that foster values, respect, make the students aware of social differences, as well as present-day socio-environmental problems; working with them on possible solutions to diminish their influence. (Cod. 152, Primary, female)

In this model, responses that focused on the active role of the student were included, prioritising awareness and social intervention, although an important part focused on environmental issues. In other words, the important thing is for students to be aware of their own system of values, of what they think and of the search for possible alternatives, emphasising educational aims such as education in democracy, participation, and the formation of critical citizens who are involved in the management of space and institutions (Morales et al. 2015; Fuster et al. 2021; Cebrián et al. 2021; Benejam 1997; Oller 2011).

4. Discussion

The student teachers who took part in this research rated social problems as being more important than environmental problems, although high scores were given to both types of problems: 9.45 points (Min–Max: 7–10, SD = 0.85) for social problems and 8.87 points out of 10 (Min–Max: 5–10, SD = 1.12) in the case of environmental ones. This assessment coincides with the role that trainee teachers give to environmental issues (Moya and Zubillaga del Río 2020; López 2021; Yavetz et al. 2014; Yates et al. 2019) which they consider a current and future social challenge, although they have little grounding in these topics (Ghosn-Chelala and Akar 2021; Miles et al. 2006) especially among those who have studied disciplines that are not closely linked to the environment (Yavetz et al. 2014).

In this research, we set two goals aimed at ascertaining the perceptions of future teachers based on their studies, but we did not find any significant differences based on this. In other words, the assessment of social and environmental problems did not differ greatly between future primary and secondary teachers despite differences in their initial training.

One of the most significant elements mentioned above was that, among future primary and secondary teachers, gender-based violence was rated the most important social problem. Given that we are working with future teachers, this assessment is important, because of the role education plays in combating gender violence, especially bearing in mind that recent Spanish studies have shown how the percentage of young men (15 to 29 years) who deny gender violence or diminish its importance has increased recently (Rodríguez et al. 2021), and that this is a problem that nonetheless affects the younger population (Instituto de la Juventud 2022). If we analyse the data by sex, although the assessment of gender violence is a little lower in the case of the male future teachers interviewed, the difference is not significant compared to female teachers.

Following the analysis of perceptions based on a list of social and environmental problems (López-Fernández and Oller 2019; López 2021), concluding that most of the problems received a high score, the future teachers were asked about how they would address these problems in their future teaching work. The model that was found to be the most prevalent was the autonomous curriculum development model, centred on students, on experiences in their immediate environment, and on practical activities, although there is difficulty in identifying environmental problems in the immediate environment (López 2021). Among the proposed activities, good environmental practices related to good recycling and waste management habits are strongly present, which also coincides with educational experiences in more degraded environments (Ghosn-Chelala and Akar 2021).

The fact that many of the proposals in this model focus on experiences with students in the surrounding environment is significant. Other research, such as Yavetz et al. (2014), finds that, despite their area of specialism, most students recognise the importance of the environment for their future work as teachers. Addressing local socio-environmental problems yields good results (Tal 2010; Oller 2011) and helps to identify global problems (Ghosn-Chelala and Akar 2021).

In the critical curriculum, content takes centre stage. It is enriched and diversified. It is no longer exclusively factual content as in the traditional model, or content subordinate to the interest of the students who have to discover it, as in the autonomous curriculum model (Pagès 1994). Although content is at the heart of this model, the responses of the participating students focus not so much on the identification of topics, but on how they are approached, making references to awareness, to the problematisation of contents or with activities for the development of critical thinking.

Students whose response can be classed within the critical model show, as in the study by, that socio-environmental problems are part of their ethical responsibilities as future teachers to foster social and environmental change. The cross-cutting nature of socio-environmental problems, which can be addressed globally or through different subjects, is repeatedly highlighted (Álvarez-García et al. 2015), as is the use of media and news for the development of critical thinking and the promotion of participation (Tal 2010). Civic engagement within the community and school is also partly present in curriculum inclusion proposals pertaining to the critical model (Yates et al. 2019; Medina and Castro 2021; Torres-Porras and Martínez-Medina 2021; López 2004).

5. Conclusions

Knowing the perceptions of future teachers in both primary and secondary education is useful for the development of Environmental Education and for the implementation of the new Spanish curriculum framework. The training about sustainable development that student teachers have received throughout the education system is important, but the role given to socio-environmental problems and their curricular inclusion in teacher training studies is also vital.

Knowing the perceptions of future teachers about social and environmental problems is necessary to subsequently analyse how these teachers will integrate socio-environmental problems into their curriculum in the future, because it is essential for teachers to be aware in this regard. The new curriculum framework, in line with European recommendations and the SDGs, opens up new possibilities for teaching and learning about socio-environmental problems, which must undoubtedly be exploited within teacher training, as teachers are important agents of change.

The participating students give more importance to social problems than environmental ones, but with a high assessment of all of them. This aspect is positive to include socio-environmental problems in the initial training of future Primary and Secondary teachers, so that they are competent in a critical curricular development that incorporates environmental issues in a holistic way, aligned with the SDGs and with the new Spanish curricular framework.

Author Contributions: Conceptualization, methodology, investigation, resources, writing—original draft preparation, writing—review and editing, R.G.-M. and R.M.-M. All authors have read and agreed to the published version of the manuscript.

Funding: This research was funded by Junta de Andalucía with the support of FEDER Funds, grant number PYC20 RE 013 UCO.

Institutional Review Board Statement: Not applicable.

Informed Consent Statement: Not applicable.

Data Availability Statement: Data are contained within the article.

Conflicts of Interest: The authors declare no conflict of interest.

References

Alcántara, Jorge, and Silvia Medina. 2019. El uso de los itinerarios didácticos (SIG) en la educación ambiental. *Enseñanza de las Ciencias: Revista de Investigación y Experiencias Didácticas* 37: 173–88. [CrossRef]

Álvarez-García, Olaya, Jaume Sureda-Negre, and Rubén Comas-Forgas. 2015. Environmental education in pre-service teacher training: A liter-ature review of existing evidence. *Journal of Teacher Education for Sustainability* 17: 72–85. [CrossRef]

Benavides-Lahnstein, Ana I., and Jim Ryder. 2020. School teachers' conceptions of environmental education: Reinterpreting a typology through a thematic analysis. *Environmental Education Research* 26: 43–60. [CrossRef]

Benejam, Pilar. 1997. Las finalidades de la educación social. In *Enseñar y Aprender Ciencias Sociales, Geografía e Historia en la Educación Secundaria*. Edited by Benejam Pilar and Joan Pagès. Barcelona: Horsori, pp. 33–52.

Boca, Gratiela D., and Sara Saraçlı. 2019. Environmental Education and Student's Perception, for Sustainability. *Sustainability* 11: 1553. [CrossRef]

Boon, Helen J. 2010. Climate change? Who knows? A comparison of secondary students and pre-service teachers. *Australian Journal of Teacher Education* 35: 104–20. [CrossRef]

Campo, Benito, and Diego García-Monteagudo. 2020. Representaciones escolares del clima en el paisaje fluvial del río Clariano. *Didáctica Geográfica* 21: 147–74. [CrossRef]

Canals, Roser, and Neus González-Monfort. 2011. El currículo de Conocimiento del Medio Social y Cultural, y la formación de competen-cias. In *Didáctica del Conocimiento del Medio Social y Cultural en la Educación Primaria: Ciencias Sociales para Aprender, Pensar y Actuar*. Edited by Santisteban Antoni and Joan Pagès. Madrid: Síntesis, pp. 41–62.

Cebrián, Daniel, Antonio Joaquin Franco, Teresa Lupión, María del Carmen Acebal, and Ángel Blanco. 2021. *Enseñanza de las Ciencias y Problemas Relevantes de la Ciudadanía*. Barcelona: Graó.

Cucurachi, María del Socorro, Juliana Mercon, and Evodia Rivera. 2017. Aportaciones de las percepciones socio-ecológicas a la Educación Ambiental. *Entreciencias: Diálogos en la Sociedad del Conocimiento* 5: 95–110.

Daskolia, Maria, Athanasios Dimos, and Panagiotis G. Kampylis. 2012. Secondary Teachers' Conceptions of Creative Thinking within the Context of Environmental Education. *International Journal of Environmental and Science Education* 7: 269–90.

Flogaitis, Evgenia, and Evagelia Agelidou. 2003. Kindergarten teachers' conceptions about nature and the environment. *Environmental Education Research* 9: 461–78. [CrossRef]

Fuster, Carlos, Diego García-Monteagudo, and Xosé Manuel Souto. 2021. La enseñanza de problemas socio-ambientales. Reflexiones para la inno-vación educativa. *REIDICS. Revista de Investigación en Didáctica de las Ciencias Sociales* 9: 79–96. [CrossRef]

Galvis-Riaño, Carlos J., Francisco J. Perales-Palacios, and Yolanda Ladino-Ospina. 2020. Conceptions about environment and environmental educa-tion by teachers from rural schools in Bogotá-Colombia. *Ambiente & Sociedade* 23: 1–20. [CrossRef]

Ghosn-Chelala, Maria, and Bassel Akar. 2021. Citizenship education for environmental sustainability in Lebanon: Public school teachers' understandings and approaches. *Environmental Education Research* 27: 366–81. [CrossRef]

Instituto de la Juventud. 2022. *La Violencia de Género en los Jóvenes*. Available online: http://www.injuve.es/observatorio/familia-pareja-e-igualdad-de-genero/la-violencia-de-genero-en-los-jovenes (accessed on 20 May 2022).

Kyridis, Argyris, Evangelia Mavrikaki, Helen Tsakiridou, J. Daikopoulos, and H. Zigouri. 2005. An analysis of attitudes of pedagogical students towards environmental education in Greece. *International Journal of Sustainability in Higher Education* 6: 54–64. [CrossRef]

López, Diego. 2004. Problemática ambiental y educación: Una reflexión desde la Geografía. *Didáctica Geográfica* 6: 15–32. Available online: https://didacticageografica.age-geografia.es/index.php/didacticageografica/article/view/175 (accessed on 15 May 2022).

López, José A. 2021. Representación social de los problemas medioambientales del profesorado en formación. Aportaciones para la educación geográfica. *Didacticae: Revista de Investigación en Didácticas Específicas* 9: 53–71. [CrossRef]

López-Fernández, José Antonio, and Montserrat Oller. 2019. Los problemas medioambientales en la formación del profesorado de educación primaria. *REIDICS. Revista de Investigación en Didáctica de las Ciencias Sociales* 4: 93–109. [CrossRef]

Martínez-Medina, Ramón, and José C. Arrebola. 2019. Analysis of Sustainability Activities in Spanish Elementary Education Textbooks. *Sustainability* 11: 5182. [CrossRef]

Massip, Sabater Mariona, Mariona Espinet Blanch, Raul Almendro Díaz, German Llerena del Castillo, and Julia Hosta Cuy. 2021. Problemas Socio ¿qué? Los problemas socioambientales y su concreción por el profesorado en formación. *REIDICS: Revista de Investigación en Didáctica de las Ciencias Sociales* 9: 115–36. [CrossRef]

Medina, Silvia, and Francisca Castro. 2021. Los valores ambientales. In *Didáctica del Medioambiente en Educación Primaria*. Edited by López-Fernández José Antonio and Jorge Alcántara-Manzanares. Madrid: Síntesis, pp. 101–22.

Miles, Rebecca, Linda Harrison, and Amy Cutter-Mackenzie. 2006. Teacher education: A diluted environmental education experience. *Australian Journal of Environmental Education* 22: 49–59. [CrossRef]

Ministerio de Educación. 1990. *Ley Orgánica 1/1990, de 3 de Octubre, de Ordenación General del Sistema Educativo*. Available online: https://www.boe.es/eli/es/lo/1990/10/03/1 (accessed on 20 May 2022).

Ministerio de Educación. 2006. *Ley Orgánica 2/2006, de 3 de Mayo, de Educación*. Available online: https://www.boe.es/eli/es/lo/2006/05/03/2 (accessed on 15 May 2022).

Ministerio de Educacíon. 2013. *Ley Orgánica 8/2013, de 9 de Diciembre, para la Mejora de la Calidad Educativa*. Available online: https://www.boe.es/eli/es/lo/2013/12/09/8/con (accessed on 20 May 2022).

Ministerio de Educación. 2020. *Ley Orgánica 3/2020, de 29 de Diciembre, por la que se Modifica la Ley Orgánica 2/2006, de 3 de Mayo, de Educación*. Available online: https://www.boe.es/diario_boe/txt.php?id=BOE-A-2020-17264 (accessed on 15 May 2022).

Ministerio de Educación. 2022. *Real Decreto 157/2022, de 1 de Marzo, por el que se Establecen la Ordenación y las Enseñanzas Mínimas de la Educación Primaria*. Available online: https://www.boe.es/buscar/act.php?id=BOE-A-2022-3296 (accessed on 20 May 2022).

Morales, Antonio J., Carlos Caurín, and Xosé M. Souto. 2015. Percepción del mundo: Mapas mentales y problemas socioambientales. *Didáctica Geográfica* 14: 91–108.

Morote, Álvaro-F., Jorge Olcina, and Maria Hernández. 2022. Teaching Atmospheric Hazards in the Climate Change Context—Environmental Didactic Proposals in the Mediterranean Region for Secondary Schools. *Environments* 9: 29. [CrossRef]

Mosothwane, Modise. 2002. Pre-service teachers' conceptions of environmental education. *Research in Education* 68: 26–40. [CrossRef]

Moya, Otero José, and Ainara Zubillaga del Río. 2020. *Un Currículo para un Mundo Sostenible #ODS2030*. Madrid: Anaya.

Munoz, François, Franz Bogner, Pierre Clement, and Graça S. Carvalho. 2009. Teachers' conceptions of nature and environment in 16 countries. *Journal of Environmental Psychology* 29: 407–13. [CrossRef]

Oller, Montserrat. 2011. Ensenyar geografia a partir de situacions problema. *Perspectiva Escolar* 358: 14–23.

Pagès, Joan. 1994. La didáctica de las Ciencias Sociales, el curriculum de historia y la formación del profesorado. SIGNOS. *Teoría y Práctica de la Educación* 5: 38–51.

Puk, Tom, and Adam Stibbards. 2010. Ecological concept development of pre-service teacher candidates: Opaque empty shells. *Intenational Journal of Environmental & Science Education* 5: 461–76.

Rodríguez, Elena, Daniel Calderón, Stribor Kuric, and Anna Sanmartín. 2021. *Barómetro Juventud y Género 2021. Identidades, Representaciones y Experiencias en una Realidad Social Compleja*. Madrid: Centro Reina Sofía Sobre Adolescencia y Juventud. Madrid: Fad. [CrossRef]

Tahsin, Kahalid. 2003. Pre-service High School Teachers' Perceptions of Three Environmental Phenomena. *Environmental Education Research* 9: 35. [CrossRef]

Tal, Tali. 2010. Pre-service teachersí reflections on awareness and knowledge following active learning in environmental education. *International Research in Geographical and Environmental Education* 19: 263–76. [CrossRef]

Torres-Porras, Jerónimo, and Ramón Martínez-Medina. 2021. Los problemas medioambientales. In *Didáctica del Medioambiente en Educación Primaria*. Edited by López-Fernández José Antonio and Jorge Alcántara-Manzanares. Madrid: Síntesis, pp. 123–50.

Van Petegem, Peter, Ann Blieck, Ingrid Imbrecht, and Tom Van Hout. 2005. Implementing environmental education in pre-service teacher training. *Environmental Education Research* 11: 161–71. [CrossRef]

Yates, Kimberly, April Reefer, Denice Robertson, Jennifer Hubbard-Sanchez, Jeanine Huss, and Melinda Wilder. 2019. Educators' perceptions of environmental education and professional development in teacher preparation programs. *Applied Environmental Education & Communication* 18: 207–18. [CrossRef]

Yavetz, Bela, Daphne Goldman, and Sara Pe'er. 2014. How do preservice teachers perceive 'environment' and its relevance to their area of teaching? *Environmental Education Research* 20: 354–71. [CrossRef]

social sciences

MDPI

Article

Did the Characteristics of Kosovar Teachers Influence the Results of Students in TIMSS 2019? Findings from the Performance of Kosovar Students in TIMSS 2019

Fadil Latifi [1] and Endrit Latifi [2,*]

1 Faculty of Social Sciences, College AAB, 10000 Pristina, Kosovo; fadil.latifi@universitetiaab.com
2 Independent Researcher, 10000 Pristina, Kosovo
* Correspondence: eflatifi10@gmail.com

Abstract: This paper aimed to explore the impact of Kosovar teachers' characteristics on the performance of fourth-grade students in TIMSS 2019. The paper addressed the characteristics of teachers, demographic, and academics, which are closely related to the overall performance of students. In this research, data from the TIMSS 2019 database were used, for which 4435 Kosovar students were tested, and 192 teachers were surveyed. The findings show that Kosovar students did not perform well in TIMSS 2019. They achieved 444 points in the field of mathematics and 413 points in science (average TIMSS 500 points) and were ranked 49th out of 56 countries participating in the test. Data from the teacher survey show that the characteristics of teachers, such as age, professional development, pre-service education, contemporary approach to teaching, etc., also played an important role in the low-scoring performance of Kosovar students in TIMSS 2019. Students who had a teacher with an average age of 30–60 years had a better performance on the test, achieving a difference of 20 points more, compared to students who had new or older teachers (under 30 years and over 60 years). The lack of professional development of teachers also had a major impact on student performance. On average, 25% of teachers in the last 2 years had attended training in certain areas, while on average 80% of them were willing to attend training programs that would improve the profile of their competencies.

Keywords: characteristics of teachers; Kosovo; performance; TIMSS 2019; professional development

Citation: Latifi, Fadil, and Endrit Latifi. 2022. Did the Characteristics of Kosovar Teachers Influence the Results of Students in TIMSS 2019? Findings from the Performance of Kosovar Students in TIMSS 2019. *Social Sciences* 11: 344. https://doi.org/10.3390/socsci11080344

Academic Editor: Alvaro Morote

Received: 4 July 2022
Accepted: 1 August 2022
Published: 3 August 2022

1. Introduction

Teacher quality is extremely important for student achievement, even though there is still no standard definition for what we mean by teacher quality (Seebruck 2015; Hanushek 2011). According to Barton and Coley (2009), among the factors that influence the creation of gaps in student achievement, in addition to educational policies, curriculum design, and the use of technology in teaching, are also the pre-service preparation of teachers and their experience. There is a high theoretical correlation for the relationship between student achievement and the quality of teachers (Prince 2002). There have been debates over time about the effects that school can have on student performance. There are opinions that schools have less influence compared to the social aspect of students (Coleman et al. 1966). There are others who think that the key factors in student achievement are both the classroom (Mosteller 1995) and the teachers (Dearling 2000). The quality of teaching creates a gap that can cause irreparable damage (Seebruck 2015). Curriculum reforms, textbook revision, and teacher professional development are interlinked and provide clear guidance on the ongoing changes in educational processes (Mullis and Martin 2012). The provision of constructive feedback to students by teachers, identifying needs, and easier approaches to child learning, are just some of the elements that influence student outcomes (Shala et al. 2021b).

On the other hand, in recent decades, professional development has been a subject of particular interest for teachers, through whom new educational practices have been disseminated (Borko et al. 2010; Ponte 2010). Teacher professional development programs are an important instrument for the implementation of educational policies at practical levels in terms of transferring new knowledge, skills, and values acquired in practice by students (Sasson and Miedijensky (2020). Such programs have allowed teachers to learn from research activities and practices (Ponte 2010)). "Professional development is defined as activities that develop an individual's skills, knowledge, expertise and other characteristics as a teacher" (OECD 2009). "How can we prepare students for jobs that have not yet been created, tackle social challenges that we can not yet imagine, and use technologies that have not yet been invented?"(OECD 2019).

Dam and Janssen (2021) have found that in recent years, many reforms have been initiated which have not had a direct impact on raising the quality of education because there has been a lack of oversight of their implementation. Rapid changes ranging from student diversity (different geographical areas where students come from) to developments in technology require teachers to be coherent and flexible in their work (Zeggelaar et al. (2020). Professional development should also be based on the motivation of teachers, their needs, and aspirations in terms of their professionalism in the field of teaching (Ungar 2020). It puts into function some of the characteristics of teachers (Zhang et al. (2021). The quality that teachers with students have to offer is essential in the implementation of the competencies and personal profile that they have built during their years of study as well as the attitude they have built towards the teaching processes in general (Karlberg and Bezzina (2020). In reflective teaching, the profile of teacher competencies should be as high and specific as possible (Jong et al. 2011). According to Mahler et al. (2018), two of the characteristics that teachers should have are motivation and certain knowledge of the given field. Research developed by Baier et al. (2019) emphasizes that the characteristics of teachers can influence quantitative change in the quality of teaching. Professional development of teachers from narrow fields, such as mathematics and pedagogical function, is essential to maximize the impact that teachers can have during their work with students (Bolyard and Baker 2021). According to Rubie-Davies (2006), young teachers are more interested in changes in education, and they have higher expectations of students. The concept of employing teachers for the teaching and learning process is a vague process, while these concepts are crucial in providing quality teaching which has an impact on student learning (Jacobs et al. 2014). Dybowski et al. (2017) have pointed out that students who are exposed to good teachers are more likely to increase and develop their competencies.

Sanders and Rivers (1996), who conducted a study observing the results of students from the second to fifth grade, observed that student achievement was related to teacher characteristics. While there may be abundant literature in international contexts which explains the link between teacher characteristics and student achievement, the reality is that there is a lack of evidence in the case of Kosovo. Conducting international large-scale assessments in the country has contributed to the development of large databases which could be used to explain which variables mediate student achievement in the case of Kosovo. However, despite available datasets, publications on the achievement of Kosovar students are very scarce, and among those existing, no study evaluates the impact of teacher characteristics on achievement. The present study aimed to mitigate this situation by researching the link between teacher characteristics and the performance of Kosovar students in the TIMSS 2019 study.

Research on the impact of teacher characteristics is particularly important in a country such as Kosovo, which has faced major reforms and challenges.

Since the declaration of Kosovo's independence in February 2008, Kosovo has faced various challenges in all fields, in particular in the field of education. Poor-quality education in both the pre-university and university systems, frequent changes in the education system, a lack of training for certain areas, supervision of the implementation of new approaches adopted during training, etc., are just some of the challenges we still face.

An unsatisfactory level of education has been introduced in recent years on the occasion of the development of several international assessments. Identifying the factors that have influenced the unsatisfactory results of Kosovar students is extremely difficult (Jepsen 2005) given that Kosovo has slow economic development and an almost unsupervised education system. Numerous researchers have analyzed these results and presented research on the unsatisfactory results of Kosovar students in PISA and TIMSS (Shala et al. 2021a), but the country's institutions have not treated these results as something worrying. This paper aimed to precisely point out the above-mentioned characteristics of teachers which we consider to have had a direct impact on the performance of Kosovar students in TIMSS 2019.

TIMSS (Trends in International Mathematics and Science Study) is an international assessment that serves to measure students' knowledge in two areas, mathematics and science, as well as to compare education systems (Mullis et al. 2020). This test is organized every four years; 2019 was the seventh evaluation cycle, in which Kosovo participated for the first time. In this research, the characteristics of teachers that were taken as reference points were: age, experience, quality of studies, their approach to teaching and learning, traditional teaching methodologies, lack of training, lack of supervision during the implementation of training, frequent curricular changes, and the inability to follow the trends in technology.

An important element within all these factors is the preparation of new teachers before service, or "academic and professional potential". The "academic potential" of the graduates in Kosovo, as a number, is not lacking when it is known that there are universities in the five cities of Kosovo, where there are also faculties of education that prepare new teachers, but in Kosovo, there is a lack of professional preparation of young people to face the demands of the 21st century, namely the developments in all areas, especially in the integration of technology in teaching. Teaching in these universities generally takes place through traditional methods and forms and does not make the student part of the construction of his own knowledge. Lectures are carried out in the full sense of the word "lectures", without constructiveness, creativity, and interactivity. Such teaching cannot prepare the teacher for the future. The same fact is shown by the results of the teachers' survey, which was cross-referenced with the performance results of fourth-grade students in the TIMSS 2019 test, for which young teachers' *(up to the age of 30)* students had weaker results, which proves that pre-service education has not achieved its mission.

Starting from the premise that "no matter how good the pre-service teacher education is, our expectations may not be realistic if we think they can meet all the challenges of education" (OECD 2009), we noticed that in Kosovo, both pre-service education and the professional development of the teachers were not in line with the needs of the students.

In recent years in Kosovo, programs for professional development (PD) of teachers have been offered, which in many cases have had as their goal the substantive aspect of teaching or even the methodical one, but the contemporary approaches that have been offered through the contents of the training have almost never been implemented as requested.

> "Effective professional development is continuous, includes training, practice, and feedback, and provides the right time and follow-up support. Successful programs involve teachers in learning activities that are similar to those they will use with their students and encourage the development of teacher-learning communities " (OECD 2009).

Middle-aged teachers (30–60 years old) usually participate in these trainings. Teachers over 60 years old are reluctant to participate in these trainings considering that they are close to retirement age (65 years old), while in certain cases young teachers (under 30) are thought to have been prepared during pre-service studies. The percentage of teachers declaring to belong to these ages was 32.1%. So, if we look at the results of students who had, for their teacher, a person under 30 or over 60 years old, which constituted approximately 1/3 of the surveyed teachers, we see that these students performed worse when they were tested (an average of 20 points less).

The participation and implementation of trainings are also legally regulated. According to the Administrative Instruction (MEST, UA-5/2017), to obtain their work license and to move from one level of license to a more advanced level, teachers need 100 h of training. This process has almost passed in formality and does not achieve its effect in practice. Differences in student outcomes based on teacher age have also been observed by TIMSS 2019 researchers in Turkey, Sezer and Çakan (2022), who relate student outcomes to teachers' age and experience.

According to Hanushek (1992), even if for only one year a child has as a teacher a person who is not effective in his work, in the future this student will find it impossible to cover these created gaps, and in this case, it is also the main variable (Prince 2002). Schools can do the most important thing for their students, and that is offering them good teachers, according to Hanushek (2011). Furthermore, Clotfelter et al. (2006) conclude that "the effect of teacher credentialing on student achievement is large and, therefore, credentialing is important enough to be considered in public policy decisions aimed at improving student achievement".

2. Materials and Methods

2.1. The Hypothesis

H1. *There is a link between the age of teachers and the points achieved by their students in the TIMSS 2019 test.*

H2. *Teachers' expectations for students' achievement in the test are higher than the real test results of students.*

H3. *Teachers do not implement the curriculum according to the parameters required by the central institutions.*

H4. *The needs of teachers for professional development are great, both in terms of mathematical and scientific content as well as for the implementation of the curriculum, the development of students' critical thinking, etc.*

2.2. Research Methods

This research was based on data from the TIMSS 2019 database, and only the data from Kosovo were used. TIMSS 2019 used quantitative methods, with the development of a test for fourth-grade students, as well as a questionnaire for the teachers. For the purposes of the research, we used the (TIMSS 2019 database) data from the teachers' survey as well as the average results from the students' tests in two fields, namely in mathematics and science. The data from the teacher questionnaire were cross-referenced in each case by comparing both demographic issues and the statements in the questionnaire with the average points achieved by the students on the test.

2.3. Respondents

In 2019, Kosovo participated for the first time in the TIMSS test. The sample work included (N) 4435 fourth-grade students and (M) 192 teachers from 145 schools. In terms of gender, 23.5% were male teachers and 76.5% were female.

2.4. Instruments

The paper used data from the 2019 International Trends in Mathematics and Science Study (TIMSS), which contained two observations for each student, one in mathematics and one in science. The main outcome variable was students' math and science test scores compared to teachers' questionnaire statements about their characteristics. Then, the teachers' statements regarding their characteristics were cross-referenced with the students' achievements in the test to see the impact of these characteristics. The test was conducted on a hard copy, even though 50% of students at the general state level were tested in computer form (Mullis et al. 2020).

3. Results

3.1. The Performance of Kosovar Students in TIMSS 2019

Since 2015, Kosovo has participated in five international assessments The published results of assessments such as PISA 2015, PISA 2018, and TIMSS 2019 show the serious situation in Kosovar education. The performance of Kosovar students in these evaluations was very low. Out of 56 participating countries, Kosovo ranked 49th in the TIMSS 2019 test. The findings show that there were many factors that influenced this result, and one of these factors was the teachers' characteristics. Some of these characteristics were related to the age of teachers, the preparation of students before the service, the professional preparation of teachers through training in different fields, the frequent changes in the education system, etc.

A total of 4435 students from 145 schools in Kosovo participated in the TIMSS 2019 test. On this occasion, 192 teachers who taught these classes were surveyed. From the data obtained from the TIMSS 2019 database and presented in Table 1, it can be noticed that the majority (F, 76.5%) of the surveyed teachers were female, while the minority (M, 23.5%) were male. Male teachers achieved better success with their students in the test in the field of mathematics (M, 453.4 points), while female teachers' were more successful with their students in science (F, 416.3 points). The data show that teachers had an average teaching experience of 16.3 years of work experience in teaching, while the most frequent frequency was 10. The majority of teachers (76.6%) were qualified and had completed bachelor studies or a similar 4-years program (240 ECTS), while a significant percentage (7.6%) completed master's studies.

Table 1. Teacher demographic data cross-referenced with student test scores.

Teacher's valid sample (N 192)	Male	Female
	23.5%	76.5%
Math (*Average scores of students on the test*)	453.3	447.0
Science (*Average scores of students on the test*)	402.1	416.3
Years of teaching (Mean)	16.3	
Years of teaching (Mode)	10	
Highest formal level of education completed by teachers		
1. No upper secondary	0.0%	
2. Upper secondary	6.1%	
3. Post-secondary, non-tertiary	0.6%	
4. Short-cycle tertiary	9.0%	
5. Bachelor's or equivalent	76.6%	
6. Master's or equivalent	7.6%	
7. Doctor or equivalent	0.0%	

The data in Table 2 show that the majority of teachers were between the ages of 30 and 59 years old (67.9% of them). When we crossed the data of the age of the teachers and the average number of points the students achieved on the test, we noticed that it was exactly this age of teachers (30–59 years old) that managed to better prepare their students, who achieved more points on the test. Data from Table 2 show that students who had a teacher between the ages of 30 and 59 had a better performance in mathematics (average 452 points) and in science (average 422 points) compared to students who had young teachers (under 30) or those over 60 years old, performing worse in mathematics (on average 25 points lower) and in science (on average 28 points lower).

Table 2. Teachers' calendar age compared to students' TIMSS 2019 scores.

	Age of Teachers					
Student scores on TIMSS 2019	Under 25 (6.4%)	25–29 (11.8%)	30–39 (28.0%)	40–49 (24.4%)	50–59 (15.5%)	60 or more (13.9%)
Students' scores in mathematics (average)	426.4	439.3	452.5	451.2	451.6	424.2
Students' scores in science (average)	394.2	406.8	420.0	423.7	421.6	388.4

Knowing that Kosovar students in total achieved 444 points in mathematics and 413 points in science, we noted that students scoring above this average had as a teacher a person who was older than 30 years and younger than 60 years. They managed to accumulate an average of 451 points in mathematics and 421 points in science. Looking at the differences between the generations of teachers (young teachers without experience, as well as older teachers), we could estimate that there were a large number of factors that affected these ages, starting from the impossibility of implementing the new educational system, quality in universities regarding the preparation of new teachers, the small amount of training for new teachers, lack of teaching experience, lack of energy needed for successful work by older teachers, older teachers' approach to educational reforms, the use of technology in teaching at scale, etc.

3.2. *How Much Influence Do the Teachers' Characteristics Have on the Students' Results?*

Table 3 combines the opinions of teachers regarding the content and methodological aspects of teaching with the results achieved by students. The results show that the percentage of teachers stating they had implemented the curriculum at a *very high* level (18.2%) was insufficient, while a more favorable percentage of teachers implemented it a high level (52.3%). The data in the table below show that teachers who had a high degree of success in implementing the curriculum achieved better results in both mathematics (450.7) and science (418.0).

Table 3. TIMSS 2019 scores according to teacher characteristics.

How Would You Characterize Each of the Following within Your School?		Students' Scores in Math and Science (TIMSS 2019)	
		Math	Science
My degree of success in implementing the school's curriculum	Very high	450.7	418.0
	High	441.2	409.8
	Medium	444.1	414.8
	Low	416.4	383.6
	Very low	-	-
I understand the curricular goals of the school	Very high	450.5	418.4
	High	440.3	408.0
	Medium	445.7	416.9
	Low	416.4	383.6
	Very low	-	-
My expectations for student achievement	Very high	437.6	403.2
	High	447.2	416.4
	Medium	446.1	418.1
	Low	-	-
	Very low	-	-

Although the percentage of teachers who understand the goals of the curriculum very well was low, this percentage still made a difference in student outcomes. Their students achieved a total average of 434.3 points in math (450.5) and science (418.4) with a difference of 34 points compared to the students of teachers who did not understand the curricular goals.

The data from Table 3 show us that about 21% of the teachers had high expectations for their students regarding their assessment with the TIMSS test. The students of these teachers achieved poorer results (on average 12 points less) compared to the students of teachers who had lower expectations. The percentage of teachers who had average expectations (3.0%) for their students was lower compared to other statements, but it was noted that these students had achieved higher points in mathematics (446.1) and science (418.1), with a mean of 439.8 points.

According to the teachers' statements regarding the students' inspiration for work, it seems that such an inspiration did not happen, because the teachers who circled the average alternative had better results with their students. Based on the teachers' statements, they felt very enthusiastic and inspired about the work they do. When we compared these two elements (enthusiasm and inspiration) as a percentage of the statement in all the participating countries, Kosovo had the highest percentage of all the participating countries (87.4% and 90.7%), despite the fact that the students did not manage to reach the average score determined by TIMSS of 500 points.

From the data collected and presented in Tables 3 and 4, in some cases, we noticed a discrepancy between the teachers' statements and the students' actual results. This made us understand that the theoretical part (the statements in the questionnaire) did not coincide with the practical part (the test results). These elements encourage us to further expand the field of research regarding the TIMSS assessment in the future.

Table 4. TIMSS 2019 scores according to teacher characteristics.

How Would You Characterize Each of the Following within Your School?		Students' Scores in Math and Science (TIMSS 2019)	
		Math	**Science**
I inspire my students to learn	Very high	442.2	408.8
	High	446.9	417.2
	Medium	451.4	428.2
	Low	450.6	426.9
	Very low	-	-
I am content with my profession as a teacher	Very high	444.5	419.2
	High	442.2	412.5
	Medium	-	-
	Low	-	-
	Very low	-	-
I am enthusiastic about my job	Very high	446.3	432.0
	High	415.5	397.5
	Medium	-	-
	Low	-	-
	Very low		
My work inspires me	Very high	445.2	414.4
	High	434.1	400.8
	Medium	-	-
	Low	-	-
	Very low	-	-

3.3. *Professional Development of Teachers Compared with Average Scores of Students in Mathematics and Science*

Post-war Kosovo (1999) has found the education system in a miserable state. From the year 2000 onwards, many international organizations have assisted with training education in various forms. Valuable help is also provided in terms of training, which is organized for teachers of different levels and in various fields. The programs offered through these training have enabled a change in teachers' approaches to teaching and learning in general. What has been lacking during these 20 years, and what is lacking today, is the systematic oversight of the entire teaching process as well as responsibility and accountability. In the statements of teachers (Table 4), it was noted that they were enthusiastic about work, adored their profession, and were very inspired, but, despite these facts, the results of international evaluations were not good. Out of 56 countries participating in the TIMS 2019 test, Kosovo ranked 49th with 444 points in mathematics and 413 points in science. Additionally, in the TIMSS 2019 test, Kosovo ranked last among the countries in the region. Teacher training plays an extraordinary role in raising the quality of teaching. If we look at the data in Table 5, it can be seen that in the last two years, teachers attended very little training in the field of mathematics and (Table 5) science (Table 6). Based on their statements, only 23.5% had attended any training in mathematics during the last two years, while the percentage was even lower for science training (15.9%). The percentage of teachers did not increase even when asked if they had attended training related to the pedagogical aspect (math, 20.5%. science, 16.0%).

Table 5. The professional development of teachers compared with the average scores of students in mathematics.

Teachers' Declarations about Professional Development	During the Past Two Years, Have You Participated in Any Professional Development (PD) Training in Math?	
	Yes	No
	Student scores in TIMSS 2019 compared to teachers' statements about PD	
Mathematics content	447.5	443.7
Mathematics/pedagogy instruction	452.9	442.8
Mathematics curriculum	445.6	444.4
Integrating technology into mathematics instruction	441.7	445.5
Improving students' critical thinking or problem-solving skills	447.2	443.1
Mathematics assessment	446.0	444.0
Addressing individual students' needs	448.4	443.2

Table 6. The professional development of teachers compared with the average scores of students in science.

Teachers' Declarations about Professional Development	During the Past Two Years, Have You Participated in Any Professional Development (PD) Training in Science?	
	Yes	No
	Student scores in TIMSS 2019 compared to teachers' statements about PD	
Science content	418.8	413.5
Science/pedagogy instruction	414.2	412.6
Science curriculum	419.3	413.9
Integrating technology into mathematics instruction	415.6	413.3
Improving students' critical thinking or problem-solving skills	419.3	415.1
Science assessment	415.7	413.4
Addressing individual students' needs	415.3	413.0

Approximate percentage values were obtained if we noticed other statements of teachers regarding curriculum training, integration of technology in mathematics and science, improving critical thinking or problem-solving skills of students, student assessment, or even addressing the individual needs of students. The highest average of training-related statements was 39.4% for teachers who had attended training to improve critical thinking skills.

If we analyze the data from Table 4 based on the training that the teachers had attended in the last 2 years and compared their professional development with the students' test scores, we can see that the students who were lucky enough to be part of classes taught by teachers stating they had attended training were more successful in the test. When teachers stated that they attended training in mathematical content, students achieved approximately 4 points more. In the case where the teachers participated in pedagogical and methodical instruction training, the students achieved 10 points more compared to their peers, who were in classes taught by teachers that did not attend pedagogical and methodical training.

There are small but substantial differences in student results, offering evidence for a correlation between the training attended by teachers and the students' results on the test.

Students of teachers who attended training for the pedagogical aspect of teaching achieved better results in mathematics (10 points more) as opposed to those of teachers who stated that they had not attended any such training.

When we analyzed the data from Table 6 based on the training that the teachers had attended in the last 2 years and compared their professional development with the students' test results, we could see that the results of the students who had the trained teacher were several points higher. When teachers stated that they attended training in science content, students achieved approximately 5 points more. In the case where the teachers participated in pedagogical and methodical instruction training in science, the students achieved 2 points more compared to their peers, who were in classes taught by teachers who did not attend pedagogical and methodical training. There were also such differences in the teachers' statements about other areas, such as evaluating and addressing students' needs.

If we look at Figure 1, the part where teachers evaluated professional development, it can be seen that they were willing to attend training in the above areas. Teachers' identification of these training needs means that they may potentially have difficulty addressing certain issues with students. Each of the statements given in the questionnaire, which required teachers to declare whether or not they needed a different kind of training, reached a very high percentage, even going up to 87% in certain cases (addressing the individual needs of students). All the facts prove that there is a close connection between the professional development of teachers and the results achieved by students. The more programs we have for the professional development of teachers, the higher the student results will be.

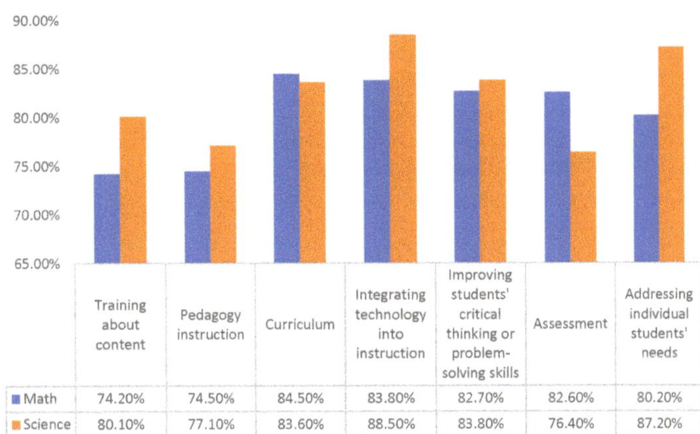

Figure 1. Teachers' needs.

4. Discussion

The results of the paper show that the characteristics of teachers are of great importance in increasing the performance of students and achieving higher results in both local and international assessments. According to Hanushek (2011), the quality of teaching and teacher characteristics are decisive factors for student learning. Ribers et al. (2021), among others, have emphasized that the practice of lifelong learning by teachers gives teachers the "right mastery". Pedagogical practices cannot be improved other than through professional development (Tang et al. 2022). Data from the literature show that for a student who spends 1 year with an ineffective teacher, this time can never be compensated for, while when spending 1 year with an active teacher, the student's benefits are doubled (Seebruck 2015). According to Grajcevci and Shala (2021), although it is known that the characteristics of teachers play a decisive role, it is still not so clear what the characteristics of a good teacher are. The analysis conducted in this study revealed that when teachers had an overall positive outlook regarding professional development, students also performed higher in mathematics and science. In addition, students tended to perform higher when teachers were at least in their 40s, which may also underline the importance of not only pedagogical knowledge but also teaching experience. Hence, future studies should focus more in depth on the impact that teacher experience has on student performance and possibly what variances in student performance this factor accounts for. Finally, in terms of professional development, TIMSS results reveal that Kosovar students benefited from any type of professional development teachers attended; however, it should be kept in mind that these improvements were not large enough to be significant, which is why they need to be reported cautiously.

Lifelong learning enables teachers to learn the best skills and practices in this area (Ribers et al. 2021). Tang et al. (2022) have emphasized that professional development is needed to improve teachers' pedagogical approaches. The experiences of teachers that are passed on to them through professional development can dramatically affect students (Didion et al. 2020; Fischer et al. 2018; Piper et al. 2018). Education is an ongoing process that does not end with graduation. Teachers are constantly improving their skills and becoming more proficient and agile in teaching (Quins University Charlote n.d.). We cannot imagine the contemporary teacher without professional and sustainable development. This development goes beyond the teaching content while also ensuring professional sustainability (Washington, 17 September 2019). Although the skills that teachers gain during their schooling are related to methodological and evaluative aspects, teacher training still plays a key role in curriculum implementation (Burgess et al. 2016).

Baumert et al. (2010) have emphasized that the teacher's pedagogical knowledge is one of the determining characteristics of a favorable student result. During the analysis of the results, it was observed that the pedagogical aspect was not one of the strengths of Kosovar teachers. They had major setbacks in understanding the curriculum as well as in its implementation in practice. The lack of basic knowledge, both from the pedagogical and methodological aspects, leads us to failure. Some of the characteristics of teachers that can have a direct impact on student achievement were the qualification of teachers, their age, teaching experience, knowledge and understanding of basic documents from the pre-university education system, lack of training in certain areas both in terms of content and pedagogy, lack of integration of technology in teaching, methodical approach regarding the development of students' critical thinking, student assessment, addressing students' individual needs, etc. Teachers' training needs were generally oriented towards the use of technology in teaching, curriculum implementation, assessment, and addressing the individual needs of students—individualized teaching. It is almost impossible to increase student success in international assessments without the professional development of teachers. The more quality training programs are offered to teachers in certain areas, the higher the student outcomes will be.

5. Conclusions

Teachers and the activities they conduct with students are very important, but even though their education in terms of duration is the same, they are different. This variability has to do with some of the characteristics they have built up over time. Researchers Jordan et al. (1997), while conducting research in Texas, observed that students who had more effective teachers achieved better results in assessments, reaching over 35 points in reading and about 50 points more in mathematics. Such effectiveness is also missing among Kosovar teachers, although this conclusion relies on data from international research. The education of pre-service teachers is lacking; seeing that teachers do not follow the trends of developments in education through professional training, we can conclude that Kosovo should take the findings so far seriously and design educational policies that foresee in-depth reforms of university and pre-university education. The ranking of Kosovar students in the TIMSS 2019 test in 49th place out of 56 participating countries (444 points in mathematics and 413 in science) is not accidental. Several factors led to this unsatisfactory result. It is very difficult to identify all the factors that caused Kosovar students to have such a low performance in this international assessment. According to Gjelaj et al. (2018), the key factor in the low quality of education is the small number of children attending preschool institutions, as we know that only 10% of children attend these institutions. Research shows that teachers are about 90% qualified, but (OECD 2009) pre-service education does not mean that teachers are ready to face all challenges. It reflects the aspirations, enthusiasm, and commitment of teachers (Ungar 2020). Research data show that the surveyed teachers have an average age of 16.3 years. Although a very young average age, the intersection of information between the calendar age of teachers and the points achieved by students in the TIMSS 2019 test proves that teachers under the age of 29 and those over 60 had lower scores by contrast with teachers aged 30–49. The difference between these two categories of teachers grouped by age was observed in the points that students accumulated on the test, and significant differences were recorded with an average of 20 points in mathematics and 22 points in science. Teachers' expectations regarding the results of their students were extremely high, but this was not reflected in the scores of their students. According to Rubie-Davies (2006), young teachers have higher expectations of their students. Teachers who had lower expectations (19.4%) for student testing were shown to be more accurate in the international ranking, while achieving better results with their students domestically. Research data show that teachers expressed that they can inspire students to learn, were very satisfied with their work (87%), were very enthusiastic at work (87.4%), and that working as teachers inspired them (90.7%). Research data show that the percentage of teachers very enthusiastic about work was over 87% *(the first in the list compared to the statements of the 56 countries participating in the test about enthusiasm)*, over 90% were inspired, and 80% had high expectations for their students. Despite the enthusiasm and inspiration of the teachers, the students' results show the opposite. Out of 56 countries participating in the test, Kosovar students achieved 444 points in mathematics and 413 points in science, ranking in 49th place. The average score set by TIMSS 2019 was 500 points. Poor results of Kosovar students in international measurements, especially in the TIMSS 2019 test, are an alarm for researchers. Based on data from the TIMSS 2019 test, the professional development of teachers is an immediate need. Only 23.5% of teachers stated that they had attended training for mathematical content, while 15.9% had attended training for content in science. Approximate percentages are also given for other areas in which teachers had attended any training. Regarding pedagogical instructions, approximately 20% had attended training in mathematics and 16% in science, and regarding the integration of technology in teaching, 16.3% had attended training in mathematics and only 9.3% in science. Teachers expressed full readiness to attend training in certain fields. In each field, the percentage of teachers who stated they would attend training ranged from 74% to 88%.

All the data presented in this paper allow us to understand that the characteristics of teachers, such as quality education before service, the age of teachers, teachers' professional development, pedagogical and methodical preparation, and knowledge and implementation of the curriculum, are some of the factors that influenced the poor results of Kosovar students in TIMSS 2019. The rise and development of teachers' characteristics is largely based on the new approaches that universities in Kosovo should offer. A new approach implies teaching based on creativity and constructivity, a type of teaching dominated by responsibility and accountability. This responsibility and accountability should also pass to other levels, especially the pre-university ones, when teachers are faced with their work with students. The data from this research, as well as the data from other research related to the field of youth education, should encourage institutions to draw up long-term policies and strategies in order to get out of this vicious circle as soon as possible.

Author Contributions: Conceptualization, F.L. and E.L.; methodology, F.L.; software, E.L.; validation, F.L. and E.L.; formal analysis, F.L.; investigation, E.L.; resources, F.L.; data curation, E.L.; writing—original draft preparation, F.L.; writing—review and editing, F.L.; visualization, E.L.; supervision, F.L.; project administration, F.L. All authors have read and agreed to the published version of the manuscript.

Funding: The authors received no financial support for the review, authorship and/or publication of this article.

Institutional Review Board Statement: Not applicable.

Informed Consent Statement: Not applicable.

Data Availability Statement: TIMSS 2019 database, https://timss2019.org/reports/achievement/ (accessed on 13 October 2021).

Acknowledgments: We would like to thanks IEA TIMSS & PIRLS, International Study Center 2019 for presenting the data of all participating states, and in this case the data of Kosovo We would like to thank Endrit Latifi for the engagement and coordination of the research.

Conflicts of Interest: The authors declare no potential conflicts of interest with respect to the research, authorship, and/or publication of this article.

References

Baier, Franziska, Ana Theresia Decker, Thamar Voss, Thilo Kleickmann, Uta Klusmann, and Mareike Kunter. 2019. What makes a good teacher? The relative importance of mathematics teachers' cognitive ability, personality, knowledge, beliefs, and motivation for instructional quality. *British Journal of Educational Psychology* 4: 767–86. [CrossRef] [PubMed]

Barton, E. Paul, and Richard J. Coley. 2009. *Parsing the Achievement Gap II*. Policy Information Report. Princeton: Educational Testing Service. Available online: https://www.ets.org/Media/Research/pdf/PICPARSINGII.pdf (accessed on 3 March 2021).

Baumert, Jürgen, Mareike Kunter, Werner Blum, Martin Brunner, Thamar Voss, Alexander Jordan, Uta Klusmann, Stefan Krauss, Michael Neubrand, and Yi-Miau Tsai. 2010. Teachers' Mathematical Knowledge, Cognitive Activation in the Classroom, and Student Progress. *American Educational Research Journal* 47: 133–80. [CrossRef]

Bolyard, Johnna, and Courtney Baker. 2021. Examining the practice of elementary mathematics specialists through narratives: Implications for professional learning and development. *Professional Development in Education* 2021: 1–15. [CrossRef]

Borko, Hilda, Jennifer Jacobs, and Karen Koellner. 2010. Contemporary Approaches to Teacher Professional Development. *International Encyclopedia of Education* 7: 548–56.

Burgess, Annete, Christine van Diggele, and Craig Mellis. 2016. Student teacher training: Participant motivation. *The Clinical Teacher* 13: 267–70. [CrossRef]

Clotfelter, T. Charles, Helen Ladd, and Jacob Vigdor. 2006. Teacher Student Matching and the Assessment of Teacher Effectiveness. *Journal of Human Resources* 41: 778–820. [CrossRef]

Coleman, S. James, Ernest Q. Campbell, Carol J. Hobson, James McPartland, Alexander M. Mood, Frederic D. Weinfeld, and Robert L. York. 1966. *Equality of Educational Opportunity*; Reprinted from *American Sociological Review* 32: 1967. Washington, DC: U.S. Government Printing Office. Available online: https://www.ssc.wisc.edu/wlsresearch/publications/files/public/Sewell-Marascuilo-Pfautz_Review.Symposium.C.et.al.pdf (accessed on 15 January 2022).

Dam, Michiel, and Fred J. J. M. Janssen. 2021. Modularity in teacher professional development—Building blocks for bridging everyday teaching practices and reform ideals centered around whole tasks. *Professional Development in Education* 2021: 1–18. [CrossRef]

Dearling, Linda Hammond. 2000. How Teacher Education Matters. *Journal of Teacher Education* 51: 166–73. Available online: https://www.researchgate.net/publication/249704714_How_Teacher_Education_Matters (accessed on 11 April 2020). [CrossRef]

Didion, Lisa, Jessica R. Toste, and Marissa J. Filderman. 2020. Teacher professional development and student reading achievement: A meta-analytic review of the effects. *Journal of Research on Educational Effectiveness* 13: 29–66. [CrossRef]

Dybowski, Christoph, Susanne Sehner, and Sigrid Harendza. 2017. Influence of motivation, self-efficacy and situational factors on the teaching quality of clinical educators. *BMC Medical Education* 17: 84. [CrossRef]

Fischer, Christian, Barry Fishman, Chris Dede, Arthur Eisenkraft, Kim Frumin, Brandon Foster, Frances Lawrenze, Abigail Jurist Levy, and Ayna McCoy. 2018. Investigating relationships between school context, teacher professional development, teaching practices, and student achievement in response to a nationwide science reform. *Teaching and Teacher Education* 72: 107–21. [CrossRef]

Gjelaj, Majlinda, Ema Rraci, and Kushtrim Bajrami. 2018. Preschool Education in Kosovo. Pristina. Available online: https://www.researchgate.net/profile/Majlinda-Gjelaj/publication/334896051_PRE-SCHOOL_EDUCATION_IN_KOSOVO/links/5d4448bc92851cd0469a6722/PRE-SCHOOL-EDUCATION-IN-KOSOVO.pdf (accessed on 4 July 2022).

Grajcevci, Albulena, and Arif Shala. 2021. A Review of Kosovo's 2015 PISA Results: Analysing the Impact of Teacher Characteristics in Student Achievement. *International Journal of Instruction* 14: 489–506. [CrossRef]

Hanushek, A. Eric. 1992. The Trade-off between Child Quantity and Quality. *Journal of Political Economy* 100: 84–117. [CrossRef]

Hanushek, A. Eric. 2011. The economic value of higher teacher quality. *Economics of Education Review* 30: 466–79. [CrossRef]

Jacobs, Johana C. G., Scheltus J. van Luijk, Francisca Galindo-Garre, Arno M. M. Muijtjens, Cees P. M. van der Vleuten, Gerda Croiset, and Fedde Scheele. 2014. Five teacher profiles in student-centred curricula based on their conceptions of learning and teaching. *BMC Medical Education* 16: 220. [CrossRef]

Jepsen, Christopher. 2005. Teacher characteristics and student achievement: Evidence from teacher surveys. *Journal of Urban Economics* 57: 302–19. [CrossRef]

Jong, Mirabelle A Schaub-de, Johanna Schönrock-Adema, Hanke Dekker, Marian Verkerk, and Janke Cohen Schotanus. 2011. Development of a student rating scale to evaluate teachers' competencies for facilitating reflective learning. *Medical Education* 45: 155–65. [CrossRef] [PubMed]

Jordan, Heather R., Robert L. Mendro, and Dash Weersinghe. 1997. Teacher Effects on Longitudinal Student Achievement: A Preliminary Report on Research on Teacher Effectiveness. In *Proceedings of the CREATE Annual Meeting, Indianapolis IN, USA, July 1997*. Kalamazoo: CREATE, Western Michigan University. Available online: https://scholar.google.com/scholar?hl=en&q=Teacher+Effects+on+Longitudinal+Student+Achievement%3A+A+Report+on+Research+in+Progress (accessed on 25 May 2021).

Karlberg, Martin, and Christopher Bezzina. 2020. The professional development needs of beginning and experienced teachers in four municipalities in Sweden. *Professional Development in Education* 12: 1–8. [CrossRef]

Mahler, Daniela, Jörg Großschedl, and Ute Harms. 2018. Does motivation matter?—The relationship between teachers' self-efficacy and enthusiasm and students' performance. *PLoS ONE* 21: e0207252. [CrossRef]

Mosteller, Frederick. 1995. The Tennessee Study of Class Size in the Early School Grades. *The Future of Children* 5: 113–27. Available online: https://www.degruyter.com/document/doi/10.1515/9781400851607.261/pdf (accessed on 11 April 2022). [CrossRef]

Mullis, Ina V. S., and Michael O. Martin. 2012. Using TIMSS and PIRLS to improve teaching and learning. *Recherches en Éducation* 14: 1–182. Available online: http://journals.openedition.org/ree/5835 (accessed on 1 February 2022). [CrossRef]

Mullis, Inva V. S., Michael O. Martin, Pierre Foy, Dana L. Kelly, and Bethany Fishbein. 2020. *TIMSS 2019 International Results in Mathematics and Science*. Newton: Boston College.

OECD. 2009. The Professional Development of Teachers. In *Creating Effective Teaching and Learning Environments: First Results from TALIS*. Paris: OECD Publishing, chp. 3. ISBN 978-92-64-05605-3.

OECD. 2019. *PISA 2018 Results (Volume I): What Students Know and Can Do*. Paris: OECD Publishing. [CrossRef]

Piper, Benjamin, Stephanie Simmons Zuilkowski, Margaret Dubeck, Evelyn Jepkemei, and Simon J. King. 2018. Identifying the essential ingredients to literacy and numeracy improvement: Teacher professional development and coaching, student textbooks, and structured teachers' guides. *World Development* 106: 324–36. [CrossRef]

Ponte, Petra. 2010. Action Research as a Tool for Teachers' Professional Development. In *International Encyclopedia of Education*, 3rd ed. Utrecht: Hogeschool Utrecht, pp. 540–47. [CrossRef]

Prince, Cynthia. 2002. The Challenge of Attracting Good Teachers and Principals to Struggling Schools. American Association of School Administrators. Available online: https://www.researchgate.net/publication/228951320_The_challenge_of_attracting_good_teachers_and_principals_to_struggling_schools (accessed on 13 March 2022).

Quins University Charlote. n.d. The Importance of Professional Development for Educators. Available online: https://online.queens.edu/resources/article/professional-development-for-educators/ (accessed on 15 July 2022).

Ribers, Bjorn, Gitte Miller Balslev, and Christine Revsbech Jensen. 2021. Education, collaboration and pedagogical phronesis: Essential dimensions in professional learning and development. *Professional Development in Education* 2021: 1–16. [CrossRef]

Rubie-Davies, Christine M. 2006. Teacher expectations and student self-perceptions: Exploring relationships. *Psychology in the Schools* 43: 537–52. [CrossRef]

Sanders, L. William, and June C. Rivers. 1996. *Cumulative and Residual Effects of Teachers on Future Student Academic Achievement*. Knoxville: University of Tennessee Value-Added Research and Assessment Center.

Sasson, Irit, and Shirley Miedijensky. 2020. Transfer skills in teacher training programs: The question of assessment. *Professional Development in Education* 2020: 1–38. [CrossRef]

Seebruck, Ryan. 2015. Teacher quality and student achievement: A multilevel analysis of teacher credentialization and student test scores in California High Schools. *McGill Sociological Review* 5: 1–18.

Sezer, Elif, and Mehtap Çakan. 2022. Role of Teacher Quality and Working Conditions in TIMSS 2019 Mathematics Achievement. *Journal of Theoretical Educational Science* 15: 395–419. [CrossRef]

Shala, Arif, Albulene Grajcevci, and Fadil Latifi. 2021a. Does Socioeconomic Status Influence Achievement? An analysis of the Performance of Kosovar Students on the 2015 and 2018 PISA Assessment. *Journal of Elementary Education* 14: 393–408.

Shala, Arif, Albulene Grajcevci, and Fadil Latifi. 2021b. Do teacher characteristics matter? Findings from the PISA performance of Kosovar students. *International Journal of Management in Education* 15: 580–97. [CrossRef]

Tang, AiBin, Wen Ye Li, and Dawei Liu. 2022. The impact of teachers' professional development in science pedagogy on students' achievement: Evidence from TIMSS 2019. *Journal of Baltic Science Education* 21: 258–74. [CrossRef]

Ungar, Orit Avidov. 2020. The professional learning expectations of teachers in different professional development periods. *Professional Development in Education*, 1–12. Available online: https://www.tandfonline.com/doi/full/10.1080/19415257.2020.1763435 (accessed on 26 July 2021). [CrossRef]

Zeggelaar, Albert, Marjan Vermeulen, and Wim M. G. Jochems. 2020. Evaluating effective professional development. *Professional Development in Education* 2020: 1–12. [CrossRef]

Zhang, Ling, Richard Allen Carter Jr., Jihong Zhang, Tiffany L. Hunt, Christopher R. Emerling, Sohyun Yang, and Fangjie Xu. 2021. Teacher perceptions of effective professional development: Insights for design. *Professional Development in Education* 2021: 1–21. [CrossRef]

social sciences

MDPI

Article

Social Sciences Teaching: Building a Holistic Approach from Student Teachers' Social Representations

Yamilé Pérez-Guilarte [1,*], Francisco Xosé Armas-Quintá [2] and Xosé Carlos Macía-Arce [3]

[1] Faculty of Educational Studies, Department of Specific Teaching Training and Research and Diagnosis Methods in Education, Universidade da Coruña, 15071 A Coruña, Spain
[2] Faculty of Teacher Training, Department of Applied Learning, Universidade de Santiago de Compostela, 27002 Lugo, Spain; francisco.armas@usc.es
[3] Faculty of Education Sciences, Department of Applied Learning, Universidade de Santiago de Compostela, 15782 Santiago de Compostela, Spain; carlos.macia@usc.es
* Correspondence: yamile.perez@udc.es

Abstract: The dynamics that today's society is facing require critical citizens capable of understanding the complexity of problems from their various dimensions. This work aimed to investigate the social representations that future primary school teachers have about socio-environmental problems, sustainability, purposes of socio-environmental education, and strategies to train in Global Citizenship Education and in Education for Sustainable Development. An exploratory case study was carried out, framed in an action-research approach in the classes of Social Sciences Teaching of the Primary Education Degree of two universities in Galicia (Spain): the University of A Coruña and the University of Santiago de Compostela. A quantitative and qualitative analysis was developed by applying a questionnaire to a sample of 200 students. The results showed that students focus on problems such as climate change or environmental pollution, leaving aside issues such as social inequalities, poverty, or gender issues. In addition, they have difficulties in recognizing the strategies by which teachers can provide their students with critical thinking that leads to social transformation. Significant differences were observed in some researched aspects according to the university of origin, the grade, and the gender of the students, with gender being the one that influenced the greatest number of questions.

Keywords: social sciences teaching; socio-environmental problems; Global Citizenship Education; Education for Sustainable Development; critical thinking; primary school

Citation: Pérez-Guilarte, Yamilé, Francisco Xosé Armas-Quintá, and Xosé Carlos Macía-Arce. 2022. Social Sciences Teaching: Building a Holistic Approach from Student Teachers' Social Representations. *Social Sciences* 11: 307. https://doi.org/10.3390/socsci11070307

Academic Editor: Alvaro Morote

Received: 17 June 2022
Accepted: 10 July 2022
Published: 14 July 2022

1. Introduction

Today, society is facing a myriad of highly complex problems, health crises, armed conflicts, media wars, and geopolitical tensions, just to mention a few. Teachers face a great challenge, that of preparing students in the development of skills that enable them to face these problems and make decisions aimed at seeking a more sustainable society. For this, it is essential that schools adopt a holistic approach to problems that allows for the formation of a critical global citizenship (González et al. 2020) that is based on the understanding of the multiple dimensions involved: social, cultural, economic, philosophical, political, climatic, territorial, etc. For example, although problems such as climate change or the depletion of non-renewable resources are worked on in schools, it is not easy for them to relate to production, distribution, and consumption models (González Reyes and Rodríguez 2020). However, problems such as gender violence, racism, or poverty, in a major way, are attributed to social problems without a connection to environmental dynamics (Massip et al. 2021).

This disjointed approach between the disciplines hinders the effective resolution of problems, since, although of different natures, they are closely linked. In this way, a problem such as climate change or extreme poverty cannot be solved without considering its link to other problems, such as the irresponsible consumption of goods and services. In this sense, the science of sustainability constitutes a backbone, an approach that permeates all disciplines (Vilches and Gil 2021). Nevertheless, education for sustainability still needs a collective reflection on these problems that avoids a biased and reductionist treatment (Barrón et al. 2010).

In the training of future Social Sciences teachers, the introduction of this interdisciplinary teaching is essential. However, previous studies show that the degree of knowledge of future teachers about socio-environmental problems is limited (Zamora-Polo et al. 2019), meaning that they are unable to train competent citizens set to face the challenges of this century (Moreno-Crespo and Moreno-Fernández 2015). Therefore, the areas of Social Studies Teaching and Experimental Sciences Teaching at universities must consider a joint approach to current problems (Massip et al. 2021; Alcántara-Manzanares and López-Fernández 2021).

For Social Sciences teaching to be oriented toward a holistic teaching–learning process, it is essential to analyze the social representations of students (Santisteban 2009; Pagès and Santisteban 2014). Hence, the present work aimed to investigate the social representations that future primary school teachers have about socio-environmental problems, sustainability, the purposes of socio-environmental education, and the strategies to train in Global Citizenship Education and in Education for Sustainable Development. The purpose of this study was that the authors, teachers in the area of Social Sciences Teaching, can introduce improvements in these subjects in the Degrees of Primary Education at two public universities in the University System of the Autonomous Community of Galicia (Spain), namely the University of A Coruña and the University of Santiago de Compostela.

1.1. A Holistic Approach of Social Sciences Teaching

In the scientific literature on Social Studies Teaching, there seems to be a consensus regarding the need to work in the classroom with real, current, complex, controversial problems that encourage students to reflect and take social action. In this sense, a group of terms have been consolidated to refer to these relevant issues that should be part of the curriculum. The term *issues-centered social studies* was introduced in the work of Evans and Saxe (1996), which had a great impact on critical social studies. In the Spanish context, it was translated as *problemas sociales relevantes* (socially relevant problems) (Pagès 1994; Benejam 1997; Santisteban 2004, 2011, 2019), while in the Francophone context, the term *questions socialement vives* (socially relevant issues) emerged (Legardez 2004; Legardez and Simonneaux 2006). Likewise, the term *controversial issues* has been used (Eulie 1966; Wellington 1986; Hess 2008) as an essential element of education and democracy for the formation of critical thinking (Santisteban 2017).

Similarly, the term *environmental problems* has gained prominence in the face of the progressive worsening of problems such as climate change, resource exploitation, water pollution, etc. There is no doubt that the relationships between causes and consequences linked to both social and environmental dimensions can be established in these problems. In this way, it is no longer possible to speak of any environmental problem that does not originate from human action in the biosphere, nor of any social problem that is not related, to a lesser or greater extent, to the management of natural resources and its needs (Massip et al. 2021). In fact, if we understood the environment as a complex system of interrelationships between all its parts, including human beings (Martínez-Medina and Torres-Porras 2021), the term *environmental* would already include the social component.

To avoid falling into the error of treating environmental problems only from their ecological dimension, some authors propose terms such as *socio-environmental problems* (Amador and Esteban 2011; Moreno-Crespo and Moreno-Fernández 2015; Ortega-Sánchez et al. 2020; Rodríguez-Marín et al. 2020; Bonilla and Garzón 2021; Martínez-Medina and Torres-Porras 2021; Massip et al. 2021) or *eco-social problems* (González Reyes and Rodríguez 2020; Santana et al. 2021). Apart from this terminological debate, Social Sciences teaching must contribute to a holistic approach to the problems of today's society. This need has recently been reflected in Spain, as a result of the new education law. In this way, Royal Decree 157/2022, of 1 March, which establishes the organization and minimum teachings of Primary Education, indicates that "the link between key competences and challenges of the 21st century is what will give meaning to learning, by bringing the school closer to real situations, issues and problems of everyday life" (Boletín Oficial del Estado 2022, p. 24402). Likewise, this Royal Decree establishes that these challenges require the following:

> Require transcending the local perspective to analyse and also commit to global problems. All this requires, on the one hand, a complex mind, capable of thinking in systemic, open terms and with a high level of uncertainty, and, on the other, the ability to empathise with relevant aspects, even if they do not affect us directly, which implies assuming the values of social justice, equity and democracy, as well as developing a critical and proactive spirit towards situations of injustice, inequality and exclusion. (Boletín Oficial del Estado 2022, p. 24402)

In this educational process based on the problems of today's society, in its understanding and intervention in the search for social and environmental justice, Education for Sustainable Development constitutes a fundamental pillar. Thus, the *Goal 4 Quality Education* of the Sustainable Development Goals (SDG) of the 2030 Agenda of the United Nations Organization, in its goal 4.7, establishes the following:

> By 2030, ensure that all learners acquire the knowledge and skills needed to promote sustainable development, including, among others, through education for sustainable development and sustainable lifestyles, human rights, gender equality, promotion of a culture of peace and non-violence, global citizenship and appreciation of cultural diversity and of culture's contribution to sustainable development. (UNDP 2022, p. 10)

Achieving these goals set out in the new educational legislation and in the SDGs depends, to a large extent, on the ability of acting teachers to concretize the curriculum (Moya and Zubillaga del Río 2020) and on the training of future teachers. Regarding the latter, some recent works show that there is still a long way to go. This is the case of the study by López-Fernández and Oller (2019), who show that the didactic proposals on environmental education designed by future primary school teachers show little concreteness in relation to reality. These focus on the development of habits such as recycling, reusing, and reducing, without questioning the current model of consumerist and capitalist society (Stiglitz 2015). In addition, the conceptual knowledge they had about the environment and the problems generated, directly and indirectly, by society was quite ambiguous.

1.2. Social Representations of Teachers in Initial Training

Professional training should start from the investigation of social representations on the role of the teacher and on the teaching of social disciplines. In this way, students have a starting point to become aware of the rationality that hides behind their social representations, so that they can assess it and, based on the results, choose to modify or maintain it (Pagès and Santisteban 2014).

Social representations are specific ways of understanding and communicating reality that influence the form of social interaction as they evolve. They represent the concepts, statements, and explanations that originate in everyday life during interpersonal communications (Moscovici 2011). The notion of representation interrelates the psychological and social approach, laying the foundations for the creation of knowledge, based on personal experiences and the influence received from society (Jodelet 1989). Therefore, knowledge can be built from the rupture of initial representations through a process consisting of the following stages (Dalongeville 2003): emergence of representations and awareness; confrontation with a situation/problem that calls the representations into question; reconstruction of new representations that reorganize the old ones and integrate the new ones; and metacognition of the processes that the subject, individually and in a group, has put into practice.

The representations belong, in part, to the field of stereotypes, prejudices, and social values, so, by studying them, a starting point can be established to be able to change the students' ideas (Santisteban 2009). Hence, social representations of future teachers are a widely researched topic (Bravo Pemjean 2002; Allieu-Mary et al. 2005; De Lourdes and Santisteban 2016; GREDICS 2018). Likewise, various investigations have shown their interest in the field of socio-environmental problems and Education for Sustainable Development (Moreno-Crespo and Moreno-Fernández 2015; Solís-Espallargas and Valderrama-Hernández 2015; Ortega-Sánchez et al. 2020; Rodríguez-Marín et al. 2020; Massip et al. 2021).

However, as appreciated in this literature review, more interdisciplinary teaching is needed to train future Social Sciences teachers from a holistic approach. This means that educational practice and research must be oriented to train competent citizens to understand the complexity of 21st century challenges. For this, it is essential that the problems of today's society be analyzed in classes, from different perspectives, to help students to develop critical thinking. In this process, analyzing student teachers' social representations is crucial to direct teaching–learning toward the construction of significant knowledge and constant improvement of teaching practice.

2. Materials and Methods

This work is framed in an action-research approach, understanding it as a self-reflection on the actions and social situations experienced by teachers, which allows them to identify their practical problems and direct actions to solve them (Elliott 2000). Action-research is constituted as a critical science that allows improved rationality and justice: of social or educational practices, of the understanding of the participants (teachers, students, directors, etc.) about it, and of the social situation in which it takes place (classrooms, schools, etc.) (Kemmis and McTaggart 1988). Specifically, an exploratory case study (Yin 1989) is carried out through inductive reasoning, allowing for the discovery of relationships and concepts in the context under study, contrasting them with others and designing an action strategy for change (Muñoz and Muñoz 2001; Bisquerra 2009).

The study is framed in the Autonomous Community of Galicia, which has the highest density of forest masses in Spain, but, at the same time, the least protected by Natura 2000, the European network of protected areas. In this sense, knowing the students' perception of Nature, society, and sustainability seemed very appropriate. This research is a continuity of previous collaborations between teachers from the area of Social Studies Teaching of two public universities in the Autonomous Community of Galicia (Spain): the University of A Coruña and the University of Santiago de Compostela. They are two out of three public universities that exist in Galicia, where students from practically all Galicia study the Primary Education Degree. Thus, this sample covers an important part of the study region.

The purpose of this research is to guide teaching practice toward the training of future primary school teachers who are competent and committed to achieving a more sustainable society. Therefore, the study was carried out from a convenience sample of 200 students in the second and third years of the Primary Education Degree from both universities.

Students from the University of Santiago de Compostela made up 70% of the sample, of which 60% were in the second year and 40% in the third. The students from the University of A Coruña, comprising 30% of the whole sample, were all in their third year. The average age of the students was 21 years. By age ranges, the distribution was as follows: 78.5% between 19 and 21 years old, 14.5% between 22 and 24 years old, 3.5% between 25 and 27 years old, 1.5% between 28 and 30 years old, and 2% were over 30 years old. In relation to gender, 66.5% identified themselves as female, 32.5% as male, and 2% as non-binary.

To investigate the social representations of future primary school teachers, a questionnaire was used as an instrument for collecting information. The one proposed by Solís-Espallargas and Valderrama-Hernández (2015) was adapted, since it is a validated instrument that includes an important part of the questions that are the object of this study. In addition, this questionnaire has been used by other authors (Ortega-Sánchez et al. 2020), allowing the results of this research to be compared with previous studies on this subject.

To respond to the objectives of this research, the questionnaire was structured in three parts: (1) social representations about socio-environmental problems, (2) social representations about socio-environmental education and Education for Global Citizenship, and (3) social representations on sustainability and Education for Sustainable Development. The adaptations to the questionnaire (Solís-Espallargas and Valderrama-Hernández 2015), which are explained below, made possible a qualitative and quantitative analysis of the data. Thus, it was made up of twelve questions, four in part 1 of the questionnaire, two in each of the parts 2 and 3, and four in the personal information part. Of these, four closed nominal questions with a single answer collected the information on the students' profile (university of study, grade, age, and gender), allowing an inferential analysis to know if they present any causal relationship with the research issues. Moreover, in this same question, typology students were also asked whether they had received training on socio-environmental topics during the degree or not.

Likewise, the scale of the three metric questions was modified. The central value was eliminated, since it was considered that the term *tending to agree* does not reflect a clear positioning of the respondent. Therefore, to better qualify the answers, the scale was made up of the following categories: totally disagree, quite disagree, slightly disagree, slightly agree, quite agree, and totally agree. Each of the metric questions encompasses several topics that can be seen in the text and graphics of Section 3. Finally, for the qualitative analysis, four open-ended questions allowed the following to be known: the subjects in which the students had training in socio-environmental problems during the degree, their capacity to define what a socio-environmental problem is, their priorities of socio-environmental problems to work on in a primary school class, and what they understand as sustainability.

To check the reliability and internal consistency of the questionnaire items, the Cronbach Alpha values were calculated through MAXQDA, a program that also allowed for the quantitative and qualitative analysis of all the research data. To calculate the Cronbach Alpha values, the questions were grouped into two groups. The first was related to the questions that investigated the degree of knowledge about socio-environmental problems and sustainability, yielding a value of 0.8. The second grouped the questions about the role of education and the strategies that teachers should follow, also obtaining a value of 0.8. Therefore, the questionnaire items have good consistency, as indicated by Oviedo and Campo-Arias (2005) for Cronbach Alpha values between 0.7 and 0.9.

The questionnaire was designed by using *Microsoft Forms* and was applied to the students of the University of Santiago de Compostela December 2021 and to those of the University of A Coruña in March 2022. In addition, it is important to point out that, although the questionnaire was in Spanish, some students responded in Galician, the co-official language in Galicia. This was taken into account when performing the statistical processing, so that similar responses in both languages were counted. All data were treated anonymously, and their use is limited to research purposes and to make improvements in this area of knowledge.

From the point of view of the methodological limitations of this study, it should be mentioned that, although the questionnaire includes open questions, the use of other research instruments, such as interviews or discussion groups, could have further enriched the results. Likewise, as it is a case study, the results must be understood in the context in which it has been developed; therefore, it cannot be generalized. However, they can be contrasted with other cases to find similarities and differences that contribute with knowledge to the area.

3. Results

This section presents the research results grouped into three subsections: (1) social representations of socio-environmental problems, (2) social representations of socio-environmental education and Education for Global Citizenship, and (3) social representations of sustainability and Education for Sustainable Development. Some textual responses from the students were included in the results, being identified by an *A*, together with the number assigned by the MAXQDA program for each response.

3.1. *Social Representations about Socio-Environmental Problems*

To begin with, the students had to indicate if they had received training in socio-environmental problems during their degree. A total of 60.5% indicated that they had not received training on this subject, while 39.5% indicated yes, fundamentally in the subjects of the area of Social Studies Teaching, as indicated by 72% of the students, and in those of Natural Sciences Teaching, as mentioned by 23% of the students.

Next, students were asked to define the concept of a socio-environmental problem. As Figure 1 shows, which includes the 100 most mentioned words, the term *sociedad* (society) was the most relevant, being mentioned by 26.5% of the students. To this we must add those who responded in Galician, indicating *sociedade*, representing 10%. The second most mentioned term was *afecta* (affects), as indicated by 24.5%, to which we must add other related terms, such as *afectan* (affect), *afectado* (affected), and *afectaciones* (affectations), which represent 4%. In third place, the term *medioambiente* (environment) appears, as mentioned by 17% of the students.

Figure 1. Word cloud of the definition of socio-environmental problems. Source: Own elaboration based on MAXQDA. Note: This figure includes terms mentioned in Spanish and Galician.

The word tree in Figure 2, centered on the most mentioned word, *sociedad* (society), shows two types of relationship that students establish between it and the *medioambiente* (environment). On the one hand, they indicate that both society and the environment are victims of socio-environmental problems since both suffer from its negative effects. It can clearly be seen in the figure how the word *afecta* (affects) unites both terms, being able to read "que afecta a la sociedad y al medioambiente" (that affects society and the environment). In this way, some definitions of socio-environmental problems were "a specific casuistry that affects both the environment and society, generating adverse effects for both" (A33) or "events that negatively affect the environment and, therefore, also have a negative effect on society" (A6). Likewise, the concern about the negative impact on health stands out, with the term *salud* (health) being mentioned by 11% of the students.

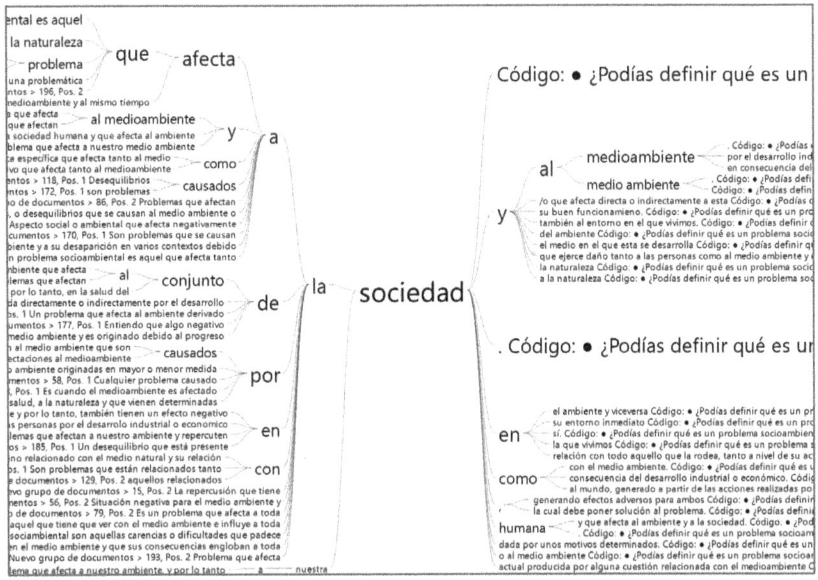

Figure 2. Word tree for the term society. Source: Own elaboration based on MAXQDA.

On the other hand, society is recognized by the students as the cause of these effects on the environment. Some statements in this regard are as follows: "difficulties in the environment caused to a greater or lesser extent by society" (A2), "problem generated by a society in its immediate environment" (A35), or "problems at the environmental level that are caused by the consumerist society in which we live" (A129). Likewise, Figure 1 shows recurring terms, such as *económico/s* (economic), *industrial* (industrial), *industria* (industry), or *desarrollo* (development) to refer to the factors causing socio-environmental problems. Some examples are the following definitions: "imbalances caused to the environment and/or the health of people that arise as a consequence of industrial or economic development" (A26), or "they are deficiencies or inconveniences to the environment produced by industrial and economic development" (A29).

It is striking that the students show their concern about the negative impact on society and the fact that it is society itself that causes it, but not about the fact that society is the only one that can solve these problems. In this sense, only one response was recorded: "it is a problem that affects our environment, and therefore our society, which must solve the problem" (A66). Likewise, there were eight students who were not able to define what a socio-environmental problem was, of which seven were second-year students at the University of Santiago de Compostela.

Figure 3 shows the most recurrent terms when the students proposed a socio-environmental problem to work on in a primary education classroom. As can be seen, *contaminación* (pollution) was the most mentioned, being proposed by 41.5% of the students, followed by *cambio climático* (climate change), as indicated by 32%. Other problems mentioned to a lesser extent were deforestation (11%), global warming (5%), and concern about water (5%) in regard to its contamination, as well as its consumption.

Figure 3. Cloud of words of socio-environmental problems to work on in the classroom. Source: Own elaboration based on MAXQDA. This figure includes terms mentioned in Spanish and Galician.

It should be noted that problems such as social inequalities or poverty had very little presence: 3.5% and 2%, respectively. Precisely, 36% of students disagreed with the fact that poverty is an environmental problem. Of these, 25% totally disagreed or quite disagreed. As can be seen in the graph in Figure 4, only 17% totally agreed that poverty is an environmental problem. Likewise, it can be seen how gender issues are not perceived as a socio-environmental problem by 42.5% of the students, with only 13.5% totally agreeing with it.

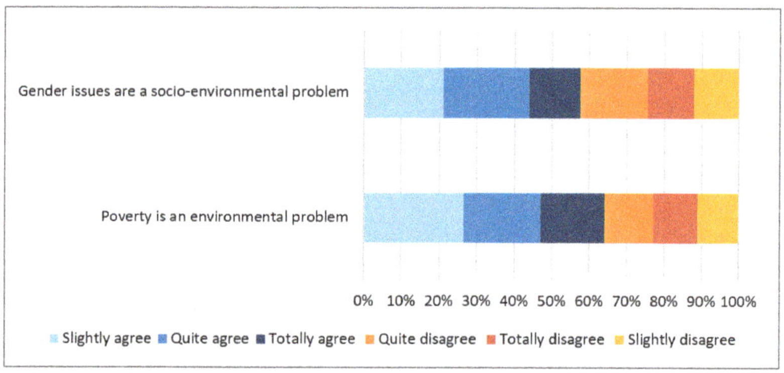

Figure 4. Social representations of poverty and gender issues. Source: Own elaboration based on MAXQDA.

3.2. Social Representations on Education for Global Citizenship and Socio-Environmental Education

There is a consensus among the students regarding the need to promote conflict resolution strategies to achieve social transformation. Thus, this idea is supported by 97% of the students, of which 75.5% totally agreed (Figure 5). However, 78.5% of future primary school teachers consider that it is inappropriate to work on political content in class. The remaining 22.5% disagreed; however, only 9% of them totally or quite disagreed. Likewise, 87.5% of the students consider that teachers should present the contents, while avoiding reflecting on their own ideology. Of the remaining students, 8% slightly disagreed, 2.5% quite disagreed, and 2% totally disagreed.

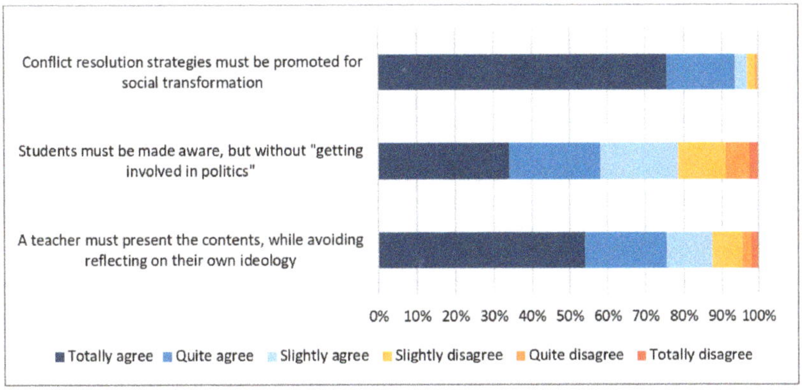

Figure 5. Social representations about the presence of politics, ideology, and conflict resolution in classes. Source: Own elaboration based on MAXQDA.

As can be seen in Figure 6, practically all the students, 98.5%, consider that the purpose of socio-environmental education is "Developing thinking and acting in students that is conducive to changing the world in the direction of making it more sustainable". Similarly, there is a consensus that it aims at "Facilitating the development of pro-environmental skills and habits such as separating garbage, respecting and caring for living beings, recycling paper, etc.". Nevertheless, 2.5% of students did not agree with the latter issue. In addition, socio-environmental education is also associated with "Getting to know Nature and learn to enjoy it, love it, and conserve it". However, 7.5% of the students disagreed with it, of which 3.5% quite disagreed, 2% slightly disagreed, and 2% totally disagreed.

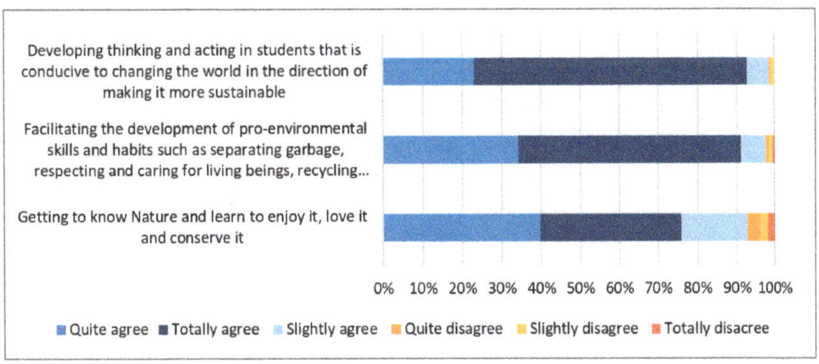

Figure 6. Social representations about the purpose of socio-environmental education. Source: Own elaboration from MAXQDA.

3.3. Social Representations about Sustainability and Education for Sustainable Development

To explain the concept of sustainability, the most used terms were *future* (future) and *futuras generaciones* (future generations), as cited by 52% of the students. Next, the term *recursos* (resources) was mentioned by 25.5%, followed by the terms *necesidades* (needs), as used by 18%; *equilibrio* (balance) by 17%; and *presente* (present) by 12%. Figure 7 allows us to understand the link established by the students between these terms. Thus, they were able to write definitions of sustainability, such as the following: "meeting the needs of society without compromising the resources we have or the following generations" (A171), "meeting needs guaranteeing a future balance" (A54), or "that covers current needs without negatively affecting the future" (A67).

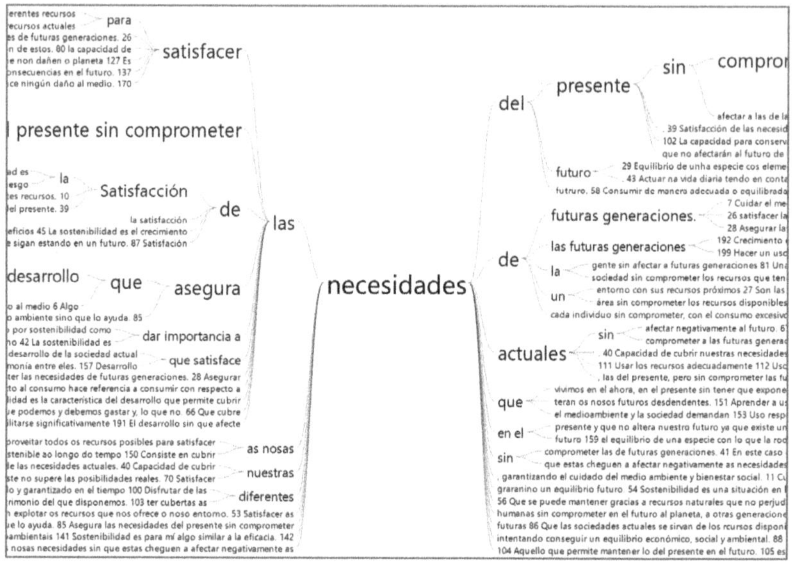

Figure 7. Word tree for the definition of sustainability. Source: Own elaboration based on MAXQDA.

The importance of protecting the environment, Nature, or natural resources was collected in 24% of the definitions. However, the need for a balance between economic and social aspects had very little representation. The social dimension was mentioned only by 3.5% of the students and economic issues by 2.5%. Of these, only 1% related the environmental, economic, and social dimensions. One defined it as follows: "satisfaction of needs trying to achieve an economic, social and environmental balance" (A88). Meanwhile, the other indicated the following: "economic and social model that does not exhaust the available resources but manages them as best as possible and with the least impact on the environment" (A35). However, these two definitions do not refer to the present and the future.

The opinions in relation to the strategies that teachers must implement to include the contents of ESD in primary education classes are shown in Figure 8. As can be seen, 97% of the students agreed, although to a different extent, in that the teacher must "Develop in the students complex thinking that entails social creativity and civil ethics", as well as "Helping to build critical ethical thinking based on the analysis of reality". A consensus was also seen in 97.5% of the future primary school teachers in the item "Ensuring the students always explain and contrast their ideas, so that they become aware of what they think and so that they mobilize their conceptions in the construction of new knowledge".

However, the other two questions asked did not have the same consensus as the previous ones. On the one hand, 89.5% of the students agreed, of which 28% totally agreed, with "Making many outings to the environment and recycling workshops". A total of 10.5% disagreed with this statement, of which 6.5% slightly disagreed, 3% quite disagreed, and 1% totally disagreed. On the other hand, the aspect of greatest disagreement was the fact that the teacher must "Try to persuade and convince the students that they have to respond to the current environmental crisis by supporting certain forms of action". A total of 78% of the students agreed, of which 21.5% totally agreed; meanwhile, 10% slightly disagreed, 6.5% quite disagreed, and 5.5% totally disagreed.

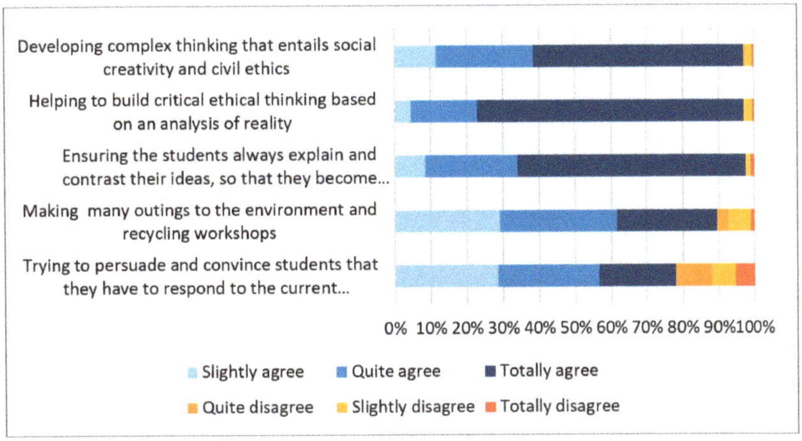

Figure 8. Social representations about strategies that teachers must implement to include the contents of ESD in primary education classes. Source: Own elaboration from MAXQDA.

3.4. *Inferential Analysis*

To find out if the university of origin, the grade, age, or gender influenced the students' responses and to what extent, the Pearson Chi-square test and the Contingency Coefficient C were used. The results indicate that age had no influence on any of the issues analyzed, while the university, grade, and gender did.

The variable "training in socio-environmental problems" depended on the university and the grade. A total of 65.7% of the students at the University of Santiago de Compostela indicated that they had not had training in socio-environmental problems; this contrasts with 48.3% at the University of A Coruña. However, this could be explained by the fact that all second-year students came from the University of Santiago de Compostela, 86% of whom indicated that they had not had training in socio-environmental problems, a percentage that is significantly different from the 41.4% of third-year students who indicated having had training. However, the significance values and the Contingency Coefficient C (Table 1) show that the grade exerts more influence ($p = 0.0000$, C = 0.418) than the university ($p = 0.0318$, C = 0.161). Therefore, the students' perception of having had training in socio-environmental problems depends more on the grade than on the center of origin.

Table 1. Pearson's Chi-square test for the variable "training in socio-environmental problems".

Dependent Variable	Bilateral Asymptotic Significance (*p*)		Contingency Coefficient C	
	University	Grade	University	Grade
Training in socio-environmental problems	0.0318	0.0000	0.161	0.418

Source: Own elaboration from MAXQDA.

Table 2 shows that there were also two other grade dependent variables. This influenced the purpose of the socio-environmental education "Getting to know Nature and learn to enjoy it, love it, and conserve it" ($p = 0.0338$, C = 0.239). Thus, 41.4% of third-year students fully agreed with this statement, compared to 29.8% of second-year students, thus evidencing a traditional vision of socio-environmental education. Differences are also observed in those who indicated that they totally disagree, with 4.8% of the second-year students and none of the third-year students. When grouping the categories into agree and disagree, no significant differences are observed between the two grades, leaning, in both cases, mostly toward agreement (91.7% of second-year students and 93.1% of third-year students).

Table 2. Pearson's Chi-square test with other items dependents of the grade.

Dependent Variable	Bilateral Asymptotic Significance (p)	Contingency Coefficient C
	Grade	
Getting to know Nature and learn to enjoy it, love it, and conserve it	0.0338	0.239
Making many outings to the environment and recycling workshops	0.0175	0.253

Source: Own elaboration from MAXQDA.

The grade was also related to the sustainability education strategy of "Making many outings to the environment and recycling workshops" (p = 0.0175, C = 0.253). In this case, the most significant difference is that 45.2% of second-year students quite agreed with it, compared to 23.3% of third-year students. As the significance values and the Contingency Coefficient C show, the dependency and the degree of correlation are stronger for this question than for the perception of the purpose of socio-environmental education "Getting to know Nature and learn to enjoy it, love it, and keep it".

Gender was related to the eight questions listed in Table 3, but there are two that stand out due to the intensity of this relationship. One of them is the purpose of socio-environmental education: "Facilitating the development of pro-environmental skills and habits such as separating garbage, respecting and caring for living beings, recycling paper, etc." (p = 0.0000, C = 0.426). The category with the highest consensus was totally in agreement, as indicated by 57% of the students. In the case of women, this behavior was higher (67.2%) than for men (36.9%) and for non-binary people (50%).

Table 3. Results of the Pearson's Chi-square test for the gender variable.

Dependent Variable	Bilateral Asymptotic Significance (p)	Contingency Coefficient C
	Gender	
Conflict resolution strategies must be promoted for social transformation	0.0016	0.333
Getting to know Nature and learn to enjoy it, love it, and conserve it	0.0056	0.332
Facilitating the development of pro-environmental skills and habits such as separating garbage, respecting and caring for living beings, recycling paper, etc.	0.0000	0.426
Developing thinking and action in students conducive to changing the world in the direction of making it more sustainable	0.0008	0.343
Making many outings to the environment and recycling workshops	0.0000	0.402
Ensuring that students always explain and contrast their ideas, so that they become aware of what they think and mobilize their conceptions in the construction of new knowledge	0.0004	0.354
Helping to build critical ethical thinking based on an analysis of reality	0.0023	0.328
Developing complex thinking that entails social creativity and civil ethics	0.0216	0.308

Source: Own elaboration from MAXQDA.

The second issue of high dependency is the sustainability education strategy "Make many outings to the environment and recycling workshops" (p = 0.0000, C = 0.402). In this case, 89.5% of the students were in favor, with 28% totally agreeing, 32.5% quite agreeing, and 29% slighting agreeing. However, differences in terms of gender are observed in these categories. For example, the percentage of women who totally agreed (35.1%) and slightly agreed (36.6%) was above the global value. This behavior is similar to the previous variable, which is understandable, since they are two variables that are thematically related.

Non-binary students were the ones who stood out the most in the category of quite agree (50%). While men exceeded the global value in the slightly disagree category (43.1%).

Another question investigated and which is related to the two previous ones is the purpose of socio-environmental education: "Getting to know Nature and learn to enjoy it, love it, and conserve it". It also shows that women's vision is too focused on knowing, caring for, and conserving Nature through school outings and recycling activities. However, women stood out in the rest of the questions focused on the development of critical thinking and social action. For example, 82.4% totally agreed with conflict resolution strategies for social transformation, when the global value was 75.5%. Likewise, 82.4% were totally in favor of the fact that the teacher should help build critical ethical thinking based on the analysis of reality; meanwhile, for the group, it was 74%.

4. Discussion

The discussion of the results is based on the complementation of the quantitative and qualitative results obtained from the questionnaire. In this way, open-ended questions allowed to deepen and confirm the results obtained in the metric questions. Moreover, the discussion is enriched by contrasting with previous research and the relevant literature related to the research issues.

One of the most surprising issues of the results is the fact that 60.5% indicated that they did not have training in socio-environmental problems; however, this result was largely influenced by second-year students. Even if the subjects in the areas of Social Sciences Teaching and Natural Sciences Teaching were the most mentioned by the students who indicated having had this training, this result could indicate that an interdisciplinary approach has not yet been consolidated in the primary education grades of the universities participating in this study. However, it should be noted that the contents included in the programs differ between grades and universities, because they are live programs that can change from one year to the next. It could even be the case of completely different programs within the same university. For example, the University of Santiago de Compostela has two faculties that teach a primary education degree, and each one has its own programs, depending on the coordinating professor and the area. This is an issue to which attention should be paid, since, as indicated by Massip et al. (2021), the construction of models that allow us to draw relationships between socio-environmental concepts is essential for the training of future teachers.

One finding that stands out regarding social representations on socio-environmental problems is that 96% of students were able to define what they are. However, as other authors have already found (Solís-Espallargas and Valderrama-Hernández 2015; Ortega-Sánchez et al. 2020; Massip et al. 2021), issues such as social inequalities, poverty, or gender issues are less appreciated as socio-environmental or environmental problems than climate change or environmental pollution, for example. A complete dissociation between society and the environment stands out, since no responses were obtained pointing out the environment as a complex system, which encompasses not only ecological aspects, but also economic, political, social, and cultural ones (Martínez-Medina and Torres-Porras 2021).

The results in relation to the treatment that the politics and ideology of teachers should have in the classes coincide with the works of Solís-Espallargas and Valderrama-Hernández (2015) and Ortega-Sánchez et al. (2020). In both cases, it can be seen how more than 43% of students strongly agree that politics should not be present in the content of the classes and that the content should avoid the teacher's ideology. In this study, values of 78.5% were obtained for the question of politics and 87.5% for ideology, but it should be noted that, in the scale used in this case, the central value was eliminated with which the students had to position themselves, and so they either agreed or disagreed with the issues raised.

These results are very worrying. On the one hand, the teachers' ideology is always present in the content they choose; it cannot be separated from it, so they must be aware of it and of the influence it has on their students (Solís-Espallargas and Valderrama-Hernández 2015). On the other hand, if you want to train critical citizens under a global citizenship education model, political education is an essential dimension (González-Monfort and Santisteban 2020). Indeed, González et al. (2020, p. 5) indicate that "Political Science allows students to understand the notions of power, systems of government and political organisation (both national and international)".

The development in the students of a way of thinking and action conducive to achieving a more sustainable world obtained the consensus of 98.5% of the students. However, when contrasted with other questions, shortcomings in knowledge are observed in this regard. The results show a vision of socio-environmental education focused on a teaching model based on care, respect, and love for Nature in a superficial way. For example, 97.5% of the students agreed, of which 57% totally agreed, that the purpose of socio-environmental education should focus on the development of pro-environmental skills and habits such as separating garbage, respecting and caring for living beings, recycling paper, etc. This vision is reaffirmed, markedly among women, in opposition to a teaching model that goes beyond this vision of citizenship on a personal or participatory basis to seek a citizenship that fights for justice (Westheimer and Kahne 2004). Again, a parallelism can be seen with the results of the works by Solís-Espallargas and Valderrama-Hernández (2015) and Ortega-Sánchez et al. (2020), in which more than 70% share this vision. This lack of a holistic vision of sustainability was evidenced in the definition issued by the students, in which the environmental dimension acquired a greater role to the detriment of other dimensions involved in the concept: social or economic (UN 1987).

These problems are also reflected in the strategies for working on Education for Sustainable Development in the classroom, where students do not manage to achieve total coherence in their answers. On the one hand, they agreed (more than 97%) on the need to develop critical thinking in students, starting from reality and leading to social creativity and civil ethics. Likewise, there is a general agreement (more than 97%) that students should be made explicit and contrast their ideas so that they become aware of what they think and manage to mobilize their conceptions for the construction of new knowledge. Women had a greater role in recognizing the importance of these issues. However, the fact that 78% agreed that the teacher must persuade and convince students to adopt certain forms of action, aligning with an educational model of imposition, is contradictory. Similar to the rest of the results, there are coincidences with the works of Solís-Espallargas and Valderrama-Hernández (2015) and Ortega-Sánchez et al. (2020), in which 50% or more of the students share this vision.

5. Conclusions

This research shows the relevance of the study of social representations in improving Social Science Teaching, in this case, in the training of future primary school teachers. The study allowed us to know the vision of the students on socio-environmental problems, sustainability, the purposes of socio-environmental education, and the strategies to train in Education for Global Citizenship and in Education for Sustainable Development.

Most of the students understand the concept of a socio-environmental problem and the double implication of society, as the affected party and as the cause. However, they do not refer to the role society has in resolving these socio-environmental conflicts. In addition, they focus on problems such as climate change or environmental pollution, leaving aside issues such as social inequalities, poverty, or gender issues. These results show the need for teachers to work with current problems from a holistic view, for example, based on the concept of sustainability. As verified in this study, the economic and social dimensions were practically absent when defining the term *sustainability*, thus focusing the attention on environmental aspects.

From an educational point of view, in general, future primary school teachers agree on the need to develop critical thinking that leads to social transformation. However, they have difficulty recognizing how it can be performed in practice, which is evidenced when they point out that political content should be excluded from the classroom, or, for example, by not being aware of the influence that teachers' ideology has on students. Likewise, they manifest a vision of socio-environmental education that is superficial, imposing, and too focused on care and respect for the environment, which makes the development of a critical teaching–learning model impossible.

To reverse this situation, above all, more awareness is needed. Teachers must find formulas to teach, but also to motivate and raise awareness. In this sense, research and teaching practice should be oriented toward a transformative education that encourages students to reflect and act in pursuing social and environmental justice. For this, innovative teaching proposals must be based on debate about current society problems both in the near environment and globally, and the search of their solution. These activities de-signed in the classes can be put into practice by student teachers during the internship period they carry out in primary schools. In this way, they can become familiar with the challenges involved in addressing problems such as armed conflicts, fake news, and hate crimes, among others. In addition, teachers can encourage students to orient their end-of-grade projects to these topics.

In summary, the results of this research provided valuable information to introduce improvements in the teaching–learning of socio-environmental problems and sustainability. In this way, future primary school teachers will be better trained in Education for Global Citizenship and Education for Sustainable Development, which will allow them to face the challenges imposed by the new primary education curriculum. In addition, this study opens the door to future research on Social Science Teaching that can contribute to the improvement of teaching practice in initial teacher training.

Author Contributions: Conceptualization, methodology, investigation, formal analysis, writing—original draft, and writing—review and editing, Y.P.-G., F.X.A.-Q. and X.C.M.-A. All authors have read and agreed to the published version of the manuscript.

Funding: This research was funded by the Erasmus+ Programme of the European Union within the action KA2—Cooperation for innovation and the exchange of good practices—Capacity Building in the field of Higher Education, grant number 609897-EPP-1-2019-1-ES-EPPKA2-CBHE-JP.

Institutional Review Board Statement: Not applicable.

Informed Consent Statement: Not applicable.

Data Availability Statement: Not applicable.

Conflicts of Interest: The authors declare no conflict of interest.

References

Alcántara-Manzanares, Jorge, and José Antonio López-Fernández. 2021. *Didáctica del Medioambiente en Educación Primaria*. Madrid: Editorial Síntesis.

Allieu-Mary, Nicole, Jacky Fontabona, and Pascal Meriaux. 2005. *Journées d'Études Didactiques de l'histoire et de la géographie. Quelles questions sont posées du point de vue de l'articulation. Recherches en didactiques—Formation aux didactiques de la géographie et de l'histoire*. Paris: INRP.

Amador, Luis V., and Macarena Esteban. 2011. Desde la educación social a la educación ambiental. Hacia una intervención educativa socioambiental. *Revista de Humanidades* 18: 147–60. [CrossRef]

Barrón, Ángela, Antonio Navarrete, and Didac Ferrer-Balas. 2010. Sostenibilización curricular en las universidades españolas. ¿Ha llegado la hora de actuar? *Revista Eureka Enseñanza y Divulgación de las Ciencias* 7: 297–315. Available online: https://revistas.uca.es/index.php/eureka/issue/view/197 (accessed on 19 April 2022).

Benejam, Pilar. 1997. Las finalidades de la educación social. In *Enseñar y Aprender Ciencias Sociales, Geografía e Historia*. Edited by Pilar Benejam and Joan Pagès. Barcelona: ICE-UB/Horsori, pp. 33–51.

Bisquerra, Rafael. 2009. *Metodología de la Investigación Cualitativa*, 2nd ed. Madrid: La Muralla.

Boletín Oficial del Estado. 2022. Royal Decree, of 1st of March, Which Establishes the Organisation and Minimum Teachings of Primary Education, 2nd of March 2022. *Boletín Oficial del Estado* 52: 24386–504. Available online: https://www.boe.es/boe/dias/2022/03/02/pdfs/BOE-A-2022-3296.pdf (accessed on 19 May 2022).

Bonilla, Yaquelin, and Isabel Garzón. 2021. El abordaje de cuestiones socioambientales para la formación eco-ciudadana en la educación básica primaria. *Revista Educación y Ciudad* 40: 199–214. [CrossRef]

Bravo Pemjean, Liliana. 2002. La Formación inicial del profesorado de secundaria en didáctica de las ciencias sociales en la Universidad Autónoma de Barcelona: Un estudio de caso. Ph.D. thesis, Universitat Autònoma de Barcelona, Barcelona, Spain. Available online: https://ddd.uab.cat/record/37526 (accessed on 19 April 2022).

Dalongeville, Ainla. 2003. Noción y práctica de la situación problema en historia. *Enseñanza de las Ciencias Sociales* 2: 3–12. Available online: https://dialnet.unirioja.es/servlet/articulo?codigo=637183 (accessed on 19 April 2022).

De Lourdes, Marta Cecilia, and Antoni Santisteban. 2016. Representaciones sociales sobre participación democrática en estudiantes de formación inicial del profesorado a partir de sus experiencias en el movimiento estudiantil. *Fronteras* 3: 35–53. Available online: https://dialnet.unirioja.es/servlet/articulo?codigo=5710256 (accessed on 6 May 2022).

Elliott, John. 2000. *El cambio educativo desde la investigación-acción*, 3rd ed. Madrid: Ediciones Morata.

Eulie, Joseph. 1966. Controversial issues and the social studies. *The Clearing House: A Journal of Educational Strategies, Issues and Ideas* 41: 89–91. [CrossRef]

Evans, Ronald W., and David Warren Saxe. 1996. *Handbook on Teaching Social Issues*. Washington, DC: National Council for the Social Studies.

González Reyes, Luis, and Víctor Manuel Rodríguez. 2020. ¿Cómo educar desde una perspectiva ecosocial? In *Un Currículo para un Mundo Sostenible #ODS2030*. Edited by José Moya Otero and Ainara Zubillaga del Río. Madrid: Anaya, pp. 119–30.

González, Gustavo, María Ballbé, and Delfín Ortega-Sánchez. 2020. Global Citizenship and Analysis of Social Facts: Results of a Study with Pre-Service Teachers. *Social Sciences* 9: 65. [CrossRef]

González-Monfort, Neus, and Antoni Santisteban. 2020. Alfabetización crítica para interpretar problemas sociales. *Íber. Didáctica de las Ciencias Sociales, Geografía e Historia* 99: 39–45. Available online: https://www.grao.com/es/producto/revista-iber-099-abril-20-interpretar-problemas-y-conflictos-contemporaneos-ib099 (accessed on 6 May 2022).

GREDICS. 2018. *Literacidad crítica y posverdad*. Barcelona: Universitat Autònoma de Barcelona, Available online: https://ddd.uab.cat/record/214742 (accessed on 6 May 2022).

Hess, Diana. 2008. Controversial issues and democratic discourse. In *Handbook of Research on Social Studies Education*. Edited by Linda S. Levstik and Cynthia A. Tyson. Hillsdale: Lawrence Erlbaum, pp. 124–37.

Jodelet, Denise. 1989. *Les Représentations Sociales*. Paris: Presses Universitaires de France (PUF).

Kemmis, Stephen, and Robin McTaggart. 1988. *Cómo planificar la investigación acción*. Barcelona: Laertes.

Legardez, Alain. 2004. Transposition didactique et rapports aux savoirs: L'exemple de questions économiques et sociales, socialement vives. *Revue Française de Pédagogie* 149: 19–27. Available online: https://www.persee.fr/doc/rfp_0556-7807_2004_num_149_1_3169 (accessed on 18 May 2022). [CrossRef]

Legardez, Alain, and Laurence Simonneaux. 2006. *L'école àl'épreuve de l'actualité. Enseigner les Questions vives*. Issy-les-Molineaux/Paris: ESF Éditeur.

López-Fernández, José Antonio, and Monstserrat Oller. 2019. Los problemas medioambientales en la formación del profesorado de educación primaria. *REIDICS. Revista de Investigación en Didáctica de las Ciencias Sociales* 4: 93–109. [CrossRef]

Martínez-Medina, Ramón, and Jerónimo Torres-Porras. 2021. El medioambiente y la educación ambiental. In *Didáctica del Medioambiente en Educación Primaria*. Edited by Jorge Alcántara-Manzanares and José Antonio López-Fernández. Madrid: Editorial Síntesis, pp. 41–76.

Massip, Mariona, Mariona Espinet, Rául Almendro, Germán Llerena, and Júlia Hosta. 2021. Problemas Socio ¿qué? Los problemas socioambientales y su concreción por el profesorado en formación. *REIDICS. Revista de Investigación en Didáctica de las Ciencias Sociales* 9: 115–36. [CrossRef]

Moreno-Crespo, Pilar, and Olga Moreno-Fernández. 2015. Problemas socioambientales: Concepciones del profesorado en formación inicial. *Andamios* 12: 73–96. [CrossRef]

Moscovici, Serge. 2011. Prólogo. In *El Discurso de lo Cotidiano y el Sentido Común*. Edited by Nicky Hayes and Wolfgang Wagner. Barcelona: Anthropos Editorial, pp. IX–XVIII.

Moya, José, and Ainara Zubillaga del Río. 2020. *Un Currículo para un Mundo Sostenible #ODS2030*. Madrid: Anaya.

Muñoz, Pilar, and Inmaculada Muñoz. 2001. Intervención en la familia: Estudio de casos. In *Modelos en investigación cualitativa en Educación Social y Animación Sociocultural*. Edited by Gloria Pérez Serrano. Madrid: Narcea, pp. 221–52.

Ortega-Sánchez, Delfín, Almudena Alonso-Centeno, and Miguel Corbí. 2020. Socio-Environmental Problematic, End-Purposes, and Strategies Relating to Education for Sustainable Development (ESD) through the Perspectives of Spanish Secondary Education Trainee Teachers. *Sustainability* 12: 5551. [CrossRef]

Oviedo, Heidi Celina, and Adalberto Campo-Arias. 2005. Aproximación al uso del coeficiente alfa de Cronbach. *Revista Colombiana de Psiquiatría* 34: 572–80. Available online: https://www.redalyc.org/pdf/806/80634409.pdf (accessed on 18 May 2022).

Pagès, Joan. 1994. La didáctica de las ciencias sociales, el currículum y la formación del profesorado. *Signos. Teoría y Práctica de la Educación* 13: 38–51.

Pagès, Joan, and Antoni Santisteban. 2014. Una mirada del pasado al futuro en la didáctica de las ciencias sociales. In *Una mirada al pasado y un proyecto. Investigación e innovación en didáctica de las ciencias sociales*. Edited by Joan Pagès and Antoni Santisteban. Barcelona: Servicio de Publicaciones de la UAB/AUPDCS, vol. 1, pp. 17–39.

Rodríguez-Marín, Fátima, María Puig Gutiérrez, Lidia López-Lozano, and Alicia Guerrero Fernández. 2020. Early Childhood Preservice Teachers' View of Socio-Environmental Problems and its Relationship to the Sustainable Development Goals. *Sustainability* 12: 7163. [CrossRef]

Santana, Diana, Antonio José Morales, and Isabel Henarejos. 2021. Educación ambiental y proyectos educativos Iniciativas escolares de transformación ecosocial. *Íber. Didáctica de las Ciencias Sociales, Geografía e Historia* 102: 20–26. Available online: https://www-grao-com.ezbusc.usc.gal/es/producto/educacion-ambiental-y-proyectos-educativos-ib102100234 (accessed on 19 May 2022).

Santisteban, Antoni. 2004. Formación de la Ciudadanía y educación política. In *La Formación de la Ciudadanía: Las TICs y los Nuevos Problemas*. Edited by María Isabel Vera and David Pérez. Alicante: AUPDCS/Universidad de Alicante, pp. 377–88.

Santisteban, Antoni. 2009. *Programa de Doctorado en Didáctica de las Ciencias Sociales*. Barcelona: Universidad Autónoma de Barcelona.

Santisteban, Antoni. 2011. Las finalidades de la enseñanza de las ciencias sociales. In *Didáctica del Conocimiento del Medio Social y Cultural en la Educación Primaria*. Edited by Antoni Santisteban and Joan Pagès. Madrid: Síntesis, pp. 43–64.

Santisteban, Antoni. 2017. Educació política i mitjans de comunicació. *Perspectiva Escolar* 395: 12–16. Available online: https://www.rosasensat.org/revista/lensenyament-de-la-politica/ (accessed on 19 May 2022).

Santisteban, Antoni. 2019. La enseñanza de las Ciencias Sociales a partir de problemas sociales o temas controvertidos: Estado de la cuestión y resultados de una investigación. *El Futuro del Pasado* 10: 57–79. [CrossRef]

Solís-Espallargas, Carmen, and Rocío Valderrama-Hernández. 2015. La educación para la sostenibilidad en la formación de profesorado. ¿Qué estamos haciendo? *Foro de Educación* 13: 165–92. [CrossRef]

Stiglitz, Joseph E. 2015. *La gran Brecha. Qué Hacer con las Sociedades Desiguales*. Barcelona: Penguin Random House Grupo Editorial.

UN. 1987. Report of the World Commission on Environment and Development: Our Common Future. United Nations Brundtland Commission. Available online: https://sustainabledevelopment.un.org/content/documents/5987our-common-future.pdf (accessed on 19 May 2022).

UNDP. 2022. Goal 4 Quality Education. Sustainable Development Goals. United Nations Development Programme. Available online: https://www.undp.org/sustainable-development-goals#quality-education (accessed on 18 April 2022).

Vilches, Amparo, and Daniel Gil. 2021. Contribución de la educación a la transición a la sostenibilidad Una perspectiva interdisciplinar. *Íber Didáctica de las Ciencias Sociales, Geografía e Historia* 102: 8–14. Available online: https://www-grao-com.ezbusc.usc.gal/es/producto/contribucion-de-la-educacion-a-la-transicion-a-la-sostenibilidad-ib102100232 (accessed on 19 May 2022).

Wellington, Jerry J. 1986. *Controversial Issues in the Curriculum*. Oxford: Basil Blackwell.

Westheimer, Joel, and Joseph Kahne. 2004. What kind of citizen? The politics of educating for democracy. *American Educational Research Journal* 41: 1–31. [CrossRef]

Yin, Robert. 1989. *Case Study Research. Design and Methods*. Los Angeles: Sage.

Zamora-Polo, Francisco, Jesús Sánchez-Martín, Mario Corrales Serrano, and Luis Espejo-Antúnez. 2019. What do university students know about Sustainable Development Goals? A realistic approach to the reception of this UN Program Amongst the Youth Population. *Sustainability* 11: 3533. [CrossRef]

social sciences

MDPI

Article

Challenges with Complex Situations in the Teaching and Learning of Social Sciences in Initial Teacher Education

Cristina Yáñez de Aldecoa [1,*] and Isabel María Gómez-Trigueros [2]

1 GRIE (Interdisciplinary Research Group on Education), University of Andorra, AD600 Sant Julià de Lòria, Andorra
2 Department of General Didactics and Specific Didactics, Education Faculty, University of Alicante, 03690 San Vicente del Raspeig, Spain; isabel.gomez@ua.es
* Correspondence: cyanez@uda.ad

Abstract: This article not only presents a paradigm shift as a methodological model for teaching heritage and social sciences (SSCC), but also offers a methodological foundation for the challenge-based learning (CBL) methodology. We present various educational innovations in social science teaching and cultural heritage education based on the use of CBL during initial teacher training at the University of Andorra. These methodological proposals take into account the TPACK model (Technological Pedagogical Content Knowledge) based on the interrelation of three types of knowledge: pedagogical content knowledge (PCK), technological content knowledge (TCK), and technological pedagogical knowledge (TPK). This set of best practices requires students to respond to a complex social challenge by designing and creating specific educational proposals for tackling content and learning through gamification. Students work on the various dimensions of the SSCC, thereby developing spatial competency, teaching competency, and competency in democratic citizenship. ICTs are included throughout in order to develop students' digital competency. As a result, students feel empowered through having acquired the different competencies and developed an awareness of the value of cultural heritage as a cornerstone of democratic citizenship.

Keywords: higher education; initial teacher training; challenge-based learning (CBL); TPCAK (technological pedagogical content knowledge); relevant social problems; cultural heritage education; education for democratic citizenship

Citation: Yáñez de Aldecoa, Cristina, and Isabel María Gómez-Trigueros. 2022. Challenges with Complex Situations in the Teaching and Learning of Social Sciences in Initial Teacher Education. *Social Sciences* 11: 295. https://doi.org/10.3390/socsci11070295

Academic Editor: Alvaro Morote

Received: 23 May 2022
Accepted: 4 July 2022
Published: 8 July 2022

1. Introduction

Universities and higher education institutions, as providers of knowledge, are closely interconnected with societal challenges and are intimately linked to the commitment to help to develop sustainable societies (Leijon et al. 2021). It is essential to promote study plans and educational practices that enable students to face complex situations and become transforming agents. In this sense, the Educational 2030 Framework for Action (FFA) examines the learning outcomes to prepare students to become active citizens. Higher Education is the setting to deliver a fair and sustainable education and social development (Diez et al. 2021).

In the first half of the 20th century, Dewey (1995, 2002) defended teaching based on the real problems of society in the development of reflective thought and in the learning of democratic participation. Dewey defends the thesis that knowledge is built from experience, in a continuous interaction with the environment. Social Science pedagogy should train students to develop critical thinking based on social justice and democracy as pillars of democratic citizenship competence. In an increasingly diverse changing society, wherein concepts are continually questioned, the critical spirit of students must be taught and trained to nurture a critical understanding based on justice, and thus influence the formation of active and committed citizens (Dam and Volman 2004; Diez et al. 2021; Pagès and Santisteban 2011).

This paper presents a line-up of Social Sciences cases of good innovative education practices in Initial Teacher Education at the University of Andorra, conducted with students of the Bachelor of Science in Education through the challenge-based learning (CBL) methodology. Once the challenge is launched, it is worth highlighting the creativity and innovation of the final results as well as the different interconnections established with other subjects, such as mathematics or language.

The main objective is to promote the use and application of technological resources and digital materials, resulting from the integration of applied technologies and cultural heritage, to bring cultural heritage closer to the classroom (Yáñez 2017; Larraz et al. 2014). Students, based on their previous research work, documentation, and analysis of educational materials on the web, design various proposals in order to meet the requirements of the challenge. These proposals usually have a high technological component, and therefore, it is an activity that helps to develop the Teachers' Digital Competence (TDC). These didactic proposals, focused on primary school students, have as a goal bringing cultural heritage into the classroom to not only make it more accessible, but also to learn and acquire knowledge from a new perspective. This type of task requires that the teacher monitors and accompanies the student throughout the whole process to ensure that corrections can be introduced as the task is being designed and developed. It is part of the Learning by Doing methodology that enhances the development of skills to respond to a complex situation and consequently learn from experience (Schank 2002; Yánez and Gómez-Trigueros 2022). The results show how, from an interdisciplinary approach and a methodology focused on challenge-based learning (CBL), specific and transversal competences can be worked and achieved during initial teacher training.

1.1. Competencies in Higher Education

Globalization has created an increasingly diverse and hyperconnected world. In order to confront the new challenges, individuals must acquire new competencies (understood as skills, attitudes, and values) in order to achieve individual goals and autonomous development, together with collective goals involving the necessary interaction with society (Yáñez and Gómez-Trigueros 2021). In today's information and communication society (ICS), there is a need for education to be reoriented towards innovation, the driving force behind change, and for teachers to act as agents of change.

Various European and international organizations, such as the European Commission (2022) and the OECD (2005), have reflected on the social, economic, and cultural transformations of the twenty-first century and have stressed the need to anticipate new demands given the ever-changing context, the impact of globalization, and the complex socio-cultural and economic relations that result from it. The traditional teaching model is being replaced by the learning environment model, based on real contexts, the use of new tools, and negotiation amongst peers around the meaning of knowledge (Basque et al. 2000).

1.2. Bachelor's Degree in Teaching and Learning (BCE) at the University of Andorra (UdA): A Competency-Based University Education Model

The English term 'dual' and the French term 'en alternance' both refer to a model of training that enables the student to combine academic and professional activity, with knowledge acquired in both the academic and professional fields. University tutors and school mentors are jointly responsible for the training and assessment of the activities carried out by the student in shared tutorials (Latorre et al. 2020; Pla 2015).

Dual training was implemented in the Bachelor's degree in Teaching and Learning (BCE) between 2014–2015 and 2018–2019. The teaching–learning (T-L) approach includes students attending theoretical classes at the university three days a week and undertaking work experience in primary schools two days a week. This T-L model enables students to learn about the theory of different subjects and to complement it with teaching practice. Students are thus better able to integrate theoretical and practical knowledge and develop competencies in a professional environment (Monné 2017).

In 2018, in response to a changing, competency-based society, the University of Andorra (UdA), the only public university in the Principality of Andorra, decided to launch a new educational model focused on competency-based learning and assessment, with competency being defined as 'an action or intervention in which a set of content (facts and concepts, procedures, attitudes and values) is incorporated and employed by the student in order to successfully respond to professional challenges' (Larraz et al. 2020, p. 2; Perrenoud 2004). The model is based on the development and assessment of both specific and cross-disciplinary competencies focused on multilingualism, democratic culture, and the values of the Sustainable Development Goals (SDGs).

The current curriculum of the BCE complies with the new educational model of the University of Andorra, with a competency-based methodology that eliminates the boundaries between individual subjects. Each semester is divided into two modules, in which students must overcome a complex challenge by means of guided work, individual work, and seminars. Each challenge is associated with learning outcomes that are derived from the specific and cross-disciplinary competencies of the degree (Saz et al. 2020).

Competency-based work benefits from low student–teacher ratios, which makes it possible for students to be closely monitored. These educational innovations focus on the importance of heritage education and dissemination, a 360° experience in which educational and cultural institutions work to safeguard and conserve heritage.

2. Theoretical Foundations

2.1. *The New Learning Environments*

Pujolàs (2003) has argued that education is experiencing a paradigm shift with the advent of what has been called twenty-first-century education (Peñalba and Leiva 2019), the aim of which is to provide the resources and tools to share knowledge. It is therefore necessary to generate new learning contexts or spaces that encourage the traditional top-down system to evolve into one that enhances the application of knowledge and focuses on competency-based learning (Monné and Yáñez 2019).

Society is constantly changing, and the learning opportunities that arise must be seized. Schools must extend beyond the curriculum and leave the classroom to deliver more comprehensive, experiential learning, taking advantage of the opportunities offered by the immediate environment. It is therefore no longer only a question of teaching students, but of teaching them to develop reliable tools, to give them a compass to find their own way in an increasingly complex, volatile, and uncertain world (OECD 2019).

The aim of the so-called 360° education is to provide cross-disciplinary teaching beyond the school walls, taking advantage of the opportunities and possibilities offered by the environment and the territory as settings for different kinds of learning, understanding territory as the geographic element (physical and human), in which the human being is contextualized. It provides a broader view by expanding the time and spaces for learning. It views the territory, the community, and cooperation with different cultural agents as an educational ecosystem full of opportunities for learning, for connecting formal and non-formal, school and non-school learning. A holistic education with fluid boundaries between formal and non-formal learning should provide a better response to a complex, uncertain society characterized by various economic, social, cultural, and technological changes. These interconnected structures of knowledge require not only competency-based learning, but also interdisciplinary work.

Gangopadhyay (2014) has explained how museums, and cultural centers in general, can help to generate rapid innovations and changes in education and learning. The author focuses on new learning settings and experiences through the application of three central strategies, involving proactive, relevant, and innovative practices (Yáñez et al. 2015).

Camacho (2018, p. 22) suggests that educational innovation should be understood as 'a dynamic process of vision, change, and creation, which requires energy and passion in order to create and implement new ideas'. In line with this approach, the various teaching innovations we present are based on the following foundations:

1. CBL (challenge-based learning);
2. Critical and reflective decision making;
3. Interdisciplinarity;
4. Cooperative work and autonomy in learning;
5. Democratic citizenship and human rights.

2.2. Challenge-Based Learning (CBL)

CBL is a learning opportunity in which students, guided by the teacher, collaborate to solve real and relevant challenges, in a context outside the classroom. The exhaustive bibliographic review carried out by Leijon et al. (2021, p. 3) establishes that the challenge-based learning (CBL) methodology "connects to entrepreneurial learning as well as self-regulated learning, organisational double-loop learning, and authentic learning". To sum up, we can trace theories of learning stemming from cognitive, but also constructive and socio-cultural perspectives, wherein learning is active, relational, practical, and situated within both the individual and the organisation (Argyris and Schön 1978; Kolb 1984; Lave and Wenger 1991). It facilitates the acquisition of basic skills in the Social Sciences, allowing to work and connect with real-world scenarios and facilitates as well the development of entrepreneurship and critical thinking, forming active and committed citizens.

Teaching based on this strategy aims to focus on activities that pose problematic situations whose resolution requires analysing, discovering, developing hypotheses, confronting, reflecting, arguing and communicating ideas to achieve student learning.

> "This didactic methodology reviews the T-L methods to determine the best practices to produce learning environments to develop and foster creativity. The results observed in the students are: critical and divergent thinking, new ideas and innovative solutions to problems".
>
> (Olivares et al. 2018, p. 232, based on Fletcher 2011)

This approach encourages working in projects or integrated tasks (Edu Trends 2016; Gaskins et al. 2015; Pérez-Sánchez et al. 2020). This is a didactic guarantee for an effective contribution to the development of key competences, as well as to the meaningful learning of curricular content. This methodology involves different actions to be followed: analyse; diagnose; observe; investigate; devise; decide; design; analyse the feasibility, sustainability and impact on the environment; implement; and evaluate.

Hence, this process fosters the students' acquisition of Social Sciences' basic skills. They allow to work and connect with real-world scenarios and also facilitates the development of entrepreneurship and critical thinking, forming active and committed citizens. Through CBL, students develop the following competencies: ability to research, analyse, and manage information; decision making to solve a complex problem; critical thinking innovation and creativity; ability to argue and reason effectively; collaborative learning; teamwork; development of democratic attitudes and values; ability to transmit to communicate; and ICT integration.

2.3. The TPACK Model

One T-L model that focuses on the correct use of technologies in the classroom is the Technological Pedagogical Content Knowledge (TPACK) model. This model suggests that teachers need to acquire pedagogical knowledge (PK), content knowledge (CK), and technological knowledge (TK) (Mishra and Koehler 2006, 2007). Furthermore, it stresses the importance of the interaction of these three elements (CK, PK, and TK) in the T-L process. The resulting network of interrelations must be understood and mobilized by teachers in order to correctly integrate ICTs in their daily activity. According to this model, teachers do not use these areas of knowledge in isolation; rather, by working with them in an interrelated way, they develop new kinds of knowledge, namely:

- *Pedagogical Content Knowledge*, or PCK, which refers to the different pedagogical skills that a teacher can employ to teach a specific discipline (or content), adapting to the available resources and the diversity of students in the classroom.
- *Technological Content Knowledge*, or TCK, which is related to the mastery of disciplinary knowledge about the subject to be taught and the ability to discern the specific technologies that will enable students to learn.
- *Technological Pedagogical Knowledge*, or TPK, which refers to the teacher's understanding of the effect of ICT tools on their students' learning process. The teacher should discern when and how to use such resources to enable students to acquire knowledge.

The TPACK model takes on board the fact that technology is here to stay. Faced with this reality, teachers must be trained in the use of technologies and acquire the skills to adapt to the changes introduced by new software and hardware.

Moreover, the TPACK model provides a new perspective on incorporating ICTs in the classroom, focusing not only on instrumental competencies but also on their interrelation with the didactic component (Gómez-Trigueros and Yáñez 2021). Educational institutions must ensure that teachers are trained in the various competencies, including teacher digital competency (TDC). Such mastery and understanding should enable them to use technology correctly in their teaching (Gómez-Trigueros et al. 2019; Gómez-Trigueros and Yáñez 2021). Similarly, the TPACK model resolves issues in initial teacher training, helping trainee teachers to develop their methodology in the use of ICTs for teaching. The model encourages trainee teachers to make use of reflective practice: it prompts a reflection on their training; helps them to develop greater self-knowledge and improve their teaching practice; enables a better understanding of how to use technological tools in education (Gómez-Trigueros and Yáñez 2021); and encourages them to focus on enabling their students to learn in a meaningful way, and not on whether or not to use technological resources in the classroom.

2.4. *Cultural Heritage Education Is Essential for Democratic Citizenship and the Construction of Historical Memory*

Cultural heritage education enables individuals to learn about and relate to the past in order to understand the present, while also acquiring ideas about space and time that are essential for the construction of a critical spirit. It also helps students to develop empathy, solidarity, cooperation, and respect for other cultures (Estepa et al. 2005). In order to promote not only knowledge but also the active participation of individuals in the different cultural expressions in their environment, it is essential to develop strategies to encourage students to become interested in and develop respect for heritage. This should contribute to the development of more plural and democratic societies (Morote and Colomer 2021; Yánez and Gómez-Trigueros 2022).

One of the strategies for combatting the population's lack of interest in cultural heritage is to actively promote the teaching of heritage in schools, specifically in the field of citizenship education. Heritage should be studied not only as a resource, but also as conceptual, procedural, and attitudinal content (Yáñez 2017). As we can see on Figure 1, the objective is to raise awareness of the importance of preserving heritage for future generations and to encourage students to appreciate it and to become actively involved in its conservation and dissemination (Casanova et al. 2018; Cuenca and Martín 2014; Cuenca and Pérez 2021).

Figure 1. Steps to raise awareness about heritage. Own source.

3. Objectives

The main objective is to present a series of best practices, understood as educational resources designed to introduce primary school pupils to heritage, history, and, ultimately, the Social Sciences.

In response to the different challenges set, students had to acquire the knowledge and skills related to the different subjects and design educational resources that in some cases had an important digital component.

In addition to involving teaching and learning through formative and evaluative methodologies, the practices presented consider four dimensions of the Social Sciences: time, space, art/culture, and democratic citizenship. They also had to promote active learning with a strong creative component and encourage the use of technologies in the classroom.

This general objective was split into the following specific objectives:

- Promote teaching innovation and good practices in heritage education.
- Increase Spatial Competence (SC) (time and space) in future trainee teachers.
- To increase the Digital Competence (DC) of future teachers in the applied use of ICT through the CBL (challenge-based learning) methodology.
- Enhance Democratic Citizenship Competence (DCC) and Sustainable Development Goals (SDGs).
- To foster the competences of enquiry, critical thinking, and autonomous work as well as the ability to work in groups.
- To promote the interconnection between education and culture.

4. Methodology

These teaching innovations were based on challenge-based learning (CBL), an approach that involves students playing an active role in their learning, in line with the constructivist model and principles defined by Savery (2015):

- Introduce the project and motivate students to respond to an educational challenge.
- Inform students about the competencies to be developed and the learning expectations.
- Organize work teams, plan the tasks, and agree on a schedule.
- Mentor and guide students' work.
- Students present and justify their responses.
- Students reflect and evaluate together, encouraging self-assessment, peer assessment, and teacher assessment.

4.1. Participants and Context

These different CBL-based teaching innovations involved trainee teachers (of children aged 0–12 years) in their second and third year of the Bachelor's degree in Teaching and Learning at the University of Andorra. A total of 101 second-year students undertaking the course in 'Education in the Cultural Heritage of Andorra' and 92 third-year students taking 'Social Science Didactics' followed this curriculum during the study (Table 1).

Both courses followed the 2015 dual curriculum, which involved project-based work clearly designed to develop competencies. Given the experience acquired in CBL and the fact that the new T-L educational model was based on responding to challenges, the aim was to transfer the experience acquired through this set of experiences to the new educational model of the Bachelor's degree.

The content of the new model was delivered in two seminars (which had the same name as the courses), which formed part of the module on the basic principles and content of the Social Sciences. This module provided tools for social science teaching and encouraged social and critical thinking, along with democratic principles and culture. The main objective of SSCC teaching is to provide students with the knowledge, skills, and competencies necessary to become active and informed citizens, capable of thinking critically, understanding and explaining the perspectives of others, making judgments, and communicating effectively.

Table 1. Total number of students participating in these educational experiences. Own source.

	Courses	
Academic year	Education in the Cultural Heritage of Andorra second-year students	Social Sciences Didactics third-year students
2014–2015	21	13
2015–2016	22	19
2016–2017	22	13
2017–2018	13	21
2018–2019	15	11
2019–2020	1	14
2020–2021	7	1
TOTAL	101	92

4.2. Design of the Challenge

The challenges were complex, real situations that may present more than one possible response. They should be close to reality, arouse the student's interest and motivation, and incorporate different disciplines. They should also encourage critical thinking so that students can make reasoned decisions, based on evidence and facts, and determine the necessary resources for the proposed response and the strategies to be used for assessment.

The challenges start with the formulation of a key question, with a distinction between the nature and the purpose of the challenge (Table 2).

Table 2. Approach to the formulation of key questions. Source adapted from Lattimer (2008).

What Are They?	What Are They for?
They help to arrive to the focus of the discipline. They pose dilemmas that overturn established truths or canons.	They allow one to construct one's own understanding of the past
They can have more than one answer.	Show that history is a developmental narrative.
Connect past with present.	Challenge to analyze and examine one's own beliefs.
Must: be open-ended; be provocative; require a high order of reasoning; lead to reasoning rather than repetition or memorization; can be revisited and generate new questions and concerns; generate debate; require evidence.	Prepare students for civic participation in a democratic society.

Most of the T-L innovations we present focus on cultural heritage and the cultural landscape (the context of time and space), which have a powerful transformative capacity. Transformative heritage sites 'empower the visitor to transform their environment through educational activities that are designed to encourage permanent dialogue between action (doing), reflection (thinking), conversation (communicating), and emotion (feeling)' (CoP 2018, p. 5). This makes it necessary to ask the following questions:

1. What do we want to teach? In order to provide interpretation models, there should be dialogue between doing, thinking, and communicating.
2. How do we want to teach it? Teaching should encourage observation, analysis, and reflection, as well as innovation to encourage dialogue.

3. Where should we teach it? The setting where the T-L process takes place is fundamental. The contexts must be real, complex, stimulating, and, above all, challenging spaces.
4. Who do we want to teach and why? Teachers must be builders of knowledge and authentic agents of transformation.

In order to work on complex situations, a challenge must have the following characteristics.

- Be real, truthful, with some social, vocational, and personal relevance.
- Generate a meaningful learning situation.
- Allow the contents to be worked on in limited contexts, preferably in the environment if possible.
- Raise guiding questions.
- To produce a cognitive conflict, a need to know.
- Provoke critical thinking and move away from rote or repetitive tasks.
- Be assumable, with a gradual complexity, with the consequent planning and hierarchization of tasks that favor autonomy as well as interaction with heterogeneous groups.
- Generate a certain degree of uncertainty, as a result of the gradual process of learning by doing, which involves not knowing what the final result will be.
- Encourage work by competences.
- Encourage interdisciplinary work.
- Encourage meaningful learning.
- Encourage interrelation between doing (experimenting), thinking (knowledge structure), and communicating (expressing and disseminating in different formats).
- Evaluate and reformulate the context. How to work with the same concepts in different contexts. What changes?
- Ensuring the transfer of knowledge to the real environment.

CBL methodology can be broken down in order to establish the phases of the process, which are shown in the following Table 3 and in Figure 2.

Table 3. CBL methodology. Own Source.

CBL Process	Guidance
1. Starting point • Defining the challenge; • Brainstorming; • Activate prior knowledge.	Explaining the challenge in detail.
2. Analysis and research • Documentation and research on the issue, the theme and the different stakeholders involved. What do we know? What primary or secondary sources do we have? Who is our audience? • Consider multiple accounts and perspectives; • Analise primary sources; • Consider context; • Claim–evidence connection; • Identify the compelling and essential questions; • Define the problem and goals; • Foresee the main desirable learning outcomes.	Independent learning; Time to do research and to internalize the problem; Teamwork to share points of view and strategies; Collective mentoring sessions to review the initial approaches.
3. Planning • Planning and organizing the different tasks; • Organize ideas; • Distribute roles. • Timeline. Sequence of actions and tasks.	Individual mentoring sessions; Individual and teamwork.

Table 3. *Cont.*

CBL Process	Guidance
4. Creativity • Design of manual and digital educational activities as a response to the challenge.	Individual mentoring sessions Individual and teamwork.
5. Simulation • Intermediate defense; • Amend and correct errors.	Feedback from: Peer intermediate assessment Teacher's assessment
6. Innovation • Apply knowledge; • Develop activities; • Use ICT tools to encourage interactions.	Individual and teamwork.
7. Implement • Expose the proposal for assessment and peer evaluation; • Implement in a real scenario; • Validate at the chosen scenario. Check the effectiveness of the implemented solution.	Individual work.
8. Defense • Expose the given answer to the challenge; • Gather suggestions for improvement.	Individual work.
9. Assessment • Evaluate and amend if necessary; • Measure learning outcomes.	Individual work.

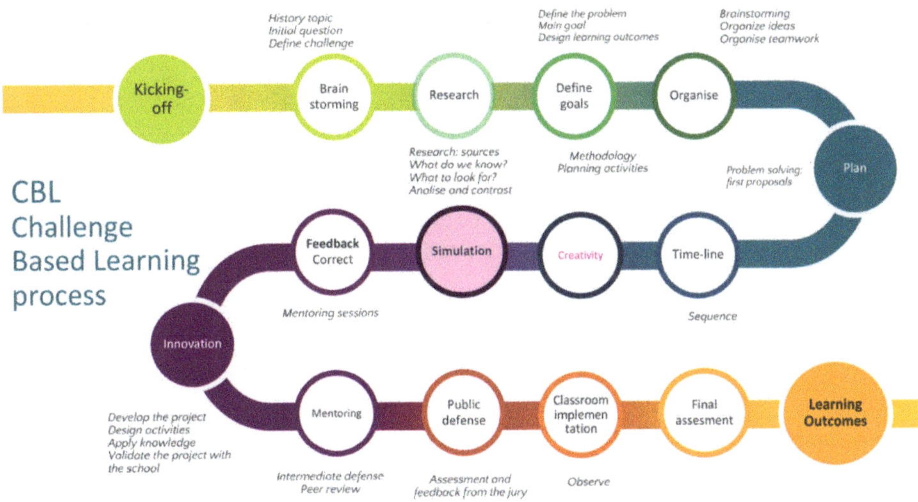

Figure 2. CBL methodology process. Own source.

4.3. *The Role of the Teacher*

The teacher uses questions to challenge students' knowledge and stimulate cognitive processes that promote critical thinking. By avoiding leading questions, the interaction takes place at a metacognitive level.

In his discussion of the tutorial process, as Barrows states from Savery and Duffy (1995, p. 11): 'The ability of the tutor to use enabling teaching skills during the small group learning process is the major determinant of the quality and the success of any educational method aimed at (1) developing students' thinking or reasoning skills (problem solving, metacognition, critical thinking) as they learn, and (2) helping them to become independent, self-directed learners (learning to learn, learning management). Tutoring is a teaching skill central to problem-based, self-directed learning' (Barrows 1992, p. 12).

5. Results

Cognitive as well as emotional, cultural, and social aspects operate in this multidimensional process. This process has led to the replacement of the traditional way in solving mechanized problems, by a new methodology to solve real, interdisciplinary, professional problems that allow to find multiple solutions (Couso et al. 2008; Pérez and Chamizo 2011).

5.1. Map of Educational Innovations

The overall objective of the activities presented was to promote the knowledge of and respect for cultural heritage as well as to generate an emotional bond with heritage that would ensure that today's trainee teachers would be true agents of transformation as active teachers in the future.

The different teaching innovations were divided into a series of categories based on the type of activity (see Table 4).

Table 4. Categories of the group activities developed. Own source.

A.	Transfer to society: activities in collaboration with social agents (schools, town councils, museums, etc.).
B.	Experiences of teaching innovation in the classroom: manipulative activities to promote creativity, enquiry and emotional education (hands-on and heart-on).
C.	Heritage education and LKT interdisciplinary activities between the subjects of Heritage Education and LKT and to promote Geographical Information Technologies (GIT).
D.	Gamification: activities to promote interest in heritage, where the basis is enquiry and play.
E.	Sensorial workshops in Cultural Heritage Education: multisensory activities to promote interest in heritage.
F.	Cultural heritage in situ: field trips to be able to know, explore, and understand how heritage is managed.
G.	Democratic citizenship and human rights: from the analysis of a historical event, activities are designed to promote the competence of democratic citizenship as well as the values of human rights.
H.	Space–Time–Heritage: activities focused on working on spatial competence.

Similarly, the innovations are associated with different dimensions of the Social Sciences: history, geography, art, heritage, and democratic citizenship. History and geography are two fundamental pillars of the SSCC, since they enable learners to develop an understanding of space and time, while offering a broad vision of the world and of society.

The following table (see Table 5) shows the map of best practices in educational innovation, grouped according to category and associated with the different Social Science dimensions. They are all interrelated and encourage meaningful and relevant learning.

Table 5. Map of teaching innovation experiences in the area of cultural heritage and Social Sciences and dimensions. Own source.

Category	Experiences of Teaching Innovation	Dimension				
		History	**Geography**	**Art**	**Heritage**	**Democratic citizenship**
A. Transfer to society	Paint me a story at the Andorran Primary School (Ordino)	x	x	x	x	x
	Story: "The Stefanoni brothers"	x	x	x	x	x
	Personal heritage and generational exchange: Aula Magna				x	x
	Workshops at the Casa de la Vall. Consell General (parliament)	x			x	x
	#Museu Carmen Thyssen Andorra Challenge			x	x	
	Training of trainers: workshops in schools					x
B. Teaching innovation in the classroom	What have we learnt? One minute paper	x	x	x	x	x
	Life stories: the heritage object	x	x	x	x	x
	Museum in a box			x	x	
	Mystery box	x	x		x	
	Basket Beat: emotional education					x
	Let's reinvent the story: "Meninas' project" & "American Gothic"			x		
C. Education in Heritage and LKT	EQuadern and Andorra's heritage			x		
	Google Earth Pro^{TM} and the discovery of cultural and natural heritage	x	x	x	x	x
D. Gamification	Wix of the Romanesque of Andorra	x	x	x	x	x
	"Pass the word": history and heritage				x	
	Smart Art Lab: Art and technology	x	x	x	x	x
	Suitcase *Escape Room* to discover heritage	x	x	x	x	x
E. Sensorial workshops in Cultural Heritage Education	Cultural heritage workshops				x	x
	Design workshops for different museums				x	x
F. Cultural Heritage in situ	Visit, analysis, and workshops in different museum facilities (archaeological sites, Electricity Museum, CAEE, Casa de la Vall, Carmen Thyssen Andorra Museum, Tobacco Museum, mapping of the Church of Santa Coloma, etc).	x	x	x	x	x
	The cultural heritage from the 7 parishes	x	x	x	x	x
	Visit to temporary exhibitions: "The 1982 floods".	x				x
G. Democratic citizenship and Human Rights	*Citizens of the world*	x	x	x	x	x
	Life story. Testimony of Norman Westby, RAF (British Armed Forces) pilot in the Second World War	x	x			x
H. Space–Time, Heritage	Timelines	x	x			
	Mental maps, the geographical center, and spatial competition		x		x	
TOTAL of dimensions distributed among activities		15	14	14	19	18

The Social Science dimensions are fairly well distributed among all the activities. The focus is primarily on heritage, with 19 of the 27 activities (70.37%), followed by democratic citizenship, with 18 of the 27 activities (66.66%). A total of 55.55% worked on history, and 51.85% worked on geography and art.

Of the 27 activities, the following Table 6 shows the distribution of the activities referred to in Table 5 divided into categories. The first two, which include activities both in and outside the classroom, have the greatest weight.

Table 6. Distribution of activities by categories. Own source.

Categories and Activities							
A.	B.	C.	D.	E.	F.	G.	H.
22.2%	22.2%	7.40%	14.80%	7.40%	11.10%	7.40%	7.40%

5.2. Educational Innovations in the Social Sciences and Heritage Education and the TPACK Model

The following table (see Table 7) associates best teaching practices with the categories in the KBT (Knowledge Base for Teachers) (Shulman 1986) and TPACK models (Mishra and Koehler 2006). The following elements are verified in the KBT model: (a) content knowledge (CK), which refers to the necessary (and acceptable) content on the subject to be taught in the classroom, along with the (scientific, philosophical, and historical) resources available in which to search for further knowledge, i.e., the teacher's awareness of the bibliography on the subject they are teaching; and (b) pedagogical knowledge (PK), which refers to the training in strategies, methodologies, and pedagogy that the teacher employs to make the subject (scientific content) understandable to their pupils. This can be described as the 'wisdom of practice' (Shulman 1986, p. 15). This foundation of the KBT relates to the good teaching practice developed by teachers in the classroom, which is learnt by experience and by observing other teachers. This 'source' is the perfect combination of teachers' knowledge and its application to enable pupils to acquire, understand, and learn such knowledge.

Table 7. Association of the activities to the model: KBT (Knowledge Base For Teachers); PCK (Pedagogical Content Knowledge); TCK (Technological Disciplinary Knowledge); and TPK (Technological Pedagogical Content).

Category	Experiences of Teaching Innovation	KBT			TPACK Model		
		CK	PK	KBT (CK + PK)	PCK	TCK	TPK
A.	Paint me a story at the Andorran Primary School (Ordino)	x	x	x			
	Story: "The Stefanoni brothers"	x	x	x			
	Personal heritage and generational exchange: Aula Magna	x	x	x			
	Workshops at the Casa de la Vall. Consell General (parliament)	x	x	x			
	#Museu Carmen Thyssen Andorra Challenge				x	x	x
	Training of trainers: workshops in schools				x	x	x
B.	What have we learnt? One minute paper	x	x	x			
	Life stories: the heritage object	x	x	x			
	Museum in a box	x	x	x			
	Mystery box				x	x	x
	Basket Beat: emotional education	x	x	x			
	Let's reinvent the story: "Meninas' project" & "American Gothic"	x	x	x			
C.	EQuadern and Andorra's heritage				x	x	x
	Google Earth Pro™ and the discovery of cultural and natural heritage				x	x	x
D.	Wix of the Romanesque of Andorra				x	x	x
	"Pass the word": history and heritage	x	x	x			
	Smart Art Lab: Art and technology				x	x	x
	Suitcase *Escape Room* to discover heritage	x	x	x			

Table 7. *Cont.*

Category	Experiences of Teaching Innovation	KBT			TPACK Model		
		CK	PK	KBT (CK + PK)	PCK	TCK	TPK
E.	Cultural heritage workshops	x	x	x			
	Design workshops for different museums	x	x	x			
F.	Visit, analysis, and workshops in different museum facilities (archaeological sites, Electricity Museum, CAEE, Casa de la Vall, Carmen Thyssen Andorra Museum, Tobacco Museum, mapping of the Church of Santa Coloma, etc).	x	x	x			
	The cultural heritage from the 7 parishes	x	x	x			
	Visit to temporary exhibitions: "The 1982 floods".	x	x	x			
G.	*Citizens of the world*				x	x	x
	Life story. Testimony of Norman Westby, RAF (British Armed Forces) pilot in the Second World War	x	x	x			
H.	Timelines				x	x	x
	Mental maps, the geographical center, and spatial competition	x	x	x			

As indicated in Section 2.3 of this article, the different dimensions of the TPACK model are all valued similarly.

Of the 27 teaching innovations, 18 (66.6%) corresponded to the KBT model (Shulman 1986) and the remaining 9 (33.3%) to the TPACK model (Mishra and Koehler 2006).

5.3. Spatial, Digital, and Democratic Citizenship Competence in Some Experiences

The following describes a content summary of some of the implemented teaching innovative experiences that gather all SSCC dimensions, as well as a high Knowledge Learning and Knowledge Technologies (LKT) and Information and Communication Technologies (ICT) component (Table 8).

Table 8. Summary of some of the teaching innovative experiences. Own source.

Teaching Innovative Experiences with a LKT and ICT Component	
Web Address Content	**Content Description**
The Romanesque Wix  https://bit.ly/3ycqeah (accessed on 1 May 2022)	Group project of 2nd year ECB students to create an online platform with activities and strategies for teaching and learning about the country's Romanesque cultural heritage. Use a digital editing platform to place all created material that will become the workspace of teachers and students. The group class was divided into three, each for a stage of primary education. A fourth group was in charge of designing the platform and uploading all the activities created by the other groups. It is evaluated according to: (a) the architecture of the information; (b) the communication strategy; and (c) the viability of the project. It ended up becoming a platform available to all teachers in the country and can be found on the university's website.
E-Notebook  https://bit.ly/3KCnqFT (accessed on 1 May 2022)	Interdisciplinary experience between two subjects: LKT and Education in the cultural heritage of Andorra. Design and creation of a digital material corresponding to a teaching unit of Primary Education in the curricular field of the cultural heritage of Andorra. The final product to be obtained consists of a space in the network where the students of a certain level will find the set of materials and different learning tools corresponding to a didactic unit. Mentoring and evaluation are shared and agreed by both teachers responsible for the subjects.

Table 8. *Cont.*

Teaching Innovative Experiences with a LKT and ICT Component	
Web Address Content	**Content Description**
Smart Art Lab  https://bit.ly/3s74atO (accessed on 1 May 2022)	Design a Digital Art Lab on a specific artist. The aim is to bring art and history closer to primary school classrooms. Designing manipulative playful pedagogical activities as well as digital activities to foster knowledge, from an innovative and creative point of view, and nurture a taste for art. Use a digital editing platform as a workspace of teachers and students. Bilingual: English and Catalan.
Citizens of the world  https://bit.ly/38Uszw8 (accessed on 1 May 2022)	Design a lesson for 6th grade students to promote social understanding of historical facts and human societies and to promote the civic competence necessary for a democratic citizenship. Fosters inquiry by scaffolding curiosity. Learning requires to examine events in history and examples of human behavior. Design a digital editing platform to upload all the designed pedagogical activities, to become a workspace for teachers and students. It is not only about research and understanding the facts, but about emotional engagement and ethical reflection.
Google Earth Pro^{TM} 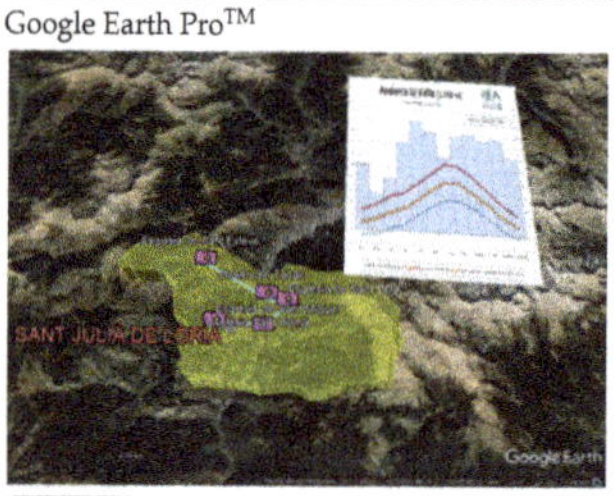 https://bit.ly/3LN42aS (accessed on 1 May 2022)	International collaborative e-learning experience between the University of Andorra and the University of Alicante, for teachers in training in the subject of Social Sciences, in the context of the 2020 lockdown. The main goal has been to use a GIT resource as an interdisciplinary strategy to bring cultural and natural heritage closer to Primary Education classrooms. An inter-university online teaching collaboration with a team learning environment

5.4. Assessment and Learning Outcomes

All the teaching innovations presented employed formative assessment to determine students' level of understanding and competency in order to plan the T-L process. The objective of the formative assessment is to guide students to achieve their objectives, so all stages of the learning process have to be assessed. Continuous assessment provides the teacher with information on the state of the T-L and allows modifications to be made in order to adapt teaching strategies so that students achieve mastery of the desired competencies. 'Students gain clear insights into how they are learning and what they can do to improve performance' (Myers 2004, p. 292).

The main objective of the assessments is to foster students' desire to learn, emphasizing autonomy and development, as well as the fundamental role of the teacher in providing continuous feedback on the work throughout the assignment. Examples of formative assessments include rubrics, games, individual and group class dynamics, projects, and presentations.

All the activities involved adapting means of assessment to the project. In the experiences involving a final product, a series of categories were established that grouped indicators based on the learning outcomes and the levels of acquisition of the competencies assigned to each project. Below, we present the categories and indicators that should be taken into account when assessing the design of educational materials in the SSCC that require a technological component (Table 9).

Table 9. Proposal of categories and indicators for the evaluation of Social Science educational materials with a technological component. Own source.

	Assessment Indicators
Architecture information	Content: coherent structure
	Presentation and design
	Variety of materials, images, photography, websites, and web resources
	Introduction and justification
	Main pedagogical and specifics goals
	Methodologies
	Activities to be developed linked to goals, methodologies, and expected learning outcomes
Evidences	Primary and secondary sources,
	Bibliographical sources and citations
Communication strategy	Speech reflection
	Oral and written communication
Feasibility	Feasibility of activities to work heritage
	Usability and accessibility
	Activities adapted to the cognitive level
Digital competence	Mechanical skills in the use of different digital and technological tools
	Didactic skills in the use of different digital and technological tools

6. Discussion

Regarding possible future lines of research that may arise from this study and the compendium of experiences, it is planned to continue this research on the competencies acquired through a CBL methodology, especially over the coming years, which will allow us to have a reading to analyze whether there is a change in the learning outcomes. Without neglecting knowledge, we think it is necessary to introduce active methodologies that allow the approaching, teaching, and learning of heritage, culture, and history in a playful and experiential way. The ideal scenario to internalize learning about heritage is field trips to see it and enjoy it in situ, but this is not always possible. That is why we must bring the heritage sites into the classroom and activate them and revive them in different possible ways through the possible multiple methodologies to create a true "heritage laboratory" in the classroom. Finally, and whenever possible, it should be with a gamification component to make it more attractive, but without losing sight of the rigor of the teaching and learning objectives. Teachers in training should experience heritage not only to know it, but to understand and experience it and become ambassadors of heritage.

This work showed, in a detailed and surgically analyzed step by step way, how to transfer the CBL and TPACK methodology to teaching activities that promote knowledge and respect for cultural heritage as well as generate an emotional bond with it in order to ensure that the trainee teacher becomes a true agent of transformation as an active teacher. The different methodologies that are hidden behind the activities, not only guarantee working the different dimensions of the SSCC, together with the bases of the TPACK model, but also this conjunction of pedagogy, knowledge, and pedagogical and technical knowledge of the methodology ensure the acquisition of KBT in future teachers.

Other works that have preceded ours confirm the need for this type of study and, above all, to be able to incorporate these methodologies in the classroom, for which it is essential that future teachers receive the best possible training (Cuenca 2013; Estepa et al.

2005; Fontal et al. 2012; Fontal and Ibáñez 2017; López-Fernández et al. 2021; Moreno-Vera et al. 2022; Morote and Colomer 2021; Yáñez et al. 2022; Yáñez and Gómez-Trigueros 2021, 2022). In this sense, it is essential to introduce an aspect that, despite not being the specific object of this article, is closely linked and is therefore undeniable. We refer to the need for all these types of activities to introduce the necessary tools to reduce the bias in the digital gender gap (Gómez-Trigueros and Yáñez 2021). There is still a pending issue in empowering women teachers in relation to technology, and we must work to make them role models for their students.

7. Conclusions

At the first International Forum on Education and Technology (FIET), possible future trends and challenges related to the role of technologies in connecting formal and non-formal learning were analysed. Some of the main discussions at the forum revolved around how to use digital technologies and how to enhance digital inclusion, digital culture, and the democratization of culture. The following recommendations emerged from the debate (Yáñez et al. 2015): (a) ensuring accessibility and connectivity worldwide; (b) promoting research on open content, data, and practices; (c) developing learning strategies to address formal and informal education, viewing the environment as an extraordinary and irreplaceable learning setting; and (d) accelerating the digitalization of culture in order to ensure access for a larger number of people.

Trainee teachers, therefore, need to develop their skills to design, assess, and present innovative T-L proposals that meet the demands of twenty-first-century schools (Monné and Yáñez 2019).

The proposals we present in this paper provide a response to these specific objectives. All of them promote awareness of heritage education by means of different strategies, and all of them are the final result of a product associated with a didactic unit. All the activities employ a process of inquiry and analysis of evidence offered by different sources, which encourages spatial competency and critical thinking. Emphasis is placed on the need for inquiry processes that allow the construction of evidence-based arguments, thus encouraging critical thinking and competency in democratic citizenship. Finally, all of them promote collaboration with different social agents, especially schools, where on many occasions it was possible to conduct a pilot during the work experience.

A large number of the activities have an important technological component. This type of educational practice initially generates a certain unease amongst students. The CBL and learning by doing methodology involves students starting from the same point, but each individual's process of inquiry, the decisions they take, and the strategies they adopt lead them to arrive at different final products, creative responses to the same challenge. The key is in the tutoring by the teacher, who must support the student at the different crossroads that arise. When students share their results in the classroom, they feel empowered and proud of their creations, especially when, as trainee teachers, they are able to carry them out in primary classrooms.

Author Contributions: Conceptualization, C.Y.d.A. and I.M.G.-T.; Formal analysis, C.Y.d.A. and I.M.G.-T.; Methodology, C.Y.d.A. and I.M.G.-T.; Writing—original draft, C.Y.d.A. and I.M.G.-T.; Writing—review & editing, C.Y.d.A. and I.M.G.-T. All authors have read and agreed to the published version of the manuscript.

Funding: This research was supported by the GRIE (Interdisciplinary Group on Education). University of Andorra. Plaça de la Germandat, 7. AD600 Sant Julià de Lòria. Andorra.

Institutional Review Board Statement: Not applicable.

Informed Consent Statement: Not applicable.

Data Availability Statement: Not applicable.

Acknowledgments: Article written as part of the work performed by the GRIE University of Andorra, Principality of Andorra. This work is connected to the innovative project Redes-I3CE Programme for research in university teaching of the Institute of Education Sciences of the University of Alicante (2021-2022 call). Ref.: 5506 "The digital teaching competence and the digital gap from the interdisciplinary and gender dimension in teacher training with TPACK" This work is led by Isabel María Gómez-Trigueros.

Conflicts of Interest: The authors declare that they have no conflict of interest.

References

Argyris, Chris, and Donald Schön. 1978. *Organizational Learning: A Theory of Action Prospective*. Reading: Addison-Westly. [CrossRef]

Barrows, Howard S. 1992. *The Tutorial Process*. Springfield: Southern Illinois University School of Medicine. Available online: https://bit.ly/3954UsU (accessed on 2 May 2022).

Basque, Josianne, Sylvie Doré, and France Henri. 2000. Facilitating a techno-pedagogical change in higher education: lessons from the SABOIR and AMETIST projects. Paper presented at XVIII Conference of the International Council for Innovation in Higher Education, Québec, QC, Canada, November 5–9. Available online: http://benhur.teluq.uquebec.ca/SPIP/jbasque/squelettes/assets/pdf/BasqueDore_Facilitating_a%20techno-pedagogical_change_in_higher_education.pdf (accessed on 3 May 2022).

Camacho, Mar. 2018. La innnovació en educació. Aspectes clau i reptes de futur. *Anuari de l'Educació de les Illes Balears* 18: 16–29. Available online: https://bit.ly/3kRngAk (accessed on 12 May 2022).

Casanova, Elisabeth, Laura Arias, and Alejandro Egea. 2018. La metodología por proyectos como oportunidad para la introducción de la historia y el patrimonio en las aulas de Educación Infantil. *Contextos Educativos* 22: 79–95. [CrossRef]

CoP (Comunitat de práctica Patrimoni i Escola). 2018. *Guia Per Avaluar el Disseny D'activitats Educatives Patrimonials*. Barcelona: Departament De Cultura Generalitat de Catalunya. Available online: https://bit.ly/3PhrVcK (accessed on 11 May 2022).

Couso, Digna, Mercè Izquierdo, and Cristian Rubilar. 2008. La resolución de problemas. In *Área y estrategias de Investigación en la Didáctica de las Ciencias Experimentales*. Curated by Cristian Merino Rubilar, Adrianna Gómez Galindo and Agustín Adúriz-Bravo. Barcelona: Universitat Autònoma de Barcelona, vol. 1, pp. 37–49. Available online: https://bit.ly/3LYr0LZ (accessed on 3 May 2022).

Cuenca, José María. 2013. La educación patrimonial: líneas de investigación actual y nuevas perspectivas. In *La Educación Patrimonial en la Escuela y el Museo: Investigación y Experiencias*. Edited by Jesús Estepa Giménez. Huelva: Universidad de Huelva. Available online: https://bit.ly/3bG2lyG (accessed on 26 April 2022).

Cuenca, José Mª., and Marta Pérez. 2021. El cuento en la enseñanza del patrimonio: Análisis de propuestas didácticas en Educación Infantil. *REIDICS. Revista de Investigación en Didáctica de las Ciencias Sociales* 8: 243–65. [CrossRef]

Cuenca, José Mª., and Myriam Martín. 2014. *Manual Para el Desarrollo de Proyectos Educativos de Museos*. Gijón: Trea.

Dam, Geer T., and Monique Volman. 2004. Critical thinking as a citizenship competence: Teaching strategies. *Learning and Instruction* 14: 359–79. [CrossRef]

Dewey, John. 1995. *Democracia y educación: Una introducción a la filosofía de la educación*. Madrid: Morata.

Dewey, John. 2002. *Cómo pensamos: Nueva exposición de la relación entre pensamiento y proceso educativo*. Barcelona: Paidós.

Diez, Rocío, Andrea Domínguez, Santiago Ponsoda, and Bárbara Ortuño. 2021. Social Science Pedagogy as a Way of Integrating Sustainable Education and Global Citizenship into the Initial Training of Pre-Primary Teachers. *European Journal of Investigation in Health, Psychology and Education* 11: 975–89. [CrossRef]

Edu Trends. 2016. *Aprendizaje Basado en Retos*. Monterrey: Instituto Tecnológico y de Estudios Superiores de Monterrey, ISBN 978-607-501-430-2. Available online: https://observatorio.tec.mx/edutrendsabr (accessed on 12 April 2022).

Estepa, Jesús, Ana M. Wamba, and Roque Jiménez. 2005. Fundamentos para una enseñanza y difusión del patrimonio desde una perspectiva integradora de las ciencias sociales y experimentales. *Investigación en la Escuela* 56: 19–26. Available online: https://bit.ly/3OFMpLw (accessed on 15 May 2022).

European Commission. 2022. *Commission Communication on a European strategy for universities. Communication from the Commission to the European Parliament, The Council, The European Economics and Social Committee and the Committee of the Regions, on a European strategy for Universities*. {SWD (2022) 6 Final}. Strasbourg: Available online: https://bit.ly/3NyvVDB (accessed on 12 May 2022).

Fletcher, Tina S. 2011. Creative Thinking in Schools: Finding the "Just Right". Challenge for Students. *Gifted Child Today* 34: 37–42. [CrossRef]

Fontal, Olaia, and Álex Ibáñez. 2017. La investigación en Educación Patrimonial. Evolución y estado actual a través del análisis de indicadores de alto impacto. *Revista de Educación* 375: 184–214. [CrossRef]

Fontal, Olaia, Paloma Ballesteros, and María Domingo. 2012. *Actas I Congreso Internacional de Educación patrimonial. Mirando a Europa: Estado de la cuestión y perspectivas de futuro*. Madrid: Instituto del Patrimonio Cultural de España (Ministerio de Educación, Cultura y Deporte) y el Observatorio de Educación Patrimonial en España. Available online: https://bit.ly/3nx1m6v (accessed on 11 May 2022).

Gangopadhyay, Paula. 2014. Time for a perfect storm! In *Building the Future of Education: Museums and the Learning Ecosystem*. Washington, DC: American Alliance of Museums, pp. 21–26. Available online: https://bit.ly/3kSdsWE (accessed on 9 May 2022).

Gaskins, Whitney, Jeffrey Johnson, Cathy Maltbie, and Anant Kukreti. 2015. Changing the Learning Environment in the College of Engineering and Applied Science Using Challenge Based Learning. *International Journal of Engineering Pedagogy (iJEP)* 5: 33–41. [CrossRef]

Gómez-Trigueros, Isabel Mª., and Cristina Yáñez. 2021. The Digital Gender Gap in Teacher Education: The TPACK Framework for the 21st Century. *European Journal of Investigation in Health, Psychology and Education* 11: 1333–49. [CrossRef]

Gómez-Trigueros, I. Mª., Mónica Ruiz-Bañuls, and Delfín Ortega-Sánchez. 2019. Digital Literacy of Teachers in Training: Moving from ICTs (Information and Communication Technologies) to LKTs (Learning and Knowledge Technologies). *Education Sciences* 9: 274. [CrossRef]

Kolb, David A. 1984. *Experiential Learning: Experience as the Source of Learning and Development*. Englewood Cliffs: Prentice-Hall.

Larraz, Virginia, Alexandra Saz, and Nadia Azzouz. 2020. El Modelo Educativo de la Universidad de Andorra: un modelo basado en competencias. Paper presented at 12th Congreso Internacional de Educación Superior "Universidad 2020", La Habana, Cuba, February 12. Available online: https://bit.ly/3FqN8fw (accessed on 18 May 2022).

Larraz, Virginia, Cristina Yáñez, Mercè Gisbert, and Cinta Espuny. 2014. An interdisciplinary study in initial teacher training. *Journal of New Approaches in Educational Research (NAER Journal)* 3: 67–74. Available online: https://bit.ly/3kPhM8S (accessed on 11 April 2022). [CrossRef]

Latorre, Marisol, Vergara Claudia, Morales María Alejandra, Orbeta Alejandra, Escobar Claudia, and Quiroga Marta. 2020. Prácticas de acompañamiento de profesores tutores en carreras de pedagogía: análisis de los dispositivos de aseguramiento y certificación de la calidad de los aprendizajes terminales. *Calidad en la educación* 53: 182–218. [CrossRef]

Lattimer, Heather. 2008. Challenging History: Essential Questions in the Social Studies Classroom. *Social Education* 72: 326–29. Available online: https://bit.ly/3snpnjs (accessed on 12 April 2022).

Lave, Jean, and Etienne Wenger. 1991. *Situated Learning: Legitimate Peripheral Participation*. Cambridge: Cambridge University Press. Available online: https://bit.ly/3MY2afA (accessed on 2 April 2022).

Leijon, Marie, Petri Gudmundsson, Patricia Staaf, and Cecilia Christersson. 2021. Challenge based learning in higher education—A systematic literature review. *Innovations in Education and Teaching International*, 1–10. [CrossRef]

López-Fernández, José A., Silvia Medina, Miguel J. López, and Roberto García-Morís. 2021. Perceptions of Heritage among Students of Early Childhood and Primary Education. *Sustainability* 13: 10636. [CrossRef]

Mishra, Punya, and Matthew J. Koehler. 2006. Technological pedagogical content knowledge: A framework for integrating technology in teacher knowledge. *Teachers College Record* 108: 1017–54. Available online: https://davidshutkin.files.wordpress.com/2014/12/mishra_koehler_tpack.pdf (accessed on 15 March 2022).

Mishra, Punya, and Matthew J. Koehler. 2007. Technological pedagogical content knowledge (TPCK): Confronting the wicked problems of teaching with technology. In *Proceedings of SITE 2007–Society for Information Technology & Teacher Education International Conference*. Edited by Roger Carlsen, Karen McFerrin, Jerry Price, Roberta Weber and Dee Anna Willis. Chesapeake: Association for the Advancement of Computing in Education (AACE), pp. 2214–26.

Monné, Alexandra. 2017. La formació dual al Bàtxelor de Ciències de l'Educació de la Universitat d'Andorra. *Butlletí de la Universitat d'Andorra* 125. Available online: https://bit.ly/3kRpuzx (accessed on 14 April 2022).

Monné, Alexandra, and Cristina Yáñez. 2019. La interculturalitat i el multilingüisme a través del joc. Paper presented at II Congrés Internacional: el joc en la historia. Diversitat lúdica a l'escola i al temps lliure, Universitat Jaume I, Castelló, Spain, November 8–10; pp. 172–18. Available online: https://bit.ly/3yqYxdU (accessed on 2 April 2022).

Moreno-Vera, Juan Ramón, José Antonio López-Fernández, and Santiago Ponsoda-López de Atalaya. 2022. La concepción del patrimonio en el profesorado en formación de Educación Infantil. *Profesorado, Revista De Currículum Y Formación Del Profesorado* 26: 439–58. [CrossRef]

Morote, Álvaro F., and Juan Carlos Colomer. 2021. Analysis of the activities based on ICT resources in Social Science textbooks (Primary Education): An approach to social-environmental issues. *Publicaciones de la Facultad de Educación y Humanidades del Campus de Melilla* 51: 87–137. [CrossRef]

Myers, John. 2004. Assessment and Evaluation in Social Studies Classrooms: A Question of Balance. In *Challenges and Prospects for Canadian Social Studies*. Edited by A. Sears and I. Wright. Vancouver: Pacific Educational Press, chp. 21. pp. 290–301. Available online: https://bit.ly/3MbTS3P (accessed on 14 April 2022).

OECD. 2005. La definición y selección de competencias clave. Resumen ejecutivo. Available online: https://bit.ly/3vUkZdV (accessed on 2 April 2022).

OECD. 2019. OECD Future of Education and Skills 2030: OECD Learning Compass 2030. Available online: https://bit.ly/3LPP3Nf (accessed on 2 April 2022).

Olivares, Silvia Lizethl, Vanessa López Cabrera, and E. Eugenio Valdez-García. 2018. Aprendizaje basado en retos: Una experiencia de innovación para enfrentar problemas de salud pública. *Educación Médica* 19: 230–37. [CrossRef]

Pagès, Joan, and Antoni Santisteban, eds. 2011. Enseñar y aprender ciencias sociales. In *La Didáctica del Conocimiento del Medio Social y Cultural en Educación Primaria. Ciencias Sociales Para Comprender, Pensar y Actuar*. Madrid: Síntesis, pp. 23–40. Available online: https://bit.ly/3FsRW49 (accessed on 11 April 2022).

Peñalba, Alicia, and Juan José Leiva. 2019. La interculturalitat en el context universitari: Necessitats en la formació inicial dels futurs professionals de l'educació. *EDUCAR. Revista Digital de la UAB* 55: 141–58. [CrossRef]

Pérez-Sánchez, Ekin, Fernando Chavarro-Miranda, and Julian D. Riano-Cruz. 2020. Challenge-based learning: A 'entrepreneurship-oriented' teaching experience. In *Management in Education*. London: SAGE Publications. [CrossRef]

Pérez, Yosajandi, and José Antonio Chamizo. 2011. Los museos: un instrumento para el Aprendizaje Basado en Problemas (ABP). *Revista Eureka sobre Enseñanza y Divulgación de las Ciencias* 8: 312–22. [CrossRef]

Perrenoud, Philippe. 2004. Évaluer des compétences. *L'éducateur*. March, pp. 8–11. Available online: https://bit.ly/37pvdcC (accessed on 14 April 2022).

Pla, Florenci. 2015. Nou bàtxelor en ciències de l'educació: modalitat dual. *Butlletí de la Universitat d'Andorra* 111: 4–6. Available online: https://bit.ly/3vTga4k (accessed on 11 April 2022).

Pujolàs, Pere. 2003. *Aprendre junts alumnes diferents*. Vic: Eumo.

Savery, John R. 2015. Overview of problem-based learning: Definitions and distinctions. In *Essential Readings in Problem-Based Learning: Exploring and Extending the Legacy of Howard S. Barrows*. Edited by Andrew Walker, Heather Leary, Cindy Hmelo-Silver and Peggy A. Ertmer. West Lafayette: Purdue University Press, pp. 5–16.

Savery, John R., and Thomas M. Duffy. 1995. Problem Based Learning: An instructional model and its constructivist framework. *Educational Technology* 35: 31–38. Available online: https://bit.ly/3wGojbJ (accessed on 11 April 2022).

Saz, Alexandra, Blanca Carrera, Sònia Gili, and Betlem Sabrià. 2020. El perfil competencial del profesorado universitario. El caso de la Universitat d'Andorra. *Enfoque Disciplinario* 5: 39–53. Available online: https://bit.ly/3kRjGpS (accessed on 12 April 2022).

Schank, Rocher C. 2002. *Designing World-Class E-Learning: How IBM, GE, Harvard Business School, and Columbia University Are Succeding at e-Learning*. Centers for Teaching Excellence-Book Library: p. 190. Available online: https://bit.ly/3lyiYZH (accessed on 2 April 2022).

Shulman, Lee S. 1986. Those who understand: Knowledge growth in teaching. *Educational Researcher* 15: 4–14. [CrossRef]

Yáñez, Cristina. 2017. Bringing cultural heritage into primary school classrooms. Designing a virtual lab. Paper presented at EDULEARN 17 Proceedings. Paper presented at 9th International Conference on Education and New Learning Technologies, Barcelona, Spain, July 3–5; pp. 4129–37. [CrossRef]

Yáñez, Cristina, and Isabel Mª. Gómez-Trigueros. 2021. Experiencia innovadora internacional online para la introducción de la geografía y el patrimonio a través de *Google Earth^{TM}* en la formación inicial del profesorado. *Revista del Congrés Internacional de Docència Universitària i Innovació (CIDUI)* 5: 1–19. Available online: https://raco.cat/index.php/RevistaCIDUI/article/view/387826 (accessed on 12 April 2022).

Yánez, Cristina, and Isabel Mª. Gómez-Trigueros. 2022. *GHPro^{TM}* y la educación en patrimonio: inmersión multidisciplinar en la formación del profesorado para reducir la brecha digital y aprender a interpretar el paisaje cultural a través del modelo TPACK. En Ponsoda, S. (coord.): *Simposio Educación Patrimonial: Propuestas y Potencialidades del Patrimonio Para la Enseñanza-aprendizaje de las Ciencias Sociales*. Rosabel Roig-Vila, and Martines, Vicens. (Coord.). In *Actas—1st International Congress: Education and Knowledge (ICON)*. Alicante: Octaedro, p. 363. Available online: https://octaedro.com/wp-content/uploads/2022/05/9788419312372.pdf (accessed on 21 April 2022).

Yáñez, Cristina, Alexandra Okada, and Ramon Palau. 2015. New learning scenarios for the 21st century related to Education, Culture and Technology. *RUSC. Universities and Knowledge Society Journal* 12: 87–102. [CrossRef]

Yáñez, Cristina, Isabel Gómez-Trigueros, Maria Luisa Rico, and Delfín Ortega. 2022. El reto de trabajar el patrimonio cultural desde una dimensión educativa a través del modelo TPACK y las TIG. In *Llibre d'actes de les XX Jornades de Xarxes d'Investigació en Docència Universitària-XARXES 2022 i V Workshop Internacional d'Innovació en Ensenyament Superior i TIC-INNOVAESTIC 2022*. Edited by Rosana Satorre, Asunción Menargues and Rocio Díez. Alicante: Institut de Ciències de l'Educació, Vicerectorat de Transformació Digital de la Universitat d'Alacante. p. 321. ISBN 978-84-09-39019-9. Available online: https://web.ua.es/es/redes-innovaestic/documentos/llibre-actes-redes-innovaestic-2022.pdf (accessed on 2 April 2022).

social sciences

MDPI

Article

Educating in History: Thinking Historically through Historical Reenactment

José-Manuel González-González *, Jesús-Gerardo Franco-Calvo and Darío Español-Solana

Argos Group, Department of Specific Didactics, Faculty of Education, Universidad de Zaragoza, Calle Pedro Cerbuna, 12, 50009 Zaragoza, Spain; jgfranco@unizar.es (J.-G.F.-C.); despanol@unizar.es (D.E.-S.)
* Correspondence: joseman@unizar.es

Abstract: This paper aimed to identify trends in the scientific literature that relate the link between two concepts: historical thinking and historical reenactment. The definition of both concepts and their commonalities were examined. Convinced that History instruction and Heritage education could improve new methods and techniques, and aware of the benefits of reenactments in active learning and participation in and outside the classroom, we came to the obvious conclusion that merging both aspects is a must and should be disseminated. We also analyzed the presence of second-order concepts in reenactment practices and how they are addressed by actors and spectators. Reenactments foster the acquisition of critical thinking by citizens through education; their quality, however, must be improved through research and didactics—didactics based on reenactment that help us value the past and the traces still present in local areas. Local and global identity and heritage, emotions, reproduction of objects, the use of sources, relevance, empathy, multiperspectives, causation, communication, the relationship between past and present, and the sustainable economy proposed by the 2030 Agenda, are all aspects that should take center stage in turning this phenomenon into a living and lasting history as an experience.

Keywords: historical thinking; historical reenactment; reenactment thinking; history teaching; sustainability; heritage education

Citation: González-González, José-Manuel, Jesús-Gerardo Franco-Calvo, and Darío Español-Solana. 2022. Educating in History: Thinking Historically through Historical Reenactment. *Social Sciences* 11: 256. https://doi.org/10.3390/socsci11060256

Academic Editor: Alvaro Morote

Received: 21 April 2022
Accepted: 6 June 2022
Published: 10 June 2022

Publisher's Note: MDPI stays neutral with regard to jurisdictional claims in published maps and institutional affiliations.

1. Introduction: Historical Reenactment: Definition, Objectives, Characteristics, and Methods

The most widespread definition of historical reenactment—or living history—is the practice of reconstructing uses, customs, material culture, and aspects of the past based on strictly scientific guidelines to achieve objectives related to cultural dissemination and education (Español-Solana 2019c). The adequacy of this tool to provide a historical approach—and, therefore, historiographical research—as a result of experimentation (Español-Solana 2021; d'Oro 2004; De Certau 1984; Stueber 2002) should also be added to the above description. Stueber's theses are based on the philosophical perspective of historical thinking, while D'Oro confronts the theoretical views on historical reenactment held by Davidson, Gadamer, and Collingwood (Retz 2018). Consequently, reconstructing the past to understand it is not only confined to its material culture (Samida 2019), but also to practices, aptitudes, and linguistic uses, among other areas, always with scientific and academic support (Español-Solana 2019b). With this practice, we aim to achieve several objectives:

- show the importance of historical reenactment as an educational tool.
- identify the relation between historical thinking and historical reenactment.
- analyze the presence of second-order concepts in reenactment practices.

Although history can neither be experienced, nor directly observed or replicated, there are alternatives to understand it and feel it as much as possible through live or computer simulations that partially overcome the initial abstraction (Cardona Gómez and Feliu Torruella 2014; Kneebone and Woods 2014).

Historical reenactment, as a kind of reconstruction of History, aimed at researching or educating, implements a process linked to the historical method, and, in most cases, it adds another element as well: knowledge transfer or dissemination. This reenactment could be surrounded by a praxis linked to experimental archaeology (Meylan 2013), to which it is inherently related (Del Barco Díaz 2010). However, historical reenactment is characterized by a univocal inclination toward dissemination and didactics (Cózar Llistó 2013), which is captured in the documentation and reconstruction process of the material culture.

Arcega Morales (2018) underscores eagerness to learn about history, liking for performing arts, and cost issues as reasons that might explain why this phenomenon is becoming increasingly popular. Tourism, financial concerns, the need to know, the search for identity roots in the community (López Cruz and Cuenca López 2014; Corral Lafuente 2019), association involvement, or aims relating to enhancing heritage (Rojas Rabaneda 2019) are some of the goals that may arise when planning activities of this type; however, they should not be considered historical reenactments if the objectives of socializing knowledge or being a research tool are not met (Español-Solana 2019b).

In this respect, the concept of historical reenactment has deteriorated as two completely different models are termed the same way (Español-Solana 2019b; Aquillué Domínguez 2019; Rojas Rabaneda 2019; Español-Solana et al. 2020). On the one hand, there are historical festivals and, on the other, historical reenactments, in the strict sense. Although the difference between both models is known in most of Europe, in Spain, in line with a trend in southern Europe, the distinction has narrowed to the point where the former has adopted the name of the latter. As a result, activities that are actually historical festivals are labeled historical reenactments, possibly in an attempt to make the contents sound more rigorous (Balbás 2019; Rojas Rabaneda 2019). This conceptual inaccuracy involves debatable stances in the opinions seen in part of the literature on the subject. For instance, Jay Anderson analyzed living history museums from the 1890s onwards and identified three types of reenactors, based on their capacity for more or less immersion (Anderson 1984).

As has been mentioned, although historicist evocations such as festivals, markets, processions, or plays may contain heritage and historical elements, these are extremely stereotyped as they are not aimed at education or research and, therefore, they lack scientific methodologies ensuring they are considered as quality scientific or educational praxis. The reason is that they respond to commercial, tourism, or other types of objectives, and do not approach historical facts based on experience or historiography, nor do they construct accurate educational discourses. The disconnect between academia and museology in these historicist manifestations must be understood as a consequence of this disparity of objectives and, therefore, of results.

Nevertheless, the experiential factor can, and must, be subjected to an epistemological examination that, in the worst scenario, would invalidate its effectiveness for research and for thinking historically (Agnew 2004, 2007; Lévesque 2009; Van der Plaetsen 2014; Retz 2018). These folkloric manifestations may also be studied on the basis of the socializing or introductory function of historical knowledge, as well as local and regional identity.

This problem has led organizations to review their programming to try to differentiate themselves from others. In recent years, historical festivals and reenactments have changed from being viewed by specialists as representations of the past aimed at entertaining tourists to becoming models seeking to socialize knowledge on history and heritage sites (Sebares Valle 2017; Español-Solana et al. 2020). This is where we can find the second model that several authors propose (Brædder 2019; Español-Solana 2019a) related to academia, research.

Although collaborations between heritage assets and historical reenactments take place occasionally, all researchers agree that such a union should be essential, as it is one of the ways of enhancing heritage (Egberts 2014; Franco-Calvo et al. 2020; Felices de la Fuente and Hernández Salmerón 2019; Solé 2019). In fact, recent studies advocate the indissolubility between the latter and the reconstruction of the past, backing their arguments with audience studies (Español-Solana and Franco-Calvo 2021a). This enhancement allows the increase of cultural tourism, which optimizes heritage and natural resources, respects local identity and traditions, and generates socio-economic benefits in the community of origin.

For this article, we have followed the PRISMA model. For this state of the art we have used scientific databases and have found these results:

- *Web of Science* (https://www.webofscience.com/wos/woscc/basic-search) (accessed on 1 December 2021):
    - 724 articles for "reenactment", 115 from History and 25 from Art;
    - 22 texts for "historical reenactment";
    - 689 results for "historical thinking", 524 articles, 225 from the field of "educational research", 177 open access;
    - 5 results for "historical", "thinking" and "reenactment" keywords related to our field.
- *ERIC* (https://eric.ed.gov/) (accessed on 9 December 2021):
    - 176 results for "reenactment", 111 journal articles, 26 related to "History instruction";
    - 17 texts for "historical reenactment";
    - 435 results for "historical thinking", 325 journal articles, 150 from the last ten years;
    - 3 results for "historical", "thinking" and "reenactment" keywords.
- *Dialnet* (https://dialnet.unirioja.es/) (accessed on 21 November 2021):
    - 63 documents for "reenactment", 29 from Humanities, 11 from Art studies, and 10 from Social Sciences;
    - 115 for "recreación histórica", 56 from Humanities, 33 from Social Sciences, and 28 from Art. 14 documents for "historical reenactment".
    - 433 results for "pensamiento histórico", 282 journal articles, 121 from Education, and with full texts. 204 results for "historical thinking", 20 with full texts in English, 9 in Portuguese.
    - 56 results for "pensamiento" "histórico" and "recreación", 0 for "historical", "thinking" and "reenactment" keywords.

We finally have collected those more relevant to write this article.

2. Discussion

2.1. *Developing Historical Thinking through Historical Reenactment*

Thinking historically is one way to achieve quality education in history instruction, especially in secondary school. As a key competence in our subject, a priority objective is for pupils to develop critical, contextualized, source- and perspective-based, empathetic, and informed thinking. In essence, it means pupils approach history with a historian's mentality and techniques, not an easy task, even for adults; however, pupils are not expected to become historians, an aspect typical of the university stage. This leap in quality, if this type of reflective thinking is attained, results in in-depth and reasoned learning, in other words, founded on sound judgment.

Thinking historically helps young people attain comprehension, quality knowledge, and competencies (Domínguez Castillo 2015; Gómez and Chapman 2016). Knowing which events were relevant, what has changed and what has not, how the past is reflected in our surroundings, and how certain events influenced the present is crucial for this process. Interpretation and analytical skills are essential in psychosocial maturity.

There are many authors who have managed the study of historical thinking, such as Shemilt (1978; History 13–16, 1980); Lee (History Teaching and Philosophy of History, 1983; 2000; 2005); Seixas (Historical Understanding among Adolescents in a Multicultural Setting, 1993; 2006); Vilar (Pensar históricamente, reflexiones y recuerdos, 1997); Wineburg (Historical Thinking, 2001; Why Learn History, 2018); Barton (Historical Thinking in the Elementary Years: A Review of Current Research, 2004; 2017); Lévesque (Thinking Historically, 2009); or Seixas and Morton (The Big Six, 2013).

As these authors explained, historical thinking comprises several second-order aspects or concepts, or metaconcepts; cognitive skills typical of our discipline rather than repetitive skills. The historical consciousness every individual has and develops is another step to be implemented in teaching-learning processes. Although it was the British, Americans, and Canadians who researched these subjects more in-depth, professionals of other nationalities (German, Dutch, and Spanish) have also conducted studies in this area (Domínguez Castillo 2015; Rodríguez-Medina et al. 2020).

"Yet to understand the history discipline as a whole, Lee (1983) stressed the importance of drawing on procedural history, or second-order historical concepts, in conjunction with substantive history when undertaking historical inquiry" (Shaw 2021, p. 85). Wineburg questioned whether knowledge should form the basis of learning, as it does in the renowned taxonomy pyramid by Bloom, who, although much followed in didactics, is not at all recommendable for the subject of history, since critical thinking may turn this foundation upside down: "Putting knowledge at the base implies that the world of ideas is fully known and that critical thinking means gathering accepted facts in order to render judgement" (Wineburg 2018, p. 92).

Nevertheless, the study of English and Spanish secondary education curricula has determined that, while the 2008 English legislation provided that almost 50% of the content must be based on second-order concepts, in the 2015 Spanish legislation it was only 5%, although in previous Spanish laws historical thinking had more presence (Gómez and Chapman 2016, pp. 437–39). A recent study including the Portuguese curriculum has also underscored that legislative references to historical thinking were far more evident in the English case than in the other two countries analyzed (Santisteban et al. 2021, p. 21). Similarly, in Canada, Haskings-Winner has stated the difficulty of applying historical thinking in classrooms: "Sometimes the process of doing history and thinking historically can get forgotten, or teachers return to what they think history education is: lectures or the dissemination of facts, and, more recently, the use of technologies such as PowerPoint and smartboards to show pictures that reinforce talking points" (Sandwell and von Heyking 2014, p. 284). In Latin America, some countries include some of these meta concepts in their curriculum (Domínguez Castillo 2015, pp. 60–62).

In Spain, several books (Domínguez Castillo 2015), articles, and doctoral theses on history didactics emphasize the interest in competence development of historical thinking; this topic has aroused the most interest among Ph.D. students, as shown by a thorough list recently analyzed (Chaparro Sainz et al. 2020, p. 95). High-impact publications have demonstrated the importance of combining consciousness and historical thinking with the ethical dimension (Cavanna et al. 2021), and countless articles have exhibited an interest in addressing these aspects (Retz 2018; Rodríguez-Medina et al. 2020).

Historical reenactment also provides whoever experiences it with an in-depth understanding of history and facilitates informed and well-understood thinking as it presents a living history (Anderson 1984) that is also scientific, more approachable, and uses images and actions that will aid said reasoning. History has to be interpreted and imagined, and it requires narrators and reenactors. The relationship with nearby familiar elements researched or experienced with this reenactment helps consolidate second-order concepts, as shown in recent studies (Shaw 2021).

The use of historical reenactment in the classroom dates back to the 1960s in the United Kingdom (Retz 2018) and to the 1980s in the United States. Several American scientists highlighted that these activities or events had more impact on learning than dramatization, since they enabled an almost full immersion and empathic understanding of the historical moment being represented (Turner 1985). If performed with the required preparation and time, reenactment ensures pupils gain a broad, in-depth knowledge based on relevant sources and on adopting a historical perspective: "Simpler reenactments can involve students and teachers intellectually. Such intellectual involvement promotes better attention, interest, concern, reaction, and evaluation. [...] This activity can be an essential power for learning history and understanding contemporary social situations" (Turner 1985, p. 221). Turner and other authors in their more recent works advise using appropriate themes for young people that are related to their identity and local or regional heritage; simple, clear events with well-known characters or phenomena that have an impact on their emotions and end satisfaction, and pay attention to the importance of the location, the facilities, the program, the atmosphere or the ambience (López Cruz and Cuenca López 2014; Carneiro et al. 2019).

Nowadays, digitalization and new technologies, such as green screens, have enabled the introduction of reenactments in the classroom (Sheffield and Swan 2012), which is highly recommended for school settings. The process becomes far more GOALeffective if we add auditory immersion to visual immersion, since historically recreating sound may serve as an educational aid, as a primary source and emotional stimulus (Carneiro et al. 2019), by listening to or playing music (Goering and Burenheide 2010).

A recently published monograph on the relationship between violent pasts and reenactments analyzes their popularity and how this relates to the capacities and skills to think or make think historically that can be found in reenactors and spectators, and on the possible link between formal and informal education that come together as a result of these initiatives (Grever and Nieuwenhuyse 2020).

> "In the 1970s, the term 'public history' gradually became acknowledged as a historical sub-discipline. Public history became institutionalized with the founding of graduate programmes, specific journals, and associations. [...]"
>
> "As the observer of the past is always situated in a present which influences his/her view of the past, historical thinking is also about building an understanding of and reflecting on the complex relationship between past, present, and future. [...] In formal history education at secondary schools, historical thinking has increasingly been adopted as the main aim of the school subject. In the history curricula of provinces, states, and countries such as British Columbia (Canada), California (United States), Flanders (Belgium), the Netherlands, Sweden, England, and Finland, historical thinking and historical reasoning occupy the centre stage (see particularly the related chapters in Metzger and Harris 2018)". (Grever and Nieuwenhuyse 2020, pp. 487–89)

Although we need to distinguish between professional and amateur reenactment (Brædder 2019)—or, as already mentioned, between reenactment and evocation or commemoration (Español-Solana et al. 2020)—we cannot ignore that some theorists oppose the idea that reenactments may be an appropriate educational tool for the subject of history, or for true knowledge of our discipline, since they often lead to disappointment, or dilute what is represented (Cook 2004; Handler and Saxton 1988). While no simulation of the past can obviously be perfect, some movies for television and theaters have paid special attention to production and language, and have even consulted and hired expert historians to make reenactments as close to actual events as possible. Nevertheless, according to Cook, they will never be the same, since many emotional, contextual, behavioral, and attitudinal issues are impossible to reproduce: "The challenge is to find a way of illustrating critical engagement with the past in a manner that captures the imagination of a lay audience—an audience that may well be eager for dramatic narrative and impatient with ambiguity and contention" (Cook 2004, p. 495).

For this reason, some authors that have interviewed professional reenactors specialized in several historical stages have found they are inspired by a passion for learning focused on others, on themselves, and on scientific material itself.

> "Against these criticisms, scholars with a more social-constructionist approach have ventured that the essentialist distinction between authentic and inauthentic is untenable [...] Reenactment may have a genuine investigative dimension in which the quest for authenticity is not solely about dramatising an already well-

known past but generating new knowledge through the activity itself (Crang 1996, pp. 419–20; Cook 2004, pp. 487–88)". (Brædder et al. 2017, pp. 172–73)

It is authenticity (Handler and Saxton 1988; Brædder et al. 2017), or its most scientific approximation, that makes these reenactments truly good learning practices and enables them to attain considerable verisimilitude or legitimacy, auctoritas or prestige. In conclusion, it involves more in-depth knowledge than what can typically be taught in the classroom (Wineburg 2018, p. 100).

2.2. *Socializing Knowledge through Historical Reenactment and Its Didactic Use*

Researchers unanimously support the idea that history should be reconstructed with the most rigor—by avoiding anachronism or presentism—, with fixed criteria and clear concepts regarding the actions and materials to be used, and without ruling out the possibility of documenting the reconstructive process based on prior research in line with scientific methods—as occurs with experimental archaeology—to serve as the foundation for the process (Egberts 2014; Del Barco Díaz 2010; Cózar Llistó 2013; Robinson and Yerbury 2015; Español-Solana 2019b; Felices de la Fuente and Hernández Salmerón 2019; Jardón Giner and Pérez Herrero 2019; Corral Lafuente 2019; Agnew and Tomann 2019).

On the one hand, collective memory, knowledge socialization, and the participatory learning of history and heritage are some of the drivers of this discipline and are still hard to find in some events and projects. Although a didactic methodology is not applied at all times, there is a will to educate based on research results (Rojas Rabaneda 2019). On the other, we refer to a didactic component that, when inserted into the training method of new teachers, through the theatrical representation of scenes from the past, offers them the chance to work on new cognitive skills in their pupils, such as stimulating their imagination and creativity, adopting their own identity within the community, or relating and promoting social attitudes and respect for heritage (De Paz Sánchez and Ferreras Listán 2010; Sandwell and von Heyking 2014); in short, adding values related to pupils' comprehensive education. Furthermore, the didactic use of dramatization is unquestionable, by adopting a global conception that can result in a complete historical reenactment. The adoption of drama techniques in the classroom will depend on pupils' interpretive skills. It will be further outlined as it develops into a series of specific forms: symbolic play, representation of roles, and theater (Motos Teruel and Navarro Amorós 2003; Kneebone and Woods 2014). It is no less true, however, that the use, or abuse, of dramatization in non-formal didactics does not encourage interactivity between reenactors and audiences, a key element in the educational process (Español-Solana 2019b).

From this point, we should explore the main methodological principles involved in inserting historical reenactment practices into formal settings (Robinson and Yerbury 2015) using historical thinking as a reference framework at all times (Lévesque 2009). This is crucial for understanding how important a role historical reenactment can play in history and heritage didactics. As Español-Solana (2019b) points out, we have to overcome the reductionism of materiality in order to convey broader paradigms linked to explaining ideas related to the historical time, such as change and continuity or causes and consequences (Barton 2017), among other second-order concepts. In fact, practice needs discourse, explanatory, and operational resources that transgress mere reenactment of material culture, since, otherwise, the educational practice we establish will have a modest impact. Along the same lines, some authors advocate the convenience of using an approach focused on experimentation and historical thinking (Agnew 2004, 2007; Lévesque 2009).

Empathy to put ourselves in the shoes of historical figures and infer their reasons and reactions (Hernàndez Cardona 2001; Retz 2018; Solé 2019) is also indispensable, as is emotion (Carneiro et al. 2019) in both reenactors and spectators (Español-Solana 2019b). The practice of drama, however, involves developing and controlling emotions, consciously exploring feelings and moods, pursuing Goleman's emotional education (Motos Teruel and Navarro Amorós 2003) and allowing the development of intrapersonal intelligence (Gardner 1993).

Teaching for comprehension also helps attain in-depth and significant knowledge (Ishee and Goldhaber 1990). To this end, new content should be linked with subjects' concepts and experiences, and their personal and social memory; this can be achieved by planning reenactment-based methods with useful undercurrents and experiential elements (Jiménez Torregrosa and Rojo Ariza 2014) that awaken pupils' curiosity and improve and expand their cognitive skills. Nevertheless, we cannot ignore that pupils are placed center stage in learning and that they will construct their own rational structures alongside their peers (Motos Teruel and Navarro Amorós 2003). All these aspects are fundamental in order to instill knowledge in pupils using active, constructivist, and discovery learning methods. In fact, historical reenactment—following the hands-on, minds-on, and hearts-on didactic principle—establishes methodologies in which pupils solve problems through research (Cardona Gómez and Feliu Torruella 2014; Rivero Gracia and Pelegrín Campo 2019; Felices de la Fuente and Hernández Salmerón 2019). All this involves thinking processes and is linked to skills such as reasoning, divergence (Hernàndez Cardona 2001), adaptation, and flexibility. The content to be taught needs to be problematized, while avoiding linking concepts or historical data that are not very attractive for learners (Español-Solana 2019b).

Another key factor in this process is creativity (Hernàndez Cardona 2001; Felices de la Fuente and Hernández Salmerón 2019), understood as interacting with the audience rather than relying on closed scripts (Español-Solana 2019b). Although it may prove controversial, some authors defend compatibility between flexibility and rigor (Motos Teruel and Navarro Amorós 2003), since both are needed for good historical reenactment. Pupils may also experience uchronic situations (Hernàndez Cardona 2001), or disciplined historical imagination (Solé 2019), which are perfectly valid in some contexts.

Cooperative work is another strength of this methodology (Robinson and Yerbury 2015). It promotes collective feeling and social interaction. Historical reenactment normally occurs in a group, working on the skill of connecting with other participants, which favors the development of interpersonal intelligence (Gardner 1993). In addition, designing and implementing a reenactment pose a challenge that encourages critical thinking (Thelen 2003).

From the perspective of heritage didactics, including methodologies based on reenactment fosters individuals' relationships with these assets regarding belonging, ownership, and identity (Fontal 2008; López Cruz and Cuenca López 2014), which, in turn, lead to more involvement and respect (Rivero Gracia and Campo 2015).

It is very positive that historical reenactment can be carried out directly on the scene, where it takes place. Heritage sites are ideal for explaining and understanding historical events and periods (Sebares Valle 2017), and contemplation and experimentation play an extremely important role in generating emotions and feelings that help us recognize the asset and identify with it (López Cruz and Cuenca López 2014; Felices de la Fuente and Hernández Salmerón 2019). This interest brings us closer to this heritage so we can overcome that image of opacity typical of cultural assets (Bardavio and González Marcén 2003).

A clear and definitive commitment must be made, by the different administrations and historical spaces, programming historical recreations as a way of socializing knowledge, as a way to improve and respect this local historical heritage. For this, we propose to flee from that mass and structured tourism that generates negative effects at all levels. It is necessary that quality mark the success or failure of an activity of this type, but rather the transmission and democratization of knowledge, in a community eager to obtain it, but seeking economic viability that implies prosperity of the territory and quality employment. For this, quality activities must be designed, with rigor and entertainment that provide the greatest cultural satisfaction to the visitor, trying to involve the local population in the whole process. But for a historical recreation project to be considered sustainable, from a tourism and social point of view, it must also take into account other aspects such as pollution, waste produced, or social discrimination, among many other factors of development settled by the 2030 Agenda. The 17 SDGs (Sustainable Development Goals) marked the way to take action just for education, equality, economic growth, improving communities, and responsible consumption. Governments, companies, and reenactors should lead these civic politics.

2.3. Some Examples of Historical Reenactment Worldwide

Besides this general approach to certain methodological factors, we must consider the advantages this cultural industry currently offers in non-formal settings, which are easy to include in any historical and heritage education program. There are now hundreds of Spanish and international historical reenactment projects combining all the necessary factors to construct an educational event, as outlined above (Figure 1). When planned in advance by teachers, they can be visited by groups of schoolchildren and form part of educational material and practice. Many of these projects are designed to offer a wide range of didactic possibilities, since they are suitable for a variety of audiences with differing educational levels. Some examples are outlined below.

Figure 1. Medieval reenactment in Aragón, Spain.

American Civil War reenactments are quite common. The Battle of Olustee (Florida), for example, is reenacted yearly with specialists and historians coming from all over the country to mark this historical event. One day is completely dedicated to students so they can interview the reenactors. Another example is the Battle of Gettysburg, which has brought together up to 30,000 reenactors. In Virginia (USA) there is a living museum called Colonial Williamsburg. It is a long-term project reproducing an eighteenth-century colonial city where visitors immerse themselves in a didactic space reconstructing all aspects of civil life in this period. It is similar to the British model reproduced at the Jorvik Viking Centre. Jorvik allows visiting the site reconstructed with areas of the old Viking settlement of York. This project has revolutionized the world's panorama of how to reconstruct—down to the last detail—a phenomenon as complex as medieval military architecture, by showing the entire process. The largest and oldest association of reenactors, The Sealed Knot, is also in England. It was founded in 1968 by a group of soldiers who dress in seventeenth-century fashion.

In the Pacific, several maritime reenactments have been held with the help of the Polynesian Voyaging Society consisting in building ships and canoes to sail the ocean in a traditional manner. These reconstructions have had a considerable impact worldwide thanks to the documentaries on the topic and to the fact that, for decades, they were conceived as real research or technological and anthropological studies, despite the controversial views they elicited about their authenticity or the consideration toward native populations (Scanlan 2017).

In Russia, the Battle of Borodino, one of the most important events of Napoleon's French invasion of Russia, which happened in 1812, is reenacted yearly on the first Sunday of September. Another famous reenactment is that of the Battle of Kulikovo, that took place in 1380, which in 2016 held twenty consecutive editions.

In Europe, there are paradigmatic models such as Biskupin in Poland, which is the reconstruction of a settlement on an Iron Age site dating from the eighth century BC. This archaeological park combines activities for the general public through historical reconstruction and experimental archaeology, while it is still studied by archaeologists and historians as part of the dissemination offering. The Eketorp site in Sweden, dating from the fifth century, is similar; in this case, it is a mixed model that reconstructs ways of life and material culture. The reconstruction of the Iberian Citadel of Calafell and the transfer and subsequent reconstruction of the town El Cabo in Andorra, both in Spain, allow walking through streets and homes as they would have been in the pre-Roman period, thereby favoring historical reenactment as a means to spread awareness of the past. These examples fall under "open-air" or "open" models, which, following an increasingly popular philosophy, are pervading many management designs for heritage assets. This approach makes visitors participate in the work of archaeologists, architects, and specialists. These open-air museums include those known as archaeodromes, such as the Archeodromo di Poggibonsi in Italy, linked to the University of Siena. Major battles have also been reenacted in Europe, such as Waterloo (Belgium) and Bailén (Spain), part of the Napoleonic Wars, or the Battle of Hastings (United Kingdom) or the Normandy landings (France), from other periods.

The European Union has financed didactic actions as part of the Socrates program, fostering communication between school and society (De Paz Sánchez and Ferreras Listán 2010, pp. 531–32). In Portugal, the work on Lindoso Castle (Solé 2019) is worth highlighting, and in France, all the educational work around the new Guédelon Castle in Treigny, where, for over 20 years, they have been building ex novo strictly using materials, techniques, and procedures dating from the medieval period (twelfth and thirteenth centuries). The project was conceived as an open-air museum rooted in education and experimental archaeology, even though it is also an incentive to create jobs and a tourist attraction beyond compare; it is now an immense living museum. Several reenactments have been held in Italy for years inspired by both the Roman past and prehistoric cultures. Many projects promoted by the Emilia-Romagna region, and present on its website, such as Antiqva Italia, have tried to encourage a connection and training between involved operators, schools, and groups to implement good reenactment practices.

In Spain (Español-Solana 2019c), the best-known project of this kind is possibly "Open for works" by the Santa Maria Cathedral Foundation, in Vitoria (https://www.catedralvitoria.eus/). This is not a historical reconstruction model *per se*, but it successfully echoes the philosophy of this new dissemination and investigative trend in heritage. There are other important projects that schools are gradually recognizing as valuable means for history and heritage education, for example, Tarraco Viva Festival, which takes place in Tarragona every year, and Emerita Lvdica, in Mérida (Figure 2). The former was devised as an attraction for Tarragona's candidacy to become a World Heritage Site and it is now the city's most important heritage management project and, undoubtedly, the best known and most imitated in Spain. It attracts almost 200,000 visitors every year and its main purpose is to spread awareness of the Roman legacy on the peninsula. In Mérida, over 18 reenactment associations gathered in 2021, involving the local population and attracting many visitors from elsewhere. These are just a few general examples that provide clues on how to tackle any educational project through organizational complexes, mostly managed by local authorities, in which historical reenactment is the primary educational technique.

Figure 2. Tarraco reenactment, Spain.

Peracense Siglo XIII, also in Spain, exhibits the link between historical reenactment and cultural heritage (Figure 3). This interesting reenactment focuses on studying, researching, and recreating the clothing, weapons, and ways of life of the former territory of the Crown of Aragon in the thirteenth century. Other activities are also held in Peracense, which fall under the umbrella of live interpretation (Solé 2019) involving face-to-face contact between reenactors and visitors.

2.4. *The Metaconcepts of History Instruction, the Key for Historical Reenactment*

There are values in history that are sometimes hard for teachers to explain and even harder for pupils to understand (Sandwell and von Heyking 2014). It is usually said that history helps understand the present and avoid repeating past "mistakes", but, in truth, society's thinking about current aims, problems, and challenges is not well-formed. This is one of the reasons why the discipline is now distancing itself from the very society it is meant to serve. Consequently, history needs to be given value "so that students find something truly valuable there that transforms how they think about the world and themselves" (Paricio Royo 2018, p. 233). The benefits of history instruction for society are more than evident with the change in methodology that helps develop critical thinking at the service of democratic citizens. This is not something people acquire naturally as they mature psychologically (Stueber 2002); instead, they need to learn a series of skills that have to be worked on in teaching (Gómez Carrasco et al. 2014; Soria López 2015). There is also an ethical, behavioral, and attitudinal learning that is extremely useful in molding personality and for community social life. The difference between this type of history and history learned by accumulating and memorizing information is significant.

Figure 3. Peracense reenactment, Spain.

Historical thinking allows individuals to approach history independently and construct their own representation of the past as it gives them a series of instruments to analyze, understand, and interpret this history (Rüsen 2005; Santisteban Fernández 2010). It is a creative process limited to interpreting the sources historians produce to create reliable historical narratives. Some concepts are key to developing these narratives: sources, historical relevance, causation, change and continuity, historical perspective, and the ethical dimension of history (Lévesque 2009; Seixas and Morton 2013; Paricio Royo 2020). Soria López (2015) considers that the conceptual proposal put forward by Santisteban, González, and Pagès—although her statement is also applicable to Seixas and Morton's—is a dynamic model consisting of elements that are all crucial and configured in relation to the others, even though it is not a sequence of steps to follow. Citizens need a reflective and critical view of the past so they can be aware of their actions and take responsibility for them. This is far from memorizing dates and historical content, a practice followed by some teachers, which does not entail any reflection or interpretation.

That is why history learning should be focused on research, inquiry, and problems with individuals needing to deal with historical facts and develop cognitive skills and procedures for understanding and explaining them (Cardona Gómez and Feliu Torruella 2014; Paricio Royo 2018). "Unfortunately, schools are not usually designed to acquaint students with such investigative processes. Usually, curriculum structures and instructional patterns are designed to transmit content (which students are expected to remember) or to teach isolated skills (which students are expected to practice)" (Barton 2017, p. 459).

Teachers/historians, let alone their pupils, no longer limit themselves to talking or writing (Plá 2005), even though this has mainly been the way of capturing thinking so far. Today, it is usual for teachers and historians to become involved in multiple practical and multimedia actions enabling pupils or the interested general public to know more and understand better by reflective thinking (Sandwell and von Heyking 2014; Franco-Calvo 2021).

Historical relevance involves reflecting on what and who should be remembered and why and, therefore, what and who should be studied. To this end, we need to analyze the social impact of the process, the importance contemporaries gave to it, how long the consequences lasted and continued, and what it reveals about our present (Seixas and Morton 2013; Rivero Gracia and Pelegrín Campo 2019). When a group decides to recreate a specific historical period, a choice has been made, determined by a variety of reasons. From then on, the group engages in an ongoing exercise of historical relevance: What do we want our reenactment to show? Which type of characters will appear in it? Which content do we want it to have and how? Should we focus on major events or everyday life? The reenactment generates a story of the past in which characters, events, and material culture all play a role in the historical narrative (Rüsen 2005).

"Given that historical reenactment is not a historical source, historical contents must be worked on using sources" (Jiménez Torregrosa and Rojo Ariza 2014, p. 38). The use of historical sources is significantly complex, since we cannot address them if we are incapable of interpreting their partiality or intentionality (Rüsen 2005). For that reason, we must be able to classify the sources based on their types, but also to analyze, evaluate, or compare them with other evidence (Barton 2017) so that we can use them to formulate hypotheses or solutions to the problems we have set ourselves. The rigor required in historical reenactments leads to using these sources to, as far as possible, document the event or activity to be recreated based on the information provided both in written documents and by archaeologists, who establish first-order evidence for reenactors in museums and cultural institutions. In general, there is a shortage of information that archaeology cannot fill. However, we must also consider that countless questions can be asked about these archaeological objects, and these questions are not exclusively concerned with the objects themselves, but also with their context, which includes the people that used them (Egea Vivancos and Arias Ferrer 2018).

Change and continuity are essential concepts shaping history in a constant back and forth, a coherent story about how the world transformed. The aim is not merely to point out changes between two points in history, but to link a narrative that establishes connections so that these changes make sense.

> "An all-encompassing, yet in-depth view of not only change processes but also of diverse and sometimes opposing forces that have led to these changes, and of resistance and continuity factors, as well as the meaning of these series of changes and the very nature of change in history". (Paricio Royo 2018, p. 236)

As Paricio Royo (2018) also points out, this involves selecting the most relevant processes, but from a dynamic stance in which changes condition future actions, thus fostering new historical transformation processes with a display of forces that try to promote or slow down said processes. But we must also explore how this change manifests itself and how it is perceived by contemporaries, while analyzing everything remaining from the past and shaping the present, straddling both realities. We cannot ignore that all this contributes to gaining historical perspective—historians' distance from events—when attempting to explain the whys and wherefores of the past, the causes and consequences, and how events were viewed by their contemporaries (Barton 2017; Paricio Royo 2020).

A historical reenactment is an effective tool for appreciating changes in everyday objects that, by using similarities and differences, allow us to place their owners in time and space (Egea Vivancos and Arias Ferrer 2018). We need historical consciousness as it will help us assess and interpret these changes and continuities while promoting understanding of concepts such as periodization or simultaneity (Rüsen 2005). Historical consciousness is the consciousness of time as it is based on the relationships we establish between the present and the past (Santisteban Fernández 2010). Acquiring historical consciousness means accepting our situation in the world at a specific moment in the development of history we find ourselves in, where, based on that dynamic conception, we accept how fleeting our existence and way of life are (Paricio Royo 2018; Retz 2018).

The main instrument used in memorizing history is the unidirectional narrative. To escape this encyclopedic view, we have to break away from that use of narrative and take advantage of it to make the past understandable, while tackling problems linked to thinking historically (Rüsen 2005; Soria López 2015). Designing a historical narrative, in this case for historical reenactment, is no easy task if our aim is to develop historical thinking. Constructing a historical narrative involves using sources to produce a story that must be structured on an explanation of causes and consequences, which naturally involves seeking those causal connections. Why do events happen and what are their consequences?

The concept of historical empathy refers to the capacity to understand the attitudes and motives of past figures (González Monfort et al. 2009; Endacott and Brooks 2018; Paricio Royo 2019). Empathic understanding allows us to imbue other ways of life, experiences, norms, or past belief systems with meaning by putting ourselves in someone else's shoes and thereby explaining their decisions. There is a debate on whether emotional involvement is necessary or not in historical empathy exercises (Doñate Campos and Sarria 2019): defenders of improvement in the capacity to understand the emotional component against those remarking that this emotional involvement is not a form of historical reasoning (Paricio Royo 2019). Consequently, some confusion arises between the ambiguity of the terms empathy and sympathy. When conducting a rational study of the past, with no affective identification with the protagonists, or even understanding of events from an emotional standpoint, establishing the concept of historical perspective-taking was deemed necessary (Barton 2017; Paricio Royo 2020). This involves recognizing our own attitudes, beliefs, and intentions in a different historical and cultural context. We need to be able to differentiate between the past and the present by establishing a temporal distance that helps us recognize the reasons behind the actions (González Monfort et al. 2009), thereby overcoming presentism. Therefore, although empathy allows us to work on historical imagination, for it to have historical value, it must go hand in hand with contextualization (Santisteban Fernández 2010), with sources and evidence as its only starting point (Barton 2017; Paricio Royo 2019), and without forgetting that it is a process, as described by Foster, Ashby, Lee, and, more recently, others (Endacott and Brooks 2018, pp. 206, 209).

Based on these arguments we can ask ourselves: Is this not the aim of a historical reenactment? Of course it is. History can be learned by developing historical empathy through role-playing and hot-seating activities (Paricio Royo 2019), simulation (Hernàndez Cardona 2001; Kneebone and Woods 2014), recreating processes, or designing a reenactment (Rivero Gracia and Campo 2015), connecting with the more human part of history (Fontal 2008; Cardona Gómez and Feliu Torruella 2014). The more precise the historical reenactment is—approaching conditions similar to those in the past—the more empathic the experience will be. Can we really put ourselves in another's shoes? Can we feel what the protagonists of the events felt? This is one of the major problems, since empathy does not distinguish between cognitive and emotional elements. The answer to the above questions would obviously be "no" as this technique does not enable us to return to the past, but rather to interpret it (Rüsen 2005; Solé 2019). Although we are momentarily replicating an event, the society we live in is different from the one at that time, which is why our responses to stimuli will be too. Feeling safe despite the possible risks our actions may entail, and the possible presence of an audience, prevent us from thinking and acting as they did in the past, which distorts our reactions. We are limited by our way of understanding the world and we relate with others based on our own system of beliefs (Jiménez Torregrosa and Rojo Ariza 2014). But, without a doubt, historical reenactment is the best way of trying to experience history as it clarifies many of the questions we may have (Español-Solana and Franco-Calvo 2021b), always from a rational standpoint, although without excluding emotions and experiences (Paricio Royo 2019).

The ethical dimension, or historical consciousness, involves the present having its origin and acquiring significance in past eras, with societies in constant flux and individuals playing a role, so the past is part of that individual (Gómez Carrasco et al. 2014; Retz 2018). Can history help us experience the present? Historical reenactment can perform an extremely important function in historical consciousness by highlighting the changes that occurred in the past, showing how contemporaries might have experienced them, and the consequences of their action or inaction, so that we can accept the historicity of the world and our own reality in the present and future possibilities.

Some historical reenactments not only focus on the ways of life of the elite or events such as battles that are represented, everyday life is also shown, by approaching mentalities, culture, and economic activities (Felices de la Fuente and Hernández Salmerón 2019).

If the basic elements of historical thinking consist of considering historical problems based on their relevance and impact on societies, with historical consciousness and historical reasoning founded on a critical use of sources (Retz 2018; Gómez Carrasco et al. 2014), duly contextualized historical reenactment can undoubtedly help develop the typical skills of historical thinking (Felices de la Fuente and Hernández Salmerón 2019). A good reenactor will implement what we can call reenactment thinking, which involves a "reenactment method" (Cózar Llistó 2013; Español-Solana and Franco-Calvo 2021b) and a way of working that leads us to develop historical thinking. Taking the steps that Jiménez Torregrosa and Rojo Ariza (2014) establish for historical reenactment as a reference, we can observe the important relationship between the key concepts established to achieve historical thinking. Consequently, historical relevance determines the contextualization steps for the episode, including the selection of the historical moment and the event to be recreated, and the choice of characters to be protagonists and secondary figures. Concepts such as change, continuity, and causation are present in the timing stages and in the choice of setting and propos within the narrative to establish, in which historical time will also play an essential role. Historical perspective and historical consciousness are connected in order to answer questions that help understand historical empathy such as: How did the protagonists feel? How would we feel and behave in similar situations? How do the decisions made impact the present?

3. Final Reflections and Conclusions

As mentioned above, the concept of historical thinking has been one of the newest and most interesting approaches in history instruction since the end of the twentieth century. Its origins are in the English-speaking world, in works by authors such as Dickinson, Lee, and Denis Shemilt. Later, it continued in the United States (Wineburg and VanSledright) and in Canada (Seixas and Lévesque), with slight variations between them. Interesting theoretical approaches are also being produced in Spain by Domínguez, Santisteban, Gómez, and Paricio, who have helped its use become more widespread.

The possibilities of historical reenactment in formal and non-formal education are immense. Its use involves applying methods that we could call new but have been around in practice for years. Historical reenactment, either observed or designed by pupils, allows them to develop a type of thinking that we have termed reenactment thinking, which, in turn, helps them develop historical and critical thinking.

This way of working on and analyzing history as a professional team leads to significant, contextualized, active, and dynamic learning, set in heritage sites, and it creates democratic citizens who opt to appreciate history and past culture for the values they contribute to society.

Consequently, as they are experiential, historical reenactments are becoming an effective didactic tool at historical monuments, and they are fostering increased interest at universities due to their accurate portrayals, alongside other characteristics such as the motivation, emotion, and the quality they achieve or their contribution to disseminating heritage. The tourism that takes place in these heritage spaces can have a huge economic, sociocultural and environmental impact, so it is necessary to establish a sustainable tourism that is compatible with the precepts established by the 2030 Agenda. We believe that historical reenactment can, perfectly, meet those criteria established with the 17 SDGs (specifically

04, Quality Education; 05, Gender Equality; 08, Decent Work and Economic Growth; 11, Sustainable Cities and Communities; & 12 Responsible Consumption and Production) and become a good practice for our communities and students.

In short, a good historical reenactor will be able to evaluate the relevance of events and the characters involved in them; manage to put themselves into past figures' shoes to understand their motives and actions; analyze the causes and consequences of the events; gain a perspective of the facts by considering alternatives to what happened—all in meticulous detail through a historical narrative created using sources and evidence and based on the research of material and intellectual culture. From these reenacted visions, the public will learn history more completely, more effectively, and more lastingly. There is no denying the contribution reenactments make to our discipline: teaching and learning History.

Author Contributions: All authors have contributed equally to the writing of the article. All authors have read and agreed to the published version of the manuscript.

Funding: Investigation Group S50_20R: ARGOS, Aragón Government, Spain. Grupo de Investigación S50_20R: ARGOS, Gobierno de Aragón, España.

Institutional Review Board Statement: Not applicable.

Informed Consent Statement: Not applicable.

Acknowledgments: Article written as part of the work performed by Argos Group, from the IUCA Institute of Research, University of Zaragoza, Spain.

Conflicts of Interest: The authors declare no conflict of interest.

References

Agnew, Vanessa. 2004. Introduction: What is reenactment? *Criticism* 46: 327–39. [CrossRef]

Agnew, Vanessa. 2007. History's affective turn: Historical reenactment and its work in the present. *Rethinking History* 11: 299–312. [CrossRef]

Agnew, Vanessa, and Juliane Tomann. 2019. Authenticity. In *The Routledge Handbook of Reenactment Studies*. Edited by Vanessa Agnew, Jonathan Lamb and J. Tomann. Abingdon: Routledge, Taylor & Francis, chp. 3. [CrossRef]

Anderson, J. 1984. *Times Machines: The World of Living History*. Nashville: Amer Assn for State & Local.

Aquillué Domínguez, Daniel. 2019. La recreación histórica y las asociaciones culturales de recreación histórica en Aragón. In *El Recreacionismo Histórico, el Patrimonio y la Arqueología Como Motores del Turismo en el Territorio*. Edited by Marisancho S. Menjón, E. Val, I. Lasobras and A. Celma. Zaragoza: Diputación Provincial de Zaragoza, pp. 45–56.

Arcega Morales, Jesús Ángel. 2018. Las recreaciones históricas en Aragón. Un teatro en auge. *Sigma* 27: 147–70. [CrossRef]

Balbás, Yeyo. 2019. Recreación histórica del alto Medievo: Esclareciendo una época oscura. *HER&MUS* 20: 70–84.

Bardavio, Antoni, and Paloma González Marcén. 2003. *Objetos en el Tiempo. Las Fuentes Materiales en la Enseñanza de las Ciencias Sociales*. Barcelona: Horsoi Editorial.

Barton, Keith. C. 2017. Shared Principles in History and Social Science Education. In *Palgrave Handbook of Research in Historical Culture and Education*. London: Palgrave Macmillan, pp. 449–67. [CrossRef]

Brædder, Anne. 2019. Expertise and amateurism. In *The Routledge Handbook of Reenactment Studies*. Edited by Vanessa Agnew, Jonathan Lamb and J. Tomann. Abingdon: Routledge, Taylor & Francis, chp. 14.

Brædder, Anne, Kim Esmark, Tove Kruse, Carsten Tage Nielsen, and Anette Warring. 2017. Doing pasts: Authenticity from the reenactors' perspective. *Rethinking History* 21: 171–92. [CrossRef]

Cardona Gómez, Gemma, and María Feliu Torruella. 2014. Arqueología, vivencia y comprensión del pasado. *Iber: Didáctica de las Ciencias Sociales, Geografía e Historia* 78: 15–25.

Carneiro, Maria João, Celeste Eusébio, Ana Caldeira, and Ana Cláudia Santos. 2019. The influence of eventscape on emotions, satisfaction and loyalty: The case of re-enactment events. *International Journal of Hospitality Management* 82: 112–24. [CrossRef]

Cavanna, Floriano, Laura Jiménez Martínez, and Noé Valtierra Pereiro. 2021. Reconstrucción histórica. Algunas experiencias en historia pública y educación reglada. In *Recreación Histórica y Didáctica del Patrimonio. Nuevos Horizontes para un Cambio de Modelo en la Difusión del Pasado*. Edited by Dario Español-Solana and Jesús G. Franco-Calvo. Gijón: Trea, pp. 30–70.

Chaparro Sainz, Álvaro, María del Mar Felices de la Fuente, and Laura Triviño Cabrera. 2020. La investigación en pensamiento histórico. Un estudio a través de las tesis doctorales de Ciencias Sociales (1995–2020). *Panta Rei. Revista digital de Historia y Didáctica de la Historia* 14: 93–147. [CrossRef]

Cook, Alexander. 2004. The Use and Abuse of Historical Reenactment: Thoughts on Recent Trends in Public History. *Criticism* 46: 487–96. [CrossRef]

Corral Lafuente, José Luis. 2019. Origen del recreacionismo histórico y el rigor como esencia de la consolidación. In *El recreacionismo Histórico, el Patrimonio y la Arqueología Como Motores del Turismo en el Territorio*. Edited by Marisancho S. Menjón, E. Val, I. Lasobras and A. Celma. Zaragoza: Diputación Provincial de Zaragoza, pp. 27–33.

Cózar Llistó, Guillermo. 2013. La recreación histórica en España. Definición, caracterización y perspectivas de aplicación. *Glyphos Revista de Arqueología* 2: 6–28.

De Certau, Michel. 1984. *The Practice of Everyday Life*. Berkley: University of California Press.

Del Barco Díaz, Miguel. 2010. La recreación histórica como medio para la divulgación de la historia. In *La Divulgación de la Historia y Otros Estudios Sobre Extremadura*. Edited by Félix Iñesta Mena. Llerena: Sociedad Extremeña de Historia, pp. 243–54.

De Paz Sánchez, José Juan, and Mario Ferreras Listán. 2010. La recreación histórica en el proceso de enseñanza-aprendizaje en las ciencias sociales: Metodología, buenas prácticas y desarrollo profesional. In *Metodología de Investigación en Didáctica de las Ciencias Sociales*. Edited by Rosa M. Ávila, M. P. Rivero and Pedro L. Domínguez. Zaragoza: Institución Fernando el Católico, pp. 525–33.

Domínguez Castillo, José. 2015. *Pensamiento Histórico y Evaluación de Competencias*. Barcelona: Graó.

Doñate Campos, Olga, and Carmen Ferrete Sarria. 2019. Vivir la Historia: Posibilidades de la empatía histórica para motivar al alumnado y lograr la comprensión efectiva de los hechos históricos. *Didáctica de las Ciencias Experimentales y Sociales* 36: 47–60. [CrossRef]

d'Oro, Giuseppina. 2004. Re-Enactment and Radical Interpretation. *History and Theory* 43: 198–208. [CrossRef]

Egberts, Linde. 2014. Battlefield of Histories. In *Companion to European Heritage Revivals*. Cham: Springer, pp. 53–70.

Egea Vivancos, Alejandro, and Laura Arias Ferrer. 2018. El desafío de enseñar a pensar históricamente a través de la arqueología. In *Y la Arqueología Llegó al Aula*. Edited by Alejandro Egea, Laura Arias and Joan Santacana. Gijón: Ediciones TREA, pp. 329–43.

Endacott, Jason L., and Sarah Brooks. 2018. Historical Empathy: Perspectives and Responding to the Past. In *The Wiley International Handbook of History Teaching and Learning*. Edited by Scott Alan Metzger and Lauren McArthur Harris. Medford: Wiley, pp. 203–25.

Español-Solana, Darío. 2019a. Los distintos modelos de recreación histórica; hacia la consolidación de una industria cultural en el siglo XXI. In *El Recreacionismo Histórico, el Patrimonio y la Arqueología Como Motores del Turismo en el Territorio*. Edited by Marisancho S. Menjón, E. Val, I. Lasobras and A. Celma. Zaragoza: Diputación Provincial de Zaragoza, pp. 11–26.

Español-Solana, Darío. 2019b. Historia para todos: Recreación histórica, didáctica y democratización del conocimiento. *HER&MUS* 20: 7–23.

Español-Solana, Darío. 2019c. New perspectives for the dissemination of medieval history: Re-enactment in southern Europe, a view from the perspective of didactics. *Imago Temporis. Medium Aevum* 13: 333–59.

Español-Solana, Darío. 2021. Historia y Cultura Militar Durante la Expansión Feudal en el valle del Ebro, Siglos XI y XII. Presupuestos Metodológicos para una Didáctica de la Guerra en la Edad Media. Ph.D. thesis, Universidad de Zaragoza, Zaragoza, Spain.

Español-Solana, Darío, and Jesús Gerardo Franco-Calvo. 2021a. Education and Heritage of Medieval Warfare. A Study on the Transmission of Knowledge by Informal Educators in Defensive Spaces. *Education Sciences* 11: 320. [CrossRef]

Español-Solana, Darío, and Jesús Gerardo Franco-Calvo. 2021b. *Recreación Histórica y Didáctica del Patrimonio. Nuevos Horizontes Para un Cambio de Modelo en la Difusión del Pasado*. Gijón: TREA.

Español-Solana, Darío, Jesús Gerardo Franco-Calvo, and José-Manuel González-González. 2020. Recreaciones y conmemoraciones históricas, diferencias y posibilidades didácticas desde Aragón (España). *Didattica della Storia* 2: 413–26.

Felices de la Fuente, María del. Mar, and Julia Hernández Salmerón. 2019. La recreación histórica como recurso didáctico: Usos y propuestas para el aula. *HER&MUS* 20: 7–23.

Fontal, Olaia. 2008. La importancia de la dimensión humana en la didáctica del patrimonio. In *La comunicación Global del patrimonio Cultural*. Edited by S. Mateos. Gijón: TREA, pp. 53–109.

Franco-Calvo, Jesús Gerardo, Antonio Hernández Pardos, and Jesús Javier Jambrina Campos. 2020. Una forma didáctica de acercarnos al patrimonio: La recreación histórica «Peracense siglo XIII». *HER&MUS* 20: 85–101.

Franco-Calvo, Jesús Gerardo. 2021. Pensar históricamente a través de la recreación histórica. El caso del castillo de Peracense. In *Recreación Histórica y Didáctica del Patrimonio. Nuevos Horizontes para un Cambio de Modelo en la Difusión del Pasado*. Edited by Darío Español-Solana and Jesús Gerardo Franco-Calvo. Gijón: Trea, pp. 175–203.

Gardner, Howard. 1993. *Multiple Intelligences: The Theory in Practice*. Nueva York: Basic Books.

Goering, Christian Z., and Bradley J. Burenheide. 2010. Exploring the Role of Music in Secondary English and History Classrooms through Personal Practical Theory. *SRATE Journal* 19: 44–51.

Gómez Carrasco, Cosme Jesús, Jorge Ortuño Molina, and Sebastián Molina Puche. 2014. Aprender a pensar históricamente. Retos para la historia en el siglo XXI. *Revista Tempo e Argumento* 6: 5–27. [CrossRef]

Gómez, Cosme Jesús, and Arthur Chapman. 2016. Pensamiento histórico y evaluación de competencias en el currículum inglés y español. Un estudio comparativo. In *Actas del VII Simposio Internacional de Didáctica de las Ciencias Sociales en el ámbito Iberoamericano*. Santiago de Compostela: Universidad de Santiago de Compostela, pp. 433–45.

González Monfort, Neus, Rodrigo Henriquez Vazquez, Joan Pagès Blanch, and Antoni Santisteban Fernández. 2009. El aprendizaje de la empatía histórica (EH) en educación secundaria: Análisis y proyecciones de una investigación sobre la enseñanza y el aprendizaje del conflicto y la convivencia en la Edad Media. In *Un progetto Educativo per la «Strategia di Lisbona*. Edited by Rosa M. Ávila Ruiz, Beatrice Borghi and Ivo Mattozzi. Bologna: Pàtron Editore, pp. 283–90.

Grever, Maria, and Karel Van Nieuwenhuyse. 2020. Popular uses of violent pasts and historical thinking. *Journal for the Study of Education and Development* 43: 483–502. [CrossRef]

Handler, Richard, and William Saxton. 1988. Dyssimulation: Reflexivity, Narrative, and the Quest for Authenticity in "Living History". *Cultural Anthropology* 3: 242–60. [CrossRef]

Hernàndez Cardona, Francisco Xavier. 2001. Los juegos de simulación y la didáctica de la historia. *Íber. Didáctica de las Ciencias Sociales, Geografía e Historia* 30: 23–36.

Ishee, Nell, and Jeanne Goldhaber. 1990. Story Re-Enactment: Let the Play Begin! *Young Children* 45: 70–75.

Jardón Giner, Paula, and Clara Isabel Pérez Herrero. 2019. La reconstrucción dramatizada en espacios arqueológicos: Interacciones en yacimientos valencianos. *HER&MUS* 20: 24–38.

Jiménez Torregrosa, Lorena, and María del Carmen Rojo Ariza. 2014. Recreación histórica y didáctica. *Iber: Didáctica de las Ciencias Sociales, Geografía e Historia* 78: 35–43.

Kneebone, Roger, and Abigail Woods. 2014. Recapturing the history of surgical practice through simulation-based re-enactment. *Medical History* 58: 106–21. [CrossRef]

Lévesque, Stephane. 2009. *Thinking Historically: Educating Students for the Twenty-First Century*. Toronto: University of Toronto Press.

López Cruz, Inmaculada, and José María Cuenca López. 2014. El patrimonio y las personas: Símbolos e identidad cultural como elementos claves para la educación. In *Reflexionar Desde las Experiencias. Una visión Complementaria Entre España, Francia y Brasil. Actas del II Congreso Internacional de Educación Patrimonial*. Edited by Olaia Fontal, Alex Ibáñez and Lorenzo Martín. Madrid: IPCE/OEPE, pp. 1161–71.

Meylan, Karine. 2013. From research to mediation. A perspective for experimental archaeology. *Experimentelle Archäologie in Europa Bilanz* 12: 171–81.

Motos Teruel, Tomás, and Antoni Navarro Amorós. 2003. El paper de la dramatització en el currículum. *Articles: Revista de Didáctica de la Llengua i de la Literatura* 29: 10–28.

Paricio Royo, Javier. 2018. El valor de la historia. Estudio de alternativas curriculares en Secundaria (1): Una visión integrada de las transformaciones (cambio/continuidad) que conducen y modelan el presente (conciencia histórica). *Clío: History and History Teaching* 4: 232–47.

Paricio Royo, Javier. 2019. El valor de la historia. Estudio de alternativas curriculares en Secundaria (2): Aprender sobre la alteridad y la naturaleza humana a través de la empatía o toma de perspectiva histórica. *Clío: History and History Teaching* 45: 330–56.

Paricio Royo, Javier. 2020. El valor de la historia. Estudio de alternativas curriculares en Secundaria (3): Aprender a pensar desde múltiples perspectivas. *Clío: History and History Teaching* 46: 202–32. [CrossRef]

Plá, Sebastián. 2005. *Aprender a Pensar Históricamente. La Escritura de la Historia en el Bachillerato*. México: Plaza y Valdés.

Retz, Tyson. 2018. *Empathy and History: Historical Understanding in Re-Enactment, Hermeneutics and Education*. New York City: Berghahn.

Rivero Gracia, María Pilar, and Julián Pelegrín Campo. 2015. Aprender historia desde la empatía. Experiencias en Aragón. *Aula de Innovación Educativa* 240: 18–22.

Rivero Gracia, María Pilar, and Julián Pelegrín Campo. 2019. ¿Qué Historia consideran relevante los futuros docentes de Educación Infantil? *Cadernos de Pesquisa* 49: 96–120.

Robinson, Jessica, and Hillary Yerbury. 2015. Re-enactment and its information practices; tensions between the individual and the collective. *Journal of Documentation* 71: 591–608. [CrossRef]

Rodríguez-Medina, Jairo, Cosme Jesús Gómez-Carrasco, Ramón López-Facal, and Pedro Miralles-Martínez. 2020. Tendencias emergentes en la producción académica de educación histórica. *Revista de Educación* 389: 211–42. [CrossRef]

Rojas Rabaneda, Antonio. 2019. La recreación histórica en Cataluña como recurso de la socialización del conocimiento. *HER&MUS* 20: 123–47.

Rüsen, Jörn. 2005. *History: Narration, Interpretation, Orientation*. Nueva York: Berghahn Books.

Samida, Stephanie. 2019. Material culture. In *The Routledge Handbook of Reenactment Studies*. Edited by Vanessa Agnew, Jonathan Lamb and J. Tomann. Abingdon: Routledge, Taylor & Francis, chp. 26.

Sandwell, Ruth, and Amy von Heyking. 2014. *Becoming a History Teacher: Sustaining Practices in Historical Thinking and Knowing*. Toronto: University of Toronto Press.

Santisteban, Antoni, Alfredo Gomes, and Edda Sant. 2021. El currículum de historia en Inglaterra, Portugal y España: Contextos diferentes y problemas comunes. *Educar em Revista* 37: 1–24. [CrossRef]

Santisteban Fernández, Antoni. 2010. La formación de competencias de pensamiento histórico. *Clío & Asociados* 14: 34–56.

Scanlan, Emma. 2017. Reimagining National Identity through Reenactment in the Pacific and Australia. *Wasafiri* 32: 60–67. [CrossRef]

Sebares Valle, Gemma. 2017. Recreación histórica y educación: El caso de Tarragona como gran espacio patrimonial. *Iber: Didáctica de las Ciencias Sociales, Geografía e Historia* 89: 66–71.

Seixas, Peter, and Tom Morton. 2013. *The Big Six Historical Thinking Concepts*. Toronto: Nelson College Indigenous.

Shaw, Emma. 2021. Historical thinking and family historians: Renovating the house of history? *Historical Encounters* 8: 83–96. [CrossRef]

Sheffield, Caroline C., and Stephen B. Swan. 2012. Digital Reenactments: Using Green Screen Technology to Recreate the Past. *Social Education* 76: 92–95.

Solé, Gloria. 2019. A «história ao vivo»: Recriação histórica de uma feira medieval no castelo de Lindoso em Portugal. *HER&MUS* 20: 102–22.

Soria López, Gabriela Margarita. 2015. El pensamiento histórico en la educación primaria: Estudio de casos a partir de narraciones históricas. *Enseñanza de las Ciencias Sociales* 14: 83–95.

Stueber, Karsten R. 2002. The psychological basis of historical explanation: Reenactment, simulation, and the fusion of horizons. *History and Theory* 41: 25–42. [CrossRef]

Thelen, David. 2003. Learning from the past: Individual experience and re-enactment. *The Indiana Magazine of History* 99: 155–65.

Turner, Thomas N. 1985. Historical Reenactment—Can it Work as a Standard Tool of the Social Studies? *The Social Studies* 76: 220–23. [CrossRef]

Van der Plaetsen, Peter. 2014. When the Past Comes to Life. In *Companion to European Heritage Revivals*. Edited by Linde Egberts and Koos Bosma. Cham: Springer, pp. 151–67.

Wineburg, Sam. 2018. *Why Learn History (When It's Already on Your Phone)*. Chicago: University of Chicago Press.

social sciences

MDPI

Article

What Do School Children Know about Climate Change? A Social Sciences Approach

Álvaro-Francisco Morote [1,*] and María Hernández [2]

1 Department of Experimental and Social Sciences Education, Faculty Teaching Training, University of Valencia, 46022 Valencia, Spain
2 Department of Regional Geographical Analysis and Physical Geography, University of Alicante, 03080 Alicante, Spain; maria.hernandez@ua.es
* Correspondence: alvaro.morote@uv.es

Abstract: One of the subject areas that is currently most prominent in the field of education (Social Science) is climate change, given its implications for raising awareness and training the present and future society. The objectives of this study, focused on school children (Primary Education—10 to 12 years old; third cycle, Secondary Education—12–16 years old; and pre-university, Baccalaureate—17–18 years old) in the Region of Valencia (Spain), are to analyse the following: the main information channels through which children receive information on climate change; the causes and consequences that they identify with respect to this phenomenon; and the main greenhouse gas that they believe is in the atmosphere. Based on the 575 students surveyed during the academic year 2020–2021, the results indicate that the three main information media are digital (TV—82.8%, Internet—56.2% and social networks—49.4%). With respect to the causes of the phenomenon identified by the students, particularly noteworthy was pollution (70.1%) and, in terms of the effects, the increase and changes in temperature (61.7%) were of particular note. Finally, with reference to greenhouse gases, the majority responded CO_2 (63.5%). This is incorrect, as the main greenhouse gas in the atmosphere is water vapour. To sum up, we can highlight the role played by schools in training the future society and the risk arising from an increase in the information received from digital media by children as they grow older, due to the danger of misinformation.

Keywords: climate change; school; media; Social Sciences; Geography

Citation: Morote, Álvaro-Francisco, and María Hernández. 2022. What Do School Children Know about Climate Change? A Social Sciences Approach. *Social Sciences* 11: 179. https://doi.org/10.3390/socsci11040179

Academic Editors: Helen F. Ladd and Nigel Parton

Received: 13 January 2022
Accepted: 11 April 2022
Published: 13 April 2022

1. Introduction

One of the main challenges faced by the society of the twenty-first century is global warming, as expressed in the Sixth Report of the Intergovernmental Panel on Climate Change (IPCC 2022). This report highlights the importance of adapting to this phenomenon due to its effects, such as the increase in temperature, the loss of climate comfort or the intensification and greater frequency of extreme weather phenomena (floods and droughts). Therefore, raising awareness and rigorously teaching about climate change is in everyone's interest, but is of particular concern for the educational environment (Jeong et al. 2021; Masters 2020; Nelles and Serrer 2020; Verlie and Blom 2021), as, in part, the future of society depends on increasing our knowledge about this serious problem (Borhaug 2021; Ferrari et al. 2019; Kurup et al. 2021; Morote and Olcina 2021a). The latter authors consider that the challenge of addressing global warming and climate change in education (Social Sciences) should be focused on by including informed decision making in classroom practices.

Regarding the research problem addressed here (teaching and climate change), for decades, different authors (Benejam 1997; Fien 1992) have described the urgent need to discuss world problems in schools (including climate change issues) through the Didactics of Social Sciences, in its different scales. Furthermore, the content of climate change is among of the most controversial that must be explained in social science classes (Ho and

Seow 2015; Morote et al. 2021b). Thus, Benejam (1997) explains that "training our students as citizens of a democratic and alternative system" (p. 47) is necessary, among other purposes, "to preserve and value the natural and cultural heritage that we have received as legacy" (p. 48). As many authors affirm, this is based on the inclusion of Relevant Social Problems (RSP) in social science classrooms, a methodology that comes from the Anglo-Saxon sphere (Fien 1992; Evans et al. 1996). Although in Francophone literature, as López and Oller (2019) explain, the same approach is described, it is called Socially Alive Issues (SAI) (Legardez 2006; Legardez and Simonneaux 2006; Tutiaux-Guillón 2011). In Spain, this approach has also been incorporated from different research (Benejam 1997; García, Francisco, and Rafael Porlán 2000; García and Alba 2003; Pagès and Santisteban 2011; Pagès 2007; López 2011; Canals and González 2011; Santisteban et al. 2014; Díaz and Felices 2017).

With respect to the interest in teaching about climate change, some authors (Morote and Olcina 2020, 2021a; Rausell et al. 2021) have pointed out that education is one of the most important non-structural factors for adapting to and addressing this phenomenon. However, these researchers also indicate that it is one of the variables that is least frequently taken into account, with priority given to structural and political mitigation measures. The importance of climate change has also been indicated by different international bodies. In its Fifth Report (IPCC 2014), the IPCC pointed out that education was one of the main actions necessary for adapting societies to this phenomenon. In addition, the United Nations (UN 2015) identifies this variable "education" as one of the most important elements in mitigating the effects of climate change (Sustainable Development Goal n°13 "Climate action"). Similarly, the European Environmental Agency (EEA 2017) indicated the importance of the vulnerability factor, highlighting education as a way to mitigate the effects of global warming. This issue has also been a point of interest with respect to educational laws, on both an international and a national (Spain) level (Morote and Olcina 2021b). Hence, in Spain (the present study case), the Climate Change Act was published (May 2021), which, for the first time, included a section on teaching: Title VIII ("Education, Research and Innovation in the fight against climate change and energy transition").

Initially, the issue of environmental and climate change constituted a controversial topic from a scientific point of view, as the extent to which human action altered the natural conditions of climate evolution was questioned (Ho and Seow 2015). Currently, the majority of researchers back these facts with evidence (Roussel and Cutter-Mackenzie-Knowles 2020) and it has now become a conceptual topic taught in the subjects of Geography and Social Sciences (Morote and Olcina 2020, 2021a). In Spain, explanations for climate change carry more weight in the educational setting due to the requirement to address these topics in classrooms, as established both for Primary Education (Royal Decree 126/2014, 28 February; Social Sciences subject) and Secondary Education and the Baccalaureate (Royal Decree 1105/2014, 26 December; Geography subject).

Another issue to consider with respect to the teaching of climate change is the influence that the media has on both teachers' and students' understanding of the issue (Morote et al. 2021a; Wu and Otsuka 2021). One of the consequences of this is the dissemination of fake news and stereotypes that have arisen in the social representation of the youngest cohorts (Ferrari et al. 2019). In fact, Morote et al. (2021a) have analysed how the information that the majority of trainee teachers receive is drawn from the Internet, television and social networks. Kažys (2018), Brisman (2018) and Lutzke et al. (2021) explain the danger that this could have in the case of information that is unreliable or lacks rigour, or when the news is false or manipulated. This fact (false information lacking in rigour) has even been transferred to the contents in Social Science textbooks, as confirmed by Morote and Olcina (2020), where there is: (1) a predominance of information with scientific errors; and (2) an excessive influence of the media and a catastrophic view of climate change. In the case of these resources (school textbooks), this problem is even more serious since, currently, in Social Sciences classes, they continue to be the main resource used (Bel et al. 2019). This is also reflected in the social attitudes of future teachers with respect to this phenomenon, in which human action is viewed as the main cause and natural disasters its main effects (Morote 2020). As this author explains, it is true that the majority of trainee teachers have received training on this subject matter, but only 13.4% of this training is based on academic studies. Therefore, as indicated by Morote et al. (2021a), the information received is similar to that presented by the media (86.2%) and constitutes a superfluous approach to the phenomenon, which fosters the creation of stereotypes.

In Spain, there is a consolidated line of study related to the teaching of climate change in the educational sphere (Caride and Meira 2019; Escoz-Roldan et al. 2020) and also in the field of the natural sciences (Calixto 2015; Domènech 2014). However, the same cannot be said for Social Science and/or Geography. It is true that these sciences have produced extensive literature on the teaching of climatology (Martínez-Fernández and Olcina 2019; Morote and Moltó 2017; Sebastiá and Tonda 2018), but not on climate change, except for some recent publications. In this respect, over the last few years, recent studies have been carried out on the teaching of this phenomenon from three perspectives: (1) the social attitudes of trainee teachers (Morote and Hernández 2020; Morote and Moreno 2021; Morote et al. 2021a); (2) the analysis of the contents of school textbooks (Morote and Olcina 2020, 2021a); and (3) didactic proposals (Morote and Olcina 2021b). In fact, the former president of the Spanish Association of Geography expressed the need for a greater dedication and interest in this subject matter similar to that of other scientific fields and on the international scene (Olcina 2017). On an international level, many works have been published on the teaching of climate change, such as those in Central and North America (McWhirter and Shealy 2018; Li et al. 2021; Sezen-Barrie and Marbach-Ad 2021), South America (Da Rocha et al. 2020), Europe (Jeong et al. 2021; Kovacs et al. 2017; Kurup et al. 2021), Africa (Anyanwu and Grange 2017) and Asia and Oceania (Ahmad and Numan 2015; Li and Liu 2021; Verlie and Blom 2021).

The objectives of this research are to analyse the following, using a case study of Primary Education (10–12 years old; third cycle), Secondary Education (12–16 years old) and Baccalaureate (17–18 years old) students of the Region of Valencia (Spain): (1) the main information channels through which students receive information on climate change; (2) the causes and consequences of this phenomenon that are identified by the students and whether these involve differences between these three educational levels; and (3) the main greenhouse gas that students believe is present in the atmosphere. With respect to the starting hypothesis, it was believed that the students would claim to receive information about this phenomenon mainly through audio-visual media (TV, Internet, social networks). Regarding the causes, it was expected that the students would respond that they are related to human action (mainly pollution), while the effects cited by the students were expected be the increase in temperature, sea level, glacier melting, etc. (second hypothesis). In terms of the main greenhouse gas present in the atmosphere, it was expected that the response of the majority of the students would be CO_2 (third hypothesis). Significant differences were expected to be found between the three educational stages. The cognitive age of the students, a priori, was expected to influence the answers (fourth hypothesis). For this reason, we would expect the responses of the Baccalaureate students to be more elaborate and self-critical and with fewer errors in their content (for example, in relation to greenhouse gases). This study, therefore, will help to reveal the social attitudes of school children with respect to climate change and whether their understanding coincides with that of their teachers and with the information that is printed in their school textbooks. These latter considerations are addressed in the Discussion section.

2. Methods

2.1. Design of the Research

This study is based on a correlational study (non-experimental). Regarding the type of didactic research, it is characterized by its presentation of a socio-critical approach (see López and Oller 2019). As numerous authors affirm (Benejam 1997; Díaz and Felices 2017), these works are based on the inclusion of Relevant Social Problems (RSP) in Social Science classrooms. This methodology originated in the Anglo-Saxon sphere (Fien 1992; Evans et al. 1996; Ochoa 1996). It has also been developed in the Francophone literature, although it is called Socially Alive Issues (SAI) in this field (Legardez 2006; Legardez and Simonneaux 2006; Tutiaux-Guillón 2011). The research design is explanatory and transversal, as the information analysed was gathered at a specific moment (academic year of 2020–2021) and refers to a case study (eight public educational centres in the Region of Valencia, Spain): four centres of Primary Education and four centres of Secondary Education and Baccalaureate.

2.2. *Context and Respondents*

With regard to the context and the respondents, the selection procedure was conducted through non-probability sampling (availability or convenience sampling). The participants in this study were Primary Education students (third cycle; fifth and sixth; 10–12 years old), Secondary Education students (first and third; 12–16 years old) and Baccalaureate students (second year of Baccalaureate; 17–18 years old). The total number of students enrolled in these years was 605: Primary Education (n = 180), Secondary Education (n = 300) and Baccalaureate (n = 125). With respect to the representativeness of the sample and taking into account the total number of students enrolled (n = 605), a minimum of 318 students would be required to obtain a representative sample so as to achieve a confidence interval of 99% and a margin of error of 5%. Finally, since the total number of respondents was 575, a representative number was achieved (see Table 1). With respect to the socio-demographic characteristics (gender and age), the figures are similar from the point of view of gender: male (45.7%; n = 263); female (53.7%; n = 309). The average age in the whole sample was 13.8 years.

Table 1. Students who took part in the research.

Educational Stages	Enrolled (n)	Respondents (n)	Average Age	Gender
Primary Education	180	176	11.0	Male (55.7%; n = 98); Female (43.8%; n = 77)
Secondary Education	300	285	13.4	Male (44.6%; n = 127); Female (55.1%; n = 157)
Baccalaureate	125	114	17.2	Male (33.8%; n = 38); Female (65.8%; n = 75)
Total	**605**	**575**	**13.8**	**Male (45.7%; n = 263);** **Female (53.7%; n = 309)**

Source: Survey results. Own elaboration.

2.3. *Questionnaire*

The instrument designed to carry out the research was based on a questionnaire to obtain the data needed to achieve the aims proposed. It was a questionnaire prepared expressly for this research, following the model of other works about social attitudes (see López and Oller 2019; Morote and Hernández 2020; Morote and Moreno 2021; Morote et al. 2021a, 2021b). In this study, this questionnaire was adapted to the non-university school stage (12-item questionnaire—see Appendix A). The questionnaire was also validated by three researchers from: the Department of Experimental and Social Sciences Education at the University of Valencia (Spain), the Department of Mathematics and Social Sciences Education at the University of Murcia (Spain) and the Department of Regional Geographic Analysis and Physical Geography of the University of Alicante (Spain). For this study (in accordance with the aims proposed), the results were obtained from parts 1 (Item 4) and 2 (Items 7, 8 and 9) (see Table 2).

Table 2. Items of the questionnaire used for this research.

Part 1. Training on Climate Change	
Item (n°)	**Response Type/Variable**
Item 4. Of the following information media, choose the top three where you receive information on climate change:	Closed-ended question: Do not know/Do not answer/Family/Social Networks/TV/Newspapers/Radio/Internet/School-High School
Part 2. Perception of Climate Change	
Item (n°)	**Response Type/Variable**
Item 7. What are the main causes of climate change?	Open question
Item 8. What are the main consequences of climate change?	Open question
Item 9. What is the main greenhouse gas in the atmosphere?	Closed-ended question: Do not know/Do not answer/Methane (CH_4)/Ozone (O_3)/Carbon Dioxide (CO_2)/Water Vapour (H_2O)/Nitrous Oxide (NOx)

Source: own elaboration.

Item 9, "What is the main greenhouse gas in the atmosphere?", was included because most of the population is unaware that this is the main greenhouse gas. The prevailing idea, influenced by the media (Morote et al. 2021a) and also replicated in the school textbooks (Morote and Olcina 2020) is that the main greenhouse gas is CO_2, which is incorrect.

In order to assess the construct validity of the questionnaire, several procedures were carried out. First, a statistical analysis of the ordinal variables (Items 11 and 12) was conducted. For these variables, it was found that there was an acceptable standard deviation, as the value obtained was between 0 > 1. Second, the construct was subjected to the Kaiser–Meyer–Olkin (KMO) validity test, which indicates whether the factor analysis of the instrument is acceptable. The KMO test gave a positive result of 0.376, which, according to other factor reliability studies, is considered to be an acceptable level (Pérez-Gil et al. 2000). Third, being a mixed questionnaire (quantitative and qualitative), the Friedman chi-squared test (Friedman's X^2) was carried out. It generated a positive value of 303.067 (p = 0.001), which indicated that there was no discrepancy between the variables. Therefore, the variables were dependent on one another (Satorra and Bentler 2010; Sharpe 2015). The results obtained through these procedures render the research reliable, as in the case of other studies on the teaching of Social Science (Moreno-Vera et al. 2020; Morote et al. 2021b).

2.4. Procedure

The questionnaire was administered in a mid-term session (first four-month period) during the second semester (2021), with a response time of 10 minutes. It should also be noted that the questionnaire was administered prior to the teaching of sessions related to climate and natural hazards so as not to influence the answers (subject of Social Science and/or Geography). Finally, the respondents' anonymity was preserved during the entire procedure and confidentiality was guaranteed in writing.

2.5. Data Analysis

For the data analysis procedure, the program SPSS v26 (IBM, New York, NY, USA) was used to carry out a statistical–inferential analysis (non-parametric tests) of the frequencies and percentages. In the data analysis, the chi-squared test was carried out for the nominal variables (Items 4, 7, 8 and 9). Furthermore, the school stage and the opinions in the open responses (qualitative information) of Items 7 and 8 were coded (see Tables 3 and 4).

Table 3. Coding of answers to Item 7.

Response Type	Code
Do not know/Do not answer	0
Human factor	1
Pollution	2
Deforestation	3
Use of plastics	4
Overexploitation of resources	5
Natural causes	6
Mistakes	7

Source: own elaboration. Note: in "mistakes", the answers in which the students confused "causes" with "consequences" were grouped.

Table 4. Coding of answers to Item 8.

Response Type	Code
Do not know/Do not answer	0
Temperature increase	1
Rising sea level	2
Melting	3
Increase of natural hazards	4
More diseases	5
Extinction of species	6
None	7
Mistakes	8

Source: own elaboration. Note: (1) for "temperature increase", the responses related to sudden climate changes or in seasons, etc., were grouped together; (2) in "mistakes", the answers in which the students confused "causes" with "consequences" were grouped.

3. Results

3.1. *Where Do School Children Receive Information about Climate Change from?*

In the first item analysed (Item 4, "Of the following information media, choose the top three where you receive information on climate change"), the sources of information about climate global warming are examined. The overall data (a total of 1725 responses) reveal that the main media are digital (TV, Internet and social networks). Of the 575 participants, 82.8% (n = 475) responded that they receive information from television, while 56.2% (n = 323) responded that they receive information from the Internet and 49.4% (n = 284) from social networks (Table 5).

Table 5. Information media from where school students receive information on climate change (Item 4).

Information Media	n	%
Do not know/Do not answer	78	13.6
Family	215	37.4
Social Networks	284	49.4
TV	475	82.6
Newspapers	44	7.7
Radio	50	8.7
Internet	323	56.2
School–High School	256	44.5

Note: the % refers to the total number of respondents (n = 575).

When analysing these media in accordance with the stage of education (Primary Education, Secondary Education and Baccalaureate), significant differences may be observed. With a total of 528 responses, the three main media for Primary Education students (10–12 years old) are: (1) TV (80.7%; n = 142); (2) school (54.0%; n = 95); and (3) Internet

(47.2%; $n = 83$) (Figure 1). In Primary Education, the Chi-Squared test revealed that the association between these two variables (information media and schooling period) was significant (Pearson's Chi-Squared = 62.243; $p = 0.001$). They were associated significantly ($p < 0.05$), so they were dependent variables.

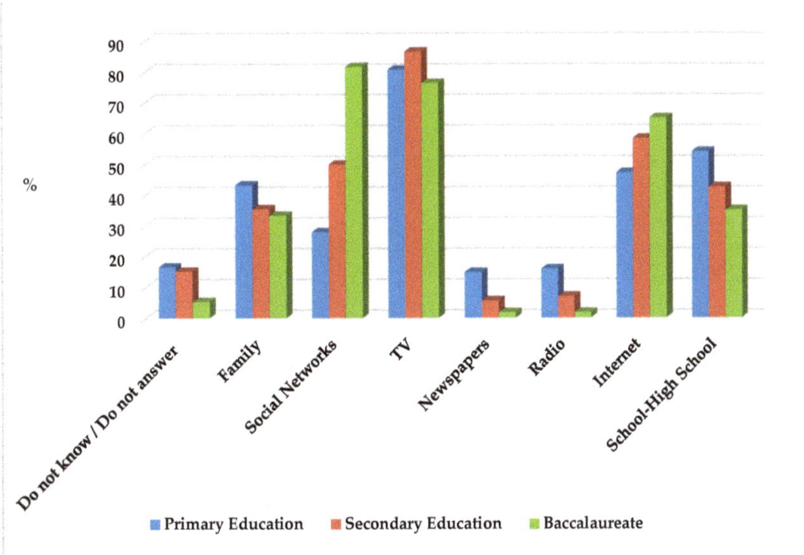

Figure 1. Channels of information through which the school children received information on climate change. Source: survey results. Own elaboration.

For the Secondary Education students (12–16 years old), with a total of 855 responses, the main media were: (1) TV (86.5%; $n = 246$); (2) Internet (58.2%; $n = 166$); and (3) social networks (49.8%; $n = 142$) (see Figure 1). The Chi-Squared test reveals that the association between these two variables (information media and schooling period) is significant (Pearson's Chi-Squared = 50.976; $p = 0.001$). There was a significant association between them ($p < 0.05$), so they were dependent variables.

With respect to the Baccalaureate students (17–18 years old), with a total of 342 responses, the main media were: (1) social networks (81.6%; $n = 93$); (2) TV (76.3%; $n = 87$); and (3) Internet (64.1%; $n = 74$). For this educational stage, the Chi-Squared test revealed that the association between these two variables (information media and school period) was significant (Pearson's Chi-Squared = 44.390; $p = 0.001$). Since they were associated significantly ($p < 0.05$), they were dependent variables.

The differences (in percentages) in the information sources used depending on the educational stage reveal that, as students progresses through the educational process, information from social networks and the Internet gain more relevance compared with the information received from school (Figure 1). In this respect, the response "school centre" is only relevant in the Primary Education stage. The responses related to schools fell from 54.0% (second position) in Primary Education to 42.5% (fourth position) in secondary education and to 35.1% among Baccalaureate students (fourth position). Meanwhile, social networks increased in relevance from 27.8% in Primary Education to 49.8% in Secondary Education and 81.6% in Baccalaureate. A similar trend can be observed for the responses related to the Internet: in Primary Education it represented 47.2% as opposed to 64.9% in the Baccalaureate. It should also be noted that Primary Education was the stage when the "family" variable had the greatest relevance (43.2%).

3.2. What Are the Causes and Consequences of Climate Change Identified by the Students?

Second, the causes (Item 7) and consequences (Item 8) of climate change according to the opinion of the school children were analysed. This serves to establish relationships between what they know and perceive about this phenomenon. With respect to the causes, if we take into account the overall data (575 responses), the results revealed that the main cause of this phenomenon was perceived thought to be pollution (70.1%; $n = 403$) (see Table 6) and, second, "no responses" (9.9%; $n = 57$). The Chi-Squared test revealed that the association between these two variables (causes and school period) was significant (Pearson's Chi-Squared= 48.516; $p = 0.001$). Since the association between them was significant ($p < 0.05$), they were dependent variables.

Table 6. "What are the main causes of climate change?" (Item 7).

Educational Stage		0	1	2	3	4	5	6	7	Total
Primary Education	*n*	24	4	108	6	11	1	9	13	176
	%	13.6%	2.3%	61.4%	3.4%	6.3%	0.6%	5.1%	7.4%	100.0%
Secondary Education	*n*	28	13	205	3	8	1	6	21	285
	%	9.8%	4.6%	71.9%	1.1%	2.8%	0.4%	2.1%	7.4%	100.0%
Baccalaureate	*n*	5	15	90	1	2	0	1	0	114
	%	4.4%	13.2%	78.9%	0.9%	1.8%	0.0%	0.9%	0.0%	100.0%
Total	*n*	**57**	**32**	**403**	**10**	**21**	**2**	**16**	**34**	**575**
	%	**9.9%**	**5.6%**	**70.1%**	**1.7%**	**3.7%**	**0.3%**	**2.8%**	**5.9%**	**100.0%**

Source: survey results. Own elaboration. Note: Do not know/Do not answer (0)/Human factor (1)/Pollution (2)/Deforestation (3)/Use of plastics (4)/Overexploitation of resources (5)/Natural causes (6)/Mistakes (7).

When these data were analysed in accordance with the educational stages, it was observed that the percentage for the item "pollution" increased as the age of the students increased, while that of "no responses" decreased. Some of the responses of the students on these causes ("pollution") were: "car pollution" (student n°57 in Primary Education); "fuel, smoke, rubbish" (student n°65 in Secondary Education); and "excessive use of transport" (student n°14 in Baccalaureate). Another distinctive factor was that, in the case of the Baccalaureate students, the second category of responses was related to "human action" (13.2%; $n = 15$) as: "consumerism" (student n°1 in Baccalaureate); "lack of awareness" (student n°68 in Baccalaureate); "low level of social awareness" (student n°104 in Baccalaureate). These responses were related to a greater sense of responsibility and a more critical perspective of the impact of society on the environment.

With respect to the consequences, if the overall data are considered (575 responses), the participants indicated that the main effects of climate change were the increase and changes in temperature (61.7%; $n = 355$) (see Table 7). The Chi-Squared test revealed that the association between these two variables (effects and school period) was significant (Pearson's Chi-Squared = 55.004; $p = 0.001$). Since the association between them was significant ($p < 0.05$), they were dependent variables. Similarly to the causes, the item "Nr/Dk" appeared (11.5%; $n = 66$). This percentage decreased as the age of the students increased and there was an increase in the percentage of responses related to the increase in temperature (in Secondary Education) and "melting" (in Baccalaureate). Among the Primary Education students, these factors represented 50.0% ($n = 88$), as opposed to the 63.9% ($n = 182$) in Secondary Education and 74.6% ($n = 85$) in Baccalaureate. Some of the opinions on temperatures were: "it is getting increasingly hotter; in Alicante, in winter, it is practically spring" (student n°2 in Primary Education); "it is 30° in winter" (student n°163 in Primary Education); "in winter it is warmer than before" (student n°176 in Primary Education); "in summer, sometimes it can get cold and in winter hot" (student n°74 in secondary education); "in summer the temperatures go above 40°" (student n°96 in secondary education); "the weather is crazy (student n°132 in Secondary Education); "strange temperatures" (student n°243 in Secondary Education); "the temperatures vary a lot from one day to the next" (student n°33 in Baccalaureate); "there is hardly a spring or autumn any more" (student n°80 in Baccalaureate).

Table 7. "What are the main consequences of climate change?" (Item 8).

Educational Stages		0	1	2	3	4	5	6	7	8	Total
Primary Education	*n*	25	88	7	9	16	3	2	7	19	176
	%	14.2%	50.0%	4.0%	5.1%	9.1%	1.7%	1.1%	4.0%	10.8%	100.0%
Secondary Education	*n*	37	182	7	24	14	2	4	3	12	285
	%	13.0%	63.9%	2.5%	8.4%	4.9%	0.7%	1.4%	1.1%	4.2%	100.0%
Baccalaureate	*n*	4	85	2	17	2	0	2	0	2	114
	%	3.5%	74.6%	1.8%	14.9%	1.8%	0.0%	1.8%	0.0%	1.8%	100.0%
Total	*n*	**66**	**355**	**16**	**50**	**32**	**5**	**8**	**10**	**33**	**575**
	%	**11.5%**	**61.7%**	**2.8%**	**8.7%**	**5.6%**	**0.9%**	**1.4%**	**1.7%**	**5.8%**	**100.0%**

Source: survey results. Own elaboration. Note: Do not know/Do not answer (0)/Temperature increase (1)/Rising sea level (2)/Melting (3)/Increase of natural hazards (4)/More diseases (5)/Extinction of species (6)/None (7)/Mistakes (8).

3.3. *What Is the Main Greenhouse Gas in the Atmosphere, According to the Students?*

Third, Item 9 ("What is the main greenhouse gas in the atmosphere?") was analysed. The overall results obtained indicate that the majority of the responses were related to carbon dioxide (CO_2) (63.5%; $n = 365$) and, second, albeit with considerably lower figures, the ozone (O_3) (14.8%; $n = 85$) (Figure 2). The Chi-Squared test revealed that the association between these two variables (greenhouse gases and school period) was significant (Pearson's Chi-Squared= 30.354; $p = 0.001$). Since the association between them was significant ($p < 0.05$), they were dependent variables.

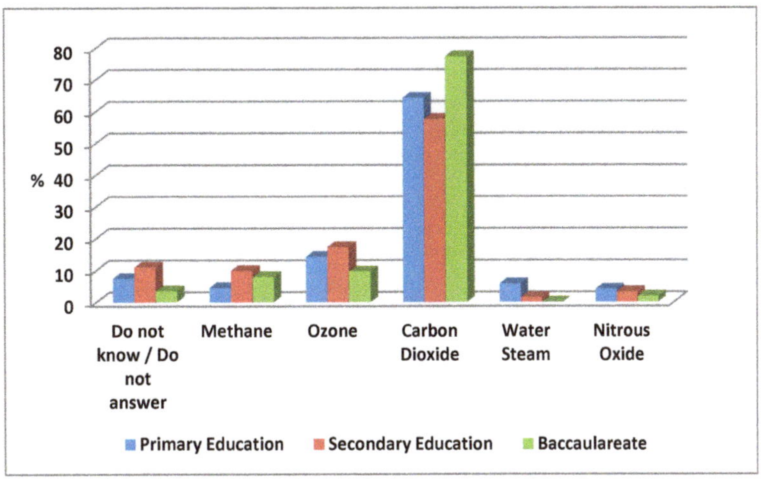

Figure 2. "What is the main greenhouse gas in the atmosphere?" (Item 9). Source: survey results. Own elaboration.

These data were directly related to the causes identified as "pollution" and were, therefore, expected. However, it should be noted that the majority of the students did not respond correctly. The question in Item 9 is related to the main greenhouse gas in the atmosphere (the correct answer is water vapour). This is different from asking about "the main greenhouse gas that is accelerating the climate change process due to anthropic causes" (the correct answer would be carbon dioxide). The correct response would be water vapour, which represents 4% of greenhouse gases, as opposed to CO_2, which accounts for 0.04%. This is important to point out as we expected that as the children progressed through the educational stages, the correct response would be higher. However, this was not the case, as no students in the Baccalaureate stage marked water vapour. By contrast, the Primary Education students (although the data were still low with respect to the main answer) are those who gave the highest percentage of correct answers (water vapour), 5.7% ($n = 10$) (see Figure 2). Moreover, a relationship could be established between the responses obtained and the information received from the information media. In this respect, we can confirm that a higher percentage of the students who received more information from school responded correctly to this question. Nevertheless, overall, the data are worrying, due to the low percentage of correct answers in all of the educational stages.

4. Discussion

The results obtained in this study highlight the importance that social networks and digital media acquire as the ages of students increase with respect to information on climate change. The first hypothesis established that "students receive information about this phenomenon mainly through audio-visual media (TV, Internet, social networks)". With respect to the media, this hypothesis was fulfilled. The response regarding "school centre" is only relevant in the Primary Education stage. In this respect, the responses related to school centres decreased from 54.0% in the primary stage to 42.5% in Secondary Sducation and to 35.1% in Baccalaureate. Meanwhile, social networks increased in relevance from 27.8% in Primary Education to 49.8% in Secondary Education and 81.6% in Baccalaureate. A similar trend can be observed for the responses related to the Internet: in Primary Education, it represented 47.2%, as opposed to 64.9% in Baccalaureate. Therefore, as students progress through the educational stages, the school centre loses its relevance and digital media gains in influence (TV, Internet and social networks), which entails risk.

Taking into account these results, we should ask whether the data obtained in this study coincide with the understanding of trainee teachers and the contents of school textbooks (the main resource used in Social Science classrooms) (Bel et al. 2019). For example, in the study by Morote (2020), the data indicated that future Primary Education teachers receive this information mainly from digital media (68.2%), distributed as follows: social networks (28.7%), television (23.1%) and Internet (16.2%). The responses related to academic studies (university training) accounted for only 13.4%. Furthermore, according to Morote et al. (2021a), 86.2% of the information about climate change received by Primary Education trainee teachers is derived from the media. In another study on Secondary Education and Baccalaureate teachers (comparing University of Valencia, UV and Murcia, UMU, Spain), the percentages of information received from digital media were 67.7% (UV) and 54.2% (UMU) (Morote and Moreno 2021). Furthermore, in Chile (University of Concepción), a predominance of the media (52.0%) was found in the social attitudes of students in three different academic fields (science and mathematics, biology and social education) (Parra et al. 2013).

Therefore, the training of future teachers is based on digital media, which leads to a low level of development in generic and transversal competencies in these topics, as acknowledged by Parra et al. (2013). Other authors also indicate the risk that could ensue if the majority of this information is drawn from media, characterised by the consumption of untruthful information that sometimes lacks scientific rigour and is overly sensationalist (Allen et al. 2018; Brisman 2018; Kažys 2018; Lutzke et al. 2021). Morote (2020) explains that the media also provide truthful information, although the majority report information that lacks rigour in order to attract audiences, using false news and/or apocalyptic headlines. This author has also researched the perceptions of this type of news by trainee teachers. The results reveal that 56.2% agree that this type of news has a manipulative objective and/or presents false information: "highly agree" (25.3%) and "agree" (30.9%). However, there is a high percentage of participants who are "indifferent" to this type of news (29.5%). Other studies have also found that the training of active teachers is deficient. A study carried out by Gallego and Castro (2020) among university lecturers (Colombia) to identify the understanding acquired in the teacher training process regarding climate change reveals that there are gaps in this training. Specifically, trainee teachers have a vague idea about the scientific model of climate change and their preparation in this area is the fruit of the popularisation of the topic, which, in many cases, is obtained from television programmes and websites. Therefore, they contribute to the transmission of implicit theories of climate change and, moreover, conceptual errors which, due to their role as teachers, will be transmitted to the future generations.

This deficient training may lead teachers to use school textbooks excessively (Morote and Souto 2020). This has been confirmed in subjects characterised by activities based on memorisation and the reproduction of content at a low cognitive level, as in the case of Geography (Kidman 2018). Furthermore, this teaching resource also lacks scientific rigour (Morote and Olcina 2020). This could mean that this topic is not taught in the classroom or is taught very loosely (Olcina 2017). In relation to climate change, the most common mistakes include the almost complete absence of discussion of the human influence (vulnerability) when referring, for example, to natural hazards (Morote and Olcina 2020). This is fundamental, as acknowledged in different reports on the effects of climate change (IPCC 2022), since it is necessary to address the vulnerability factor to adapt to this phenomenon (EEA 2017).

The second hypothesis (causes: "it is expected that the students will respond that they are related to human action –mainly pollution –"; consequences: "would be the increase in temperature, sea level, glacier melting, etc."), was confirmed. The results revealed that the phenomenon of climate change is seen by students as an issue of "pollution" and an "increase in temperatures". Furthermore, we observed that these responses became more important as the children grew older. This may have been due to the greater influence of digital information media (Internet, social networks, and TV) as the age of the students increased. In these media, there is a predominance of news identifying pollution as the cause and the increase in temperature as the effect.

Different studies carried out from the social attitudes of trainee teachers generated results similar to those of this study. For the case of Secondary Education teachers, Morote and Moreno (2021) indicate that the causes identified are human action (pollution, deforestation, and overexploitation of resources) and that the consequences are related to catastrophic effects (natural disasters). A similar result was obtained by Escoz-Roldan et al. (2020), who analysed the social attitudes of undergraduate students with respect to the risk of climate change and its relationship with water in three Spanish cities. In total, 85.0% of the respondents believed that climate change was "mainly due to human causes". Furthermore, the study conducted in two secondary schools in the United Kingdom by Kurup et al. (2021) revealed a strong understanding of the causes and effects of global warming. This was corroborated, in turn, by Chang and Pascua (2016) for Asia, who found that students' understanding of the causes of climate change was limited to the belief that the recorded changes are solely due to anthropogenic reasons. Gaps in training in terms of content can also be observed in students' difficulty in establishing relationships between the different elements and processes that intervene in the climate change process beyond associating them with processes that they consider to be "good" or "bad". In other words, they also lack the specific vocabulary to explain these processes. Furthermore, Wu and Otsuka (2021) found that, based on a sample of 657 Secondary Education students in Shanghai (China), erroneous concepts and a biased comprehension of climate change persist. The authors highlight the need to expand climate literacy and education from a conceptual and geographical perspective. If we compare this study with other works, the students always maintain the same discourse pattern: the greenhouse effect is harmful for society, it is caused by human action and global warming has disastrous effects (Da Silva and Boveloni 2009).

These gaps are associated both with the sources from which the information is obtained (the media) and the school textbooks, in which there is a predominance of catastrophic messages and, often, images taken out of context (Morote and Olcina 2020). Although this lack of scientific rigour is the dominant feature, it should also be noted that publications are emerging that address the current process of climate change simply and rigorously. An example is the study by Nelles and Serrer (2020) which, with abundant and clear graphic material, explains the different elements that comprise this complex process affecting the entire environment (the earth's climate, climate change, oceans, extreme events, ecosystems and human beings). Scharmacher-Schreiber and Stephanie (2020), using a question–answer method (is the climate becoming warmer and warmer? Can a difference of one degree be felt?), have elaborated materials on climate change directed at basic educational levels.

Furthermore, in the training on natural risks as effects of climate change, it is important to focus on how these processes are perceived by children. As indicated by Zhong et al. (2021), in a study on disaster education and its impact on children's perceptions of flooding in China, this enables us to explore its impact effectiveness and pathways. In turn, this furthers knowledge of the management of the consequences (effects) of climate change. In other words, it is important to know the effects of climate change and, using this knowledge, to implement actions aimed at managing these risks and to learn how society adapts through education, based on social representations and the perceptions of students.

The third hypothesis was also confirmed: with respect to the main greenhouse gas in the atmosphere, "it is expected that the response of the majority of the students will be CO_2". The results generated from analysing the sources from which the children obtained their information on climate change or their causes and consequences were reiterated. Chang and Pascua (2016), through semi-structured interviews with secondary students from Singapore, found "that the students' knowledge of climate change is composed of incomplete and incorrect elements" (p. 84). This was corroborated by analysing the role played by gases in the greenhouse effect. For example, the students considered that CFCs are the major greenhouse gases (19.7%); 6.6% even believed that "climate change is linked with tectonic activities (tsunami, earthquakes)". Both of these assertions are incorrect. In the first case, this was because water vapour was not taken into account; there were even students who did not consider it as an element that affects climate change. Furthermore, an even more serious mistake was relating geological processes with atmospheric processes, as Olcina (2017) has observed in Secondary Education textbooks.

This conceptual confusion coincides with the results obtained in this study, where water vapour was an "unknown" element for the respondents. These results are similar to those obtained from analysing the contents of textbooks. For example, the aforementioned study by Morote and Olcina (2020) confirms that only three out of every ten books refer to water vapour as a greenhouse gas. However, none of the textbooks indicate that it is the principal greenhouse gas in the atmosphere. Nevertheless, this fact, as expressed by the authors, is not completely negative, as 33.3% of textbooks at least mention it. Moreover, it should be explained that CO_2 is the main greenhouse gas that is increasing due to anthropic reasons (0.04%), but its presence in the atmosphere continues to be much lower than that of water vapour (4.0%).

The fourth hypothesis established that "significant differences are expected to be found between the three educational stages. The cognitive age of the students, a priori, should influence the answers. For this reason, we would expect the responses of the Baccalaureate students to be more elaborate, self-critical and with less error in their content (for example, in relation to the greenhouse gases)". This was partially confirmed. We can confirm that as the age of the students increased, the number of errors related to their responses on greenhouse gases also rose. In the case of the Baccalaureate stage, none of the students responded correctly. Again, the influence of the media could be the cause of this error.

Teaching the topic of climate change based on a holistic and integrated approach would constitute an opportunity to promote critical thought in society and among citizens. Only in this way will the current and future society be able to interpret its surrounding territory. In this respect, we can refer to the initiative in the United Kingdom that has recently incorporated teachers specialising in climate change into primary and secondary education centres (Ecoinventos 2019). Furthermore, it is also worth highlighting case study analyses that examine studies on natural risks (Aspin 2018), such as global warming (Greenwood 2018), carried out for several decades by the Geographical Association (United Kingdom) in Primary Education. The acquisition of the competencies required to face challenges in the future should be one of the objectives of future teachers. This will enable them to understand people's perspectives, social processes and human–environment interactions and to connect local and global geographical concepts. These concepts include climate change in terms of space, place, scale, physical and human processes, environmental and cultural diversity and interdependence (Chang and Wi 2018). This would enable future teachers to educate a new generation of critical thinkers about climate change, supported by a good curriculum (Kagawa and Selby 2012).

5. Conclusions

The research carried out revealed that school children mainly obtain their information from digital media (TV, Internet, and social networks). One element that is particularly noteworthy is the relevance of schools in these contents: it was found that as the students progressed through the educational stages, the information received from their schools become less relevant compared with information from the media. This is a risk for the critical-thinking training of citizens, due to the need to confirm information and news from false sources, or news that lacks scientific rigour (Lutzke et al. 2021). Therefore, there is a risk that students take this information to be reliable. Furthermore, it is worth mentioning the work being carried out from the earliest educational stages (particularly primary education) in many schools and the importance given to "family" by these students. However, the relevance of digital media is also significant in very early ages (TV and Internet).

This research goes a step further and, taking into account previous studies, highlights that this knowledge on climate change is shared by trainee teachers and even the content included in school textbooks. Therefore, we are faced with a vicious circle, as the teacher has gaps in his or her training that are passed on to the student who, in turn, will do the same in the future. Taking these training deficiencies and the need to break this vicious circle into account, future research needs to continue using surveys, but also, in particular, to analyse what is really taught in classrooms. The main limitation of this study is that what and how teachers teach climate change in Social Science classes was not analysed. This constitutes a challenge for future research. To achieve this, it is necessary to interview active teachers. In this way, the whole of the educational sphere will be completed, including schools, training at university level and even the resources used in the classroom which, in the case of Social Sciences, continues to be predominant.

As educators, we should underline the importance of raising awareness among the youngest cohort about climate change. Undoubtedly, the education factor is one of the most important pillars for the present and future adaptation to this phenomenon and its associated risks. The accurate communication of climate change to society is a key element in mitigation and adaptation policies and, as argued by Romero and Olcina (2021), "the academic world and the public powers must continue transferring the greatest amount of information possible based on proven facts" (p. 329). Therefore, teacher training should be improved with: (1) an explanation of the main issues in climate change, based on information and data provided by different scientific studies and comparing it with, for example, news stories that appear in the media or even the information provided by school textbooks; and (2) the fostering of "IOL" proposals ("Imagination + Originality + Local") (Morote and Olcina 2021a). An improvement in training is required as teaching is a huge responsibility. Finally, there is a need, each day, to reinforce concepts that students find difficult to understand. Educators must be committed to implementing methodologies that continuously enhance teaching practices in order to ensure a sound geographical education, generating critically thinking citizens who know how to interpret the environment in which they live.

Author Contributions: Conceptualization, Á.-F.M. and M.H.; methodology, Á.-F.M. and M.H.; formal review, Á.-F.M. and M.H.; and investigation, Á.-F.M. and M.H. All authors have read and agreed to the published version of the manuscript.

Funding: This research received no external funding.

Institutional Review Board Statement: Not applicable.

Informed Consent Statement: Not applicable.

Data Availability Statement: Not applicable.

Conflicts of Interest: The author declares no conflict of interest.

Appendix A

Table A1. Items of the questionnaire.

General Information	
Item (n°)	**Response Type/Variable**
Item 1. What course are you studying? Item 2. How old are you? Item 3. Gender	Open question. Open question. Closed-ended question: Male/Female/Other
Part 1. Training on Climate Change	
Item (n°)	**Response Type/Variable**
Item 4. Of the following information media, choose the top three where you receive information on climate change:	Closed-ended question: Do not know/Do not answer/Family/Social Networks/TV/Newspapers/Radio/Internet/School-High School
Item 5. What is climate change?	Open question
Part 2. Perception of Climate Change	
Item (n°)	**Response Type/Variable**
Item 6. Do you think that climate is changing? (mark with an X). Answer from 1 to 5, with 1 being least agree and 5 being most agree:	Likert scale: 1/2/3/4/5
Item 7. What are the main causes of climate change?	Open question
Item 8. What are the main consequences of climate change?	Open question
Item 9. What is the main greenhouse gas in the atmosphere?	Closed-ended question: Do not know/Do not answer/Methane (CH_4)/Ozone (O_3)/Carbon Dioxide (CO_2)/Water Vapour (H_2O)/Nitrous Oxide (NOx)
Part 3. Solutions and Proposals for Adapting to Global Warming	
Item (n°)	**Response Type/Variable**
Item 10. What do you do in your daily life to solve climate change?	Open question.
Item 11. Will climate change a problem for humanity in the future? (mark with an X). Please answer from 1 to 5, with 1 being less agree and 5 being more agree.	Likert scale: 1/2/3/4/5
Item 12. Will climate change pose a problem for humanity in the future? (mark with an X). Please answer from 1 to 5, with 1 being less agree and 5 being more agree.	Likert scale: 1/2/3/4/5

Source: Own elaboration.

References

Ahmad, Saima, and Sharker Md Numan. 2015. Potentiality of disaster management education through open and distance learning system in Bangladesh open university. *Turkish Online Journal of Distance Education* 16: 249–60. [CrossRef]

Allen, David E., Michel McAleer, and David McHardy-Reid. 2018. Fake news and indifference to scientific fact: President Trump's confused tweets on global warming, climate change and weather. *Scientometrics* 117: 625–29. [CrossRef]

Anyanwu, Raymond, and Lesley Le Grange. 2017. The influence of teacher variables on climate change science literacy of Geography teachers in the Western Cape, South Africa. *International Research in Geographical and Environmental Education* 26: 193–206. [CrossRef]

Aspin, Victoria. 2018. A week of rain .... *Primary Geography* 96: 24–25.

Bel, Juan Carlos, Juan Carlos Colomer, and Rafael Valls. 2019. Alfabetización visual y desarrollo del pensamiento histórico: Actividades con imágenes en manuales escolares. *Educación XX1* 22: 353–74. [CrossRef]

Benejam, Pilar. 1997. La selección y secuenciación de los contenidos sociales. In *Enseñar y Aprender Ciencias Sociales, Geografía e Historia*. Edited by Pilar Benejam and Joan Pagès. Barcelona: ICE Universidad de Barcelona/Horsori, pp. 71–95.

Borhaug, Frédérique. 2021. Missing links between intercultural education and anthropogenic climate change? *Intercultural Education* 32: 386–400. [CrossRef]

Brisman, Avi. 2018. Representing the "invisible crime" of climate change in an age of post-truth. *Theoretical Criminology* 22: 468–91. [CrossRef]

Calixto, Raúl. 2015. Propuesta en educación ambiental para la enseñanza del cambio climático. *Revista Electrónica Diálogos Educativos* 15: 54–68.

Canals, Roser, and Neus González. 2011. El currículo de Conocimiento del Medio Social y Cultural, y la formación de competencias. In *Didáctica del Conocimiento del Medio Social y Cultural en la Educación Primaria. Ciencias Sociales para Aprender, pensar y Actuar*. Edited by Antoni Santisteban and Joan Pagès. Madrid: Ed. Síntesis, pp. 41–62.

Caride, José Antonio, and Pablo Ángel Meira. 2019. Educación, ética y cambio climático. *Innovación Educativa* 29: 61–76. [CrossRef]

Chang, Chew-Hung, and Andy Wi. 2018. Why the World Needs Geography Knowledge in Global Understanding: An Evaluation from a Climate Change Perspective. In *Geography Education for Global Understanding, International Perspectives on Geographical Education*. Edited by A. Ali Demirci, R. Miguel González and S. Bednarz. Cham: Springer, pp. 29–42. [CrossRef]

Chang, Chew-Hung, and Liberty Pascua. 2016. Singapore students' misconceptions of climate change. *International Research in Geographical and Environmental Education* 25: 84–96. [CrossRef]

Da Rocha, Vanessa Tibola, Luciana Brandli, Jana Mazuti, Leila Dal Moro, Liane Dalla Gasperina, and Rosa Maria Kalil. 2020. Teacher's Approach on Climate Change Education a Case Study. *World Sustainability Series* 617–42. [CrossRef]

Da Silva, Henrique, and Daliane Boveloni. 2009. Los temas "cambios climáticos" y "calentamiento global" en los libros de texto: La falta de la mirada geológica. *Enseñanza de las Ciencias de la Tierra* 17: 190–95.

Díaz, Naira, and María del Mar Felices. 2017. Problemas sociales relevantes en el aula de primaria: La "cartografía de la controversia" como método. *REIDICS, Revista de Investigación en Didáctica de las Ciencias Sociales* 1: 24–38.

Domènech, Jordi. 2014. Contextos de indagación y controversias socio-científicas para la enseñanza del cambio climático. *Enseñanza de las Ciencias de la Tierra* 22: 287–96.

Ecoinventos. 2019. Reino Unido Incorpora a sus Colegios a Profesores Especialistas en el Cambio Climático. Available online: https://ecoinventos.com/reino-unido-incorpora-a-sus-colegios-a-profesores-especialistas-en-el-cambio-climatico/?fbclid=IwAR1bsdJD3MwOIFc0g4wYK9JRMZmVp3xPSeos9U-57iWDI3JNWk0GAkCH1WI (accessed on 8 August 2021).

Escoz-Roldan, Amor, José Gutiérrez-Pérez, and Pablo Ángel Meira. 2020. Water and climate change, two key objectives in the agenda 2030: Assessment of climate literacy levels and social representations in academics from three climate contexts. *Water* 12: 92. [CrossRef]

European Environment Agency (EEA). 2017. Climate Change, Impacts and Vulnerability in Europe 2016. An Indicator-Based Repor. Available online: https://www.eea.europa.eu/publications/climate-change-impacts-and-vulnerability-2016 (accessed on 1 November 2021).

Evans, Ronald W., Fred M. Newmann, and David W. Saxe. 1996. Defining issues-centered education. In *Handbook on Teaching Social Issues*. Edited by Ronald W. Evans and David W. Saxe. Washington: National Council for the Social Studies, pp. 2–5.

Ferrari, Enzo, Anne-Marie Ballegeer, Miguel Ángel Fuertes, Pablo Herrero, Laura Delgado, Diego Corrochano, Santiago Andrés-Sánchez, Kylian Marc Bisquert, Antonio Garcia-Vinuesa, Pablo Ángel Meira, and et al. 2019. Improvement on Social Representation of Climate Change through a Knowledge-Based MOOC in Spanish. *Sustainability* 11: 6317. [CrossRef]

Fien, John. 1992. Geografía, sociedad y vida cotidiana. *Documents d'Anàlisi Geogràfica* 21: 73–92.

Gallego, Adriana Patricia, and Jhonn Edgar Castro. 2020. Estudio de las representaciones sociales de los docentes sobre el cambio climático antropogénico. *Revista Científica* 38: 229–42. [CrossRef]

García, Francisco F., and Nicolás De Alba. 2003. El patrimonio urbano como ámbito para tratamiento escolar de problemas sociales y ambientales. In *El patrimonio y la didáctica de las ciencias sociales*. Edited by Ernesto Ballesteros, Cristina Fernández, José Antonio Molina and Pilar Moreno. Cuenca: Asociación Universitaria de Profesorado de Didáctica en las Ciencias Sociales (AUPDCS), pp. 81–89.

García, Francisco, and Rafael Porlán 2000. El proyecto IRES (Investigación y renovación escolar). Biblio 3W. Revista Bibliográfica de Geografía y Ciencias Sociales. 205. Available online: http://www.ub.edu/geocrit/b3w-205.htm (accessed on 1 March 2022).

Greenwood, Henry. 2018. A climate change assembley. *Primary Geography* 96: 22–23.

Ho, Li-Ching, and Tricia Seow. 2015. Teaching controversial issues in Geography: Climate change education in Singaporean Schools. *Theory and Research in Social Education* 43: 314–44. [CrossRef]

Intergovernmental Panel on Climate Change (IPCC). 2014. Climate Change 2013 and Climate Change 2014. 3 vols. Available online: http://www.ipcc.ch/ (accessed on 13 April 2021).

Intergovernmental Panel on Climate Change (IPCC). 2022. Climate Change 2021. The Physical Science Basis. Available online: https://www.ipcc.ch/report/ar6/wg1/downloads/report/IPCC_AR6_WGI_Full_Report.pdf (accessed on 17 February 2022).

Jeong, Jin Su, David González-Gómez, María Carmen Conde-Núñez, José Samuel Sánchez-Cepeda, and Félix Yllana-Prieto. 2021. Improving climate change awareness of preservice teachers (Psts) through a university science learning environment. *Education Sciences* 11: 78. [CrossRef]

Kagawa, Fumiyo, and David Selby. 2012. Ready for the storm: Education for disaster risk reduction and climate change adaptation and mitigation. *Journal of Education for Sustainable Development* 6: 207–17. [CrossRef]

Kažys, Justa. 2018. Climate change information on internet by different Baltic Sea Region languages: Risks of disinformation & misinterpretation. *Journal of Security and Sustainability Issues* 7: 685–95. [CrossRef]

Kidman, Gillian. 2018. School geography: What interests students, what interests teacher? *International Research in Geographical and Environmental Education* 27: 311–25. [CrossRef]

Kovacs, Alexandra, Horatiu Ştefănie, Carmelia Botezan, Iulia Crăciun, and Alexandru Ozunu. 2017. Assesment of natural hazards in european countries with impact on young people. Paper presented at the International Multidisciplinary Scientific GeoConference Surveying Geology and Mining Ecology Management, SGEM 2017, Albena, Bulgaria, June 29; vol. 17, pp. 73–80.

Kurup, Premnadh M., Ralph Levinson, Xia Li, and Ralph Levinson. 2021. Informed-Decision Regarding Global Warming and Climate Change Among High School Students in the United Kingdom. *Canadian Journal of Science, Mathematics and Technology Education* 21: 166–85. [CrossRef]

Legardez, Alain, and Laurence Simonneaux. 2006. *L'école à l'Épreuve de l'Actualité. Enseigner les Questions Vives*. París: Esf.

Legardez, Alain. 2006. L'enseignement des questions sociales et historiques, socialement vives. *Le cartable de Clio* 3: 245–53.

Li, Christine Jie, Martha C. Monroe, Annie Oxarart, and Tracy Ritchie. 2021. Building teachers' self-efficacy in teaching about climate change through educative curriculum and professional development. *Applied Environmental Education and Communication* 20: 34–48. [CrossRef]

Li, Yuh-Yuh, and Shu-Chiu Liu. 2021. Examining Taiwanese students' views on climate change and the teaching of climate change in the context of higher education. *Research in Science and Technological Education* 1–14. Available online: https://www.tandfonline.com/doi/abs/10.1080/02635143.2020.1830268?journalCode=crst20 (accessed on 12 January 2022).

López, José Antonio, and Montserrat Oller. 2019. Los problemas medioambientales en la formación del profesorado de educación primaria. *REIDICS: Revista de Investigación en Didáctica de las Ciencias Sociales* 4: 93–109.

López, Ramón. 2011. Conflictos sociales candentes en el aula. In *Les Questions Socialment Vives i l'Ensenyament de les Ciències Socials*. Edited by Joan Pagès and Antoni Santisteban. Barcelona: Servei de Publicacions de la UniversitatAutònoma de Barcelona, pp. 65–76.

Lutzke, Lauren, Caitlin Drummond, and Joseph Arvai. 2021. Priming critical thinking: Simple interventions limit the influence of fake news about climate change on Facebook. *Environmental Change-Human and Policy Dimensions* 5: 101964. [CrossRef]

Martínez-Fernández, Luis Carlos, and Jorge Olcina. 2019. La enseñanza escolar del tiempo atmosférico y del clima en España: Currículo educativo y propuestas didácticas. *Anales de Geografía de la Universidad Complutense* 39: 125–48. [CrossRef]

Masters, Mathilda. 2020. *123 Curiosidades que Todo el Mundo Debería Conocer Sobre el Clima*. Geoplaneta: Barcelona.

McWhirter, Nathan, and Tripp Shealy. 2018. Case-based flipped classroom approach to teach sustainable infrastructure and decision-making. *International Journal of Construction Education and Research* 16: 1–21. [CrossRef]

Moreno-Vera, Juan Ramón, Santiago Ponsoda, José Antonio López-Fernández, and Rubén Blanes-Mora. 2020. Holistic or traditional conceptions of heritage among early-childhood and primary trainee teachers. *Sustanability* 12: 8921. [CrossRef]

Morote, Álvaro-Francisco, and Enrique Moltó. 2017. El Museo del Clima de Beniarrés (Alicante). Propuesta de un recurso didáctico para la enseñanza de la Climatología. *Didáctica de las Ciencias Experimentales y Sociales* 32: 109–31. [CrossRef]

Morote, Álvaro-Francisco, and Jorge Olcina. 2020. El estudio del cambio climático en la Educación Primaria: Una exploración a partir de los manuales escolares de Ciencias Sociales de la Comunidad Valenciana. *Cuadernos Geográficos* 59: 158–77.

Morote, Álvaro-Francisco, and Jorge Olcina. 2021a. Cambio climático y sostenibilidad en la Educación Primaria. Problemática y soluciones que proponen los manuales escolares de Ciencias Sociales. *Sostenibilidad: Económica, Social y Ambiental* 3: 25–43. [CrossRef]

Morote, Álvaro-Francisco, and Jorge Olcina. 2021b. Riesgos atmosféricos y cambio climático: Propuestas didácticas para la región mediterránea en la enseñanza secundaria. *Investigaciones Geográficas* 76: 195–220. [CrossRef]

Morote, Álvaro-Francisco, and Juan Ramón Moreno. 2021. La percepción de los futuros docentes de Educación Secundaria sobre las implicaciones territoriales del cambio climático en destinos turísticos del litoral mediterráneo. *Grand Tour. Revista de Investigaciones Turísticas* 23: 261–82.

Morote, Álvaro-Francisco, and María Hernández. 2020. Social Representations of Flooding of Future Teachers of Primary Education (Social Sciences): A Geographical Approach in the Spanish Mediterranean Region. *Sustainability* 12: 6065. [CrossRef]

Morote, Álvaro-Francisco, and Xosé Manuel Souto. 2020. Educar para convivir con el riesgo de inundación. *Estudios Geográficos* 81: 1–14. [CrossRef]

Morote, Álvaro-Francisco, Benito Campo, and Juan Carlos Colomer. 2021a. Percepción del cambio climático en alumnado de 4° del Grado en Educación Primaria (Universidad de Valencia, España) a partir de la información de los medios de comunicación. *Revista Electrónica Interuniversitaria de Formación del Profesorado* 24: 131–44. [CrossRef]

Morote, Álvaro-Francisco, María Hernández, and Jorge Olcina. 2021b. Are Future School Teachers Qualified to Teach Flood Risk? An Approach from the Geography Discipline in the Context of Climate Change. *Sustainability* 13: 8560. [CrossRef]

Morote, Álvaro-Francisco. 2020. El papel de los medios de comunicación y las redes sociales en la sensibilización y educación del cambio climático. In *Fundación Nueva Cultura del Agua (FNCA), XI Congreso Ibérico de Gestión y Planificación del Agua*. Zaragoza: Fundación Nueva Cultura del Agua (FNCA), pp. 933–43.

Nelles, David, and Christian Serrer. 2020. *El pequeño Manual del Cambio Climático*. Grijalbo: Barcelona.

Ochoa, Anna S. 1996. Building a rationale for issue-centered education. In *Handbook on Teaching Social Issues*. Edited by Ronald Evans and David Saxe. Washington: National Council for the Social Studies, pp. 6–12.

Olcina, Jorge. 2017. La enseñanza del tiempo atmosférico y del clima en los niveles educativos no universitarios. Propuestas didácticas. In *Enseñanza y Aprendizaje de la Geografía para el Siglo XXI*. Edited by Rafael Sebastiá and Emilia María Tonda. Alicante: University of Alicante, pp. 119–48.

Pagès, Joan, and Antoni Santisteban. 2011. *Les Qüestionssocialment Vives i l'Ensenyament de les Ciències Socials*. Barcelona: Servei de Publicacions de la Universitat Autònoma de Barcelona, pp. 121–28.

Pagès, Joan. 2007. La enseñanza de las ciencias sociales y la educación para la ciudadanía. *Didáctica Geográfica* 9: 205–14.

Parra, Elizabeth, Claudia Castillo, and Miriam Vallejo. 2013. Representaciones sociales sobre desarrollo sostenible y cambio climático en estudiantes universitarios. *Perspectivas de la Comunicación* 6: 108–19.

Pérez-Gil, José Antonio, Salvador Moscoso, and Rafael Moreno. 2000. Validez de constructo: El uso del análisis factorial exploratorio-confirmatorio para obtener evidencia de validez. *Psicothema* 12: 442–46.

Rausell, Helena, Álvaro-Francisco Morote, Rafael Valls, and Moisés Domingos. 2021. Social representations and teaching of cross-cutting topics for history and geography teachers in training: Gender and climate change. In *Handbook of Research on Teacher Education in History and Geography*. Edited by Cosme Jesús Gómez-Carrasco, Pedro Martínez-Miralles and Ramón López-Facal. Berlín: Peter Lang Publishing, pp. 221–40.

Romero, Joan, and Jorge Olcina. 2021. *Cambio Climático en el Mediterráneo: Procesos, Riesgos y Políticas*. Valencia: Tirant Humanidades.

Roussel, David, and Amy Cutter-Mackenzie-Knowles. 2020. A systematic review of climate change education: Giving children and Young people a voice and a hand in redressing climate change. *Children's Geography* 18: 191–208. [CrossRef]

Santisteban, Antoni, Neus González, Joan Pagès, and Montserrat Oller. 2014. La introducción de temas controvertidos en el currículo de ciencias sociales: Investigación e innovación en la práctica. In *Historia e identidades culturales. V Simpósio Internacional de Didáctica de LasCiencias Sociales en el Ámbito Iberoamericano*. Edited by Joaquim Prats, Barca Isabel and López Ramón. Barcelona: Universitat de Barcelona, pp. 310–22.

Satorra, Albert, and Peter M. Bentler. 2010. Ensuring positiveness of the scaled difference Chi-Square test statistic. *Psychometrika* 75: 243–48. [CrossRef]

Scharmacher-Schreiber, Kristina, and Marian Stephanie. 2020. *¿Cuánto Calor es 1 Grado Más? ¿Qué pasa con el Cambio Climático?* Salamanca: Lóguez Ediciones.

Sebastiá, Rafael, and Emilia María Tonda. 2018. Enseñar y aprender el tiempo atmosférico y clima. In *Reflexiones sobre educación geográfica: Revisión disciplinar e innovación didáctica*. Edited by Alfonso García de la Vega. Madrid: Asociación Española de Geografía, pp. 153–76.

Sezen-Barrie, Asli, and Gili Marbach-Ad. 2021. Cultural-Historical Analysis of Feedback from Experts to Novice Science Teachers on Climate Change Lessons. *International Journal of Science Education* 43: 497–528. [CrossRef]

Sharpe, Donald. 2015. Chi-Square test is statistically significant: Now what? *Practical Assessment, Research, and Evaluation* 20: 8.

Tutiaux-Guillón, Nicole. 2011. ¿Les qüestions socialment vives, un repte per a la història i la geografía escolars? In *Les Qüestions Socialment Vives i l'Ensenyament de les Ciències Socials*. Edited by Joan Pagès and Antoni Santisteban. Barcelona: Servei de Publicacions de la UniversitatAutònoma de Barcelona, pp. 25–39.

United Nations (UN). 2015. Sustainable Development Goals. UNDP, Sustainable Development Agenda. Available online: https://www.undp.org/content/undp/es/home/sustainable-developmentgoals/resources.html (accessed on 1 December 2021).

Verlie, Blanche, and Simone Miranda Blom. 2021. Education in a changing climate: Reconceptualising school and classroom climate through the fiery atmos-fears of Australia's Black summer. *Childrens Geographies* 1–15. [CrossRef]

Wu, Jing, and Yoshiki Otsuka. 2021. Exploring the climate literacy of high school students for better climate change education. *International Journal of Global Warming* 23: 151–68. [CrossRef]

Zhong, Shuang, Qiu Cheng, Shuwei Zhang, Cunrui Huang, and Zhe Wang. 2021. An impact assessment of disaster education on children's flood risk perceptions in China: Policy implications for adaptation to climate extremes. *Science of the Total Environment* 757: 143761. [CrossRef]

social sciences MDPI

Article

Who Are the Protagonists of History? Exploratory Study on Historical Relevance after Completing Compulsory Secondary Education in Spain

Pilar Rivero [1,2,3,*], Iñaki Navarro-Neri [1,2,3] and Borja Aso [1,2,3]

1 Department of Specific Didactics, Faculty of Education, University of Zaragoza, 50009 Zaragoza, Spain; inakinavarro@unizar.es (I.N.-N.); basom@unizar.es (B.A.)
2 Research Group ARGOS, Faculty of Education, University of Zaragoza, 50009 Zaragoza, Spain
3 University Institute of Research in Environmental Sciences of Aragon (IUCA), University of Zaragoza, 50009 Zaragoza, Spain
* Correspondence: privero@unizar.es

Abstract: Which historical figures do Spanish students think are relevant after completing their compulsory schooling in the subject of history? The main objective of this research is to discover which four historical figures students choose as the most important in human history after completing their compulsory schooling, and the type of reasons they give to justify their answers. By means of a mixed study, this research involved asking 165 students, aged 16 to 17, which four figures they think are the most relevant in history and why. Based on this statement, a quantitative and qualitative analysis was conducted to explore the students' discourses justifying their choices using the categories proposed by Partington, Hunt and Lévesque. The results show a clear tendency in students to choose key figures in Spain's history, such as Christopher Columbus or Francisco Franco, or a prominent person, such as Adolf Hitler, as the most relevant historical figures, who appear in the curriculum in their last year of compulsory secondary education. Lastly, the argumentative approach regarding the level of historical significance focuses primarily on the impact of their actions on a large number of people during their historical moment, with no critical reflection on their long-term impact or consequences.

Keywords: historical relevance; education for democratic citizenship; historical narrative; historical thinking

Citation: Rivero, Pilar, Iñaki Navarro-Neri, and Borja Aso. 2022. Who Are the Protagonists of History? Exploratory Study on Historical Relevance after Completing Compulsory Secondary Education in Spain. *Social Sciences* 11: 175. https://doi.org/10.3390/socsci11040175

Academic Editor: Alvaro Morote

Received: 3 March 2022
Accepted: 2 April 2022
Published: 11 April 2022

1. Introduction

One of the main purposes of social science teaching is that, through the development of historical thinking skills, citizens can distinguish what is really relevant from what is irrelevant in the mid and long term, which is extremely important in a society as immediate and with such an overdose of information and fake news as ours is at the present time. This is why in teaching history, when its aim is to enable the exercise of responsible citizenship, exercises in reflection on historical relevance have been common in the English and Latin American context for some time now.

Completing an exercise that makes us consider which figures are historically relevant involves constructing our own historical narrative, which places us within the collective at the same time. This activity is not new; in fact, it is as old as the use of history as a means of constructing any collective identity throughout the ages.

As a first socialization space, school becomes the setting where students begin the exercise of constructing their own historical narrative to understand themselves and the world around them. However, now that we have mass media and culture has been democratized, school and family environments are no longer the only valid spaces for constructing a historical narrative.

Historical relevance is not merely concerned with how people and/or events are related in the past, but also with how these events and/or people relate to the present (Seixas 1998). As Seixas and Peck (2004) mention, historical significance and relevance involve an exercise of reflection on what must be remembered and why and consider a series of factors, such as how many people were affected, how long it lasted, and whether the reflection helps us understand the present (Lévesque 2008; Seixas and Morton 2012).

In other words, they are contents, processes, and a historiographical context that turn a mere event into a historical event (Cercadillo 2001). In short, historical relevance or significance is but a construct, since it is given when a certain figure, event, or process occupies a significant place in a historical narrative, which involves accepting that this relevance or significance given to someone, or something, varies depending on the temporal and social context (Seixas and Morton 2012).

The development of historical thinking, including reflection on historical significance, is an important line of work in the teaching of social sciences in Spain, linked to the work by Canadian researchers (Lévesque 2005, 2008; Seixas 1998, 2004, 2011; Seixas and Morton 2012), British researchers (Cooper and Chapman 2009), and American researchers (Barton 2011; VanSledright 2011, 2014; Wineburg 2001) with their postulates on meta-concepts or second-order concepts in history.

This is a consolidated line of research with a number of studies in the past 20 years focusing on what students consider historically relevant during their academic education, either in their first educational stages of primary and secondary school (Cercadillo 2002, 2004, 2006; Ibagón-Martín and Miralles 2021; Ortuño et al. 2016), or in their university education stage (Egea-Vivancos and Arias-Ferrer 2018; Rivero and Pelegrín 2019; Sáiz-Serrano and López-Facal 2015).

Learning to think historically, or what is known in Ibero-American teaching of social sciences as the development of historical thinking or historical literacy (Sáiz-Serrano and López-Facal 2014) "is a way of understanding history involving one's own critical interpretation of past events, which in turn enables meaning to be given to these events" (Ortuño et al. 2016, p. 11).

To perform this exercise of reviewing and interpreting the past, Seixas and Morton (2012) developed six basic principles, i.e., meta-concepts: working with sources, continuity and change, cause and consequence, historical perspective, the moral dimension, and historical significance or relevance. These are six meta-concepts or second-order concepts that summarized the idea posited by multiple authors on the need for students to develop a series of skills beyond mere conceptual knowledge or learning by rote to enable them to interpret the past (Barca 2000; Clark 2011; Lévesque 2008; Peck and Seixas 2008).

Educating students historically, or developing their historical thinking, first involves accepting that this knowledge is neither natural nor intuitive, as it requires an ongoing educational process (Sáiz-Serrano and López-Facal 2015). Second, it encompasses the exercise of critically interpreting the past by exploring what is known as substantive or first-order knowledge, for example, historical figures, dates, concepts, and data (Lévesque 2008; VanSledright 2011; Wineburg 2001). When explored from an empirical perspective—thinking, reasoning, arguing, and writing about history—the exercise of critically interpreting the past also forms part of historical literacy (Monte-Sano 2010; Wineburg et al. 2013).

The concept of a historical narrative as a means to convey meanings is closely tied to the first forays of history as a scientific discipline (López-Rodríguez 2015, 2020). Its beginnings served the then incipient nation states throughout the eighteenth and nineteenth centuries as a tool for constructing national identities (Berger 2007; Berger and Lorenz 2008; López and Márquez 2018; Lowenthal 1998; VanSledright 2008), replete with emotional and identity-related elements (Barton and Levstik 2004; Rüsen 2005).

Narrative thinking is a basic mental operation that organizes and gives meaning to the historical past (Carretero and Atorresi 2008; Ricoeur 1987; Rüsen 2005). When students are asked to write a summary, we can observe not only what they know but also how they represent and organize this knowledge (Sáiz-Serrano and Gómez-Carrasco 2016). As Rüsen

(2005) says, narrative is an instrument for constructing historical consciousness, which is also replete with emotional and identity-related elements (Barton and Levstik 2004; Rüsen 2004).

Consequently, historical narrative is a vitally important tool for learning about history (Pagès and Santisteban 2010), since it facilitates the development of some of the skills typical of historical thinking, such as interpreting sources or causal and intentional explanations (Santisteban 2017).

The latest bibliometric analyses highlight the increase in empirical studies published in journals with a renowned impact (Chaparro Sainz et al. 2020; Gómez-Carrasco et al. 2019). Numerous lines of research have been opened on the development of historical thinking in the past two decades, such as historical significance or relevance (Arias-Ferrer and Egea-Vivancos 2019; Egea-Vivancos and Arias-Ferrer 2018), narrative competence (Sáiz-Serrano and López-Facal 2015; Sant et al. 2014), historical empathy (Calderón-López and Arias-Ferrer 2017; Carril-Merino et al. 2020; Molina-Puche and Egea 2018); and historical consciousness (Miguel-Revilla and Sánchez-Agustí 2018; Sáiz-Serrano and López-Facal 2014). These lines of research in Spain are in consonance with the numerous publications on history education in other countries, such as the works of Counsell et al. (2016), Carretero et al. (2017), Metzger and Harris (2018), Apostolidou and Solé (2019), or Solé and Gago (2021), among others.

Also worth highlighting are studies on the potential of several didactic tools for developing historical thinking, such as the use of textbooks (Gómez-Carrasco et al. 2019), new technologies (Colomer et al. 2018), social media (Navarro-Neri et al. 2021) or the link between heritage and teaching history (Egea-Vivancos and Arias-Ferrer 2018; Egea-Vivancos et al. 2018; Rivero et al. 2018).

This study is, therefore, part of the line of research found in several countries in recent years on the development of historical thinking in students by authors including Barton and Levstik (2004), Lévesque (2005), Phillips (2002), Barca and Schmidt (2013), Sung (2020), and Alvén (2021), and in Spain by Cercadillo (2001, 2006), Rivero and Pelegrín (2019), Arias-Ferrer et al. (2019), Ibagón-Martín and Miralles (2021), Ibagón-Martín et al. (2021), among others.

Based on theoretical references and the aforementioned studies, this research uses mixed analysis to learn which figures students choose and the level of historical significance they bestow to justify that choice. The aim of this study is to consolidate the line of research referring to the development of historical thinking in students and it contributes a specific qualitative approach, such as the use of the criteria established by Partington (1980), Hunt (2000), and Lévesque (2008).

Considering all the above, the aim of this research is to explore which historical figures students believe are relevant after completing their compulsory schooling, and which approach they use for historical significance when justifying their answer. Therefore, in order to fulfil the aim of this investigation, the research questions structuring this study are: Which figures do Spanish students think are historically relevant after completing their compulsory schooling? Which approach, referring to the level of historical significance (Hunt 2000; Lévesque 2008; Partington 1980), do they use to justify their answer?

2. Materials and Methods

2.1. Sample

To answer this research questions, we used the responses of $n = 165$ students aged 16 to 17 who agreed to reply to the questionnaire anonymously. The questionnaire was completed at the start of the academic year after they had completed all their compulsory schooling years in Spain. As the number of boys and girls was similar (80 boys and 85 girls), and there was no significant difference in the answers to the questionnaire, the students were taken as a single group ($n = 165$). All participants were informed about the research following the general recommendations of the ethics committees.

2.2. Questionnaire and Method

The information was collected using an ad hoc-designed questionnaire for anonymous completion by the students during an estimated time of 20 min. The aim was for them to reflect on the following statement and write their answer:

Choose four historical figures you think are very important and answer these questions: What did this historical figure do? Why is he/she important for you? Reflect on whether he/she affected many people, caused a significant change in people's lives, had consequences that extended over a period, serves to make us realize something important, etc.

Previously, the researchers tested one class in each of the schools participating in the study, with 50 students in total. Both schools are located in neighborhoods characterized by sociocultural diversity with a considerable number of immigrant families, mostly from Latin America and North Africa, together with families from Eastern Europe. The schools participating in the research are representative of the city where the study was made.

As a result of this pilot test, the researchers changed two points: First, the number of characters to be provided by each participant was reduced, because it was found that the last answers were left blank, or the explanations were too short. Second, the researchers included a reminder of what each category of analysis consisted of in the design, because it was observed that in many cases, especially with regard to the last answers, there were explanations and evaluations that dealt with issues far removed from the main topic. The tests were administered by the same teachers who subsequently collected the data for the study, and they did so a year before the data collection that constitutes the research. The information collection instrument was previously validated by 10 researchers in education who suggested adjustments concerning the fulfilment of ethical criteria in the research.

This research used a mixed method (Tasshakori et al. 2021). It complements the quantitative and qualitative approaches already advocated for in the teaching of social sciences by authors, such as Grant (2001), Barca (2005), Barca and Schmidt (2013), Barton (2005), Sáiz-Serrano and López-Facal (2014), or Solé and Gago (2021), among others.

The qualitative analysis was performed using a categorization system based on the proposals made about historical significance by Partington (1980), Lévesque (2008), and Hunt (2000), that is, talking about an approach centered on importance when the argument focuses on the figures' value in their era; using an approach based on profundity to refer to the impact (superficial or deep) the figures had in their era; an approach centered on durability to talk about whether the figures and their actions have permanence over time; an approach mainly focusing on highlighting the number of people impacted by the figures in question; and, lastly, an approach based on the chosen figures' relevance, i.e., on the value he/she currently has or is given. This categorization is the most common in research and teaching proposals in the Anglo-Saxon sphere, as can be seen in the work proposals on historical relevance proposed by the Canadian research group Historical Thinking Project (Center of Historical Thinking Consciousness 2016).

The results and the statistical summary of this research was constructed using instruments provided by the IBM SPSS STATISTICS 26 for Windows. In the end, to make the statistical analysis more agile, historical figures that were not mentioned by at least 5% of the students were disregarded as analysis units.

3. Results

3.1. Which Historical Figures Do Students Think Are the Most Important after Completing Their Compulsory Schooling?

The first datapoint worth highlighting is that out of the 165 students participating in the study and replying with at least one historical figure (600 responses in total), 121 different figures were counted, out of which only 15 were mentioned by over 5% of the students (Table 1).

Table 1. Which figures do students think are historically relevant after completing their compulsory schooling?

Answers							
Historical Figure	*n*	**Percentage**	**Percentage of Cases**	**Historical Figure**	*n*	**Percentage**	**Percentage of Cases**
Adolf Hitler	100	15.0%	60%	Benito Mussolini	11	1.8%	7.3%
Christopher Columbus	88	13.8%	55.3%	Miguel de Cervantes	10	1.7%	6.7%
Francisco Franco	38	6.3%	25.3%	Socrates	10	1.7%	6.7%
Napoleon	32	5.3%	21.3%	Clara Campoamor	9	1.5%	6%
Isaac Newton	20	3.3%	13.3%	Catholic Monarchs	9	1.5%	6%
Albert Einstein	18	3.0%	12%	Karl Marx	9	1.5%	6%
Stalin	13	2.2%	8.7%	Gandhi	8	1.3%	5.3%
Marie Curie	12	2.0%	8%	TOTAL:	600	100%	400%

As can be seen in Table 1, two historical figures stand out from the others, as they were mentioned by over 50% of the students: Adolf Hitler (60%) and Christopher Columbus (55.3%). Far behind these percentages are choices including Francisco Franco (25.3%) and Napoleon (21.3%). Only two more figures exceed 10%: Isaac Newton (13.3%) and Albert Einstein (12%).

The first aspect of note is that the most repeated options seem to fall into one of two thematic categories: contemporary history (e.g., Adolf Hitler, Napoleon, Stalin, and Mussolini) or Spanish history (Christopher Columbus, Francisco Franco, Miguel de Cervantes, Clara Campoamor, and the Catholic Monarchs). The third thematic category encompassing a larger number of historical figures is science (Isaac Newton, Albert Einstein, and Marie Curie).

A possible explanation for these categories is that they are the last subjects students take before completing their compulsory schooling, since in Spain students in the fourth and final year of compulsory secondary education only work on contemporary universal history. Furthermore, in the first year of the Spanish baccalaureate, the students participating in this research studied the subject of scientific culture, which explains the continuous appearance of essential figures in the history of science, such as the ones mentioned above.

An in-depth analysis of the results reveals who is the students' first choice when selecting a historically relevant figure (Table 2).

Table 2. First historical figure chosen by students.

Historical Figure 1			
Figure	**Frequency**	**Percentage**	**Valid Percentage**
Christopher Columbus	39	23.6%	23.6%
Adolf Hitler	37	22.4%	22.4%
Francisco Franco	13	7.9%	7.9%
Napoleon	8	4.8%	4.8%
Clara Campoamor	5	3.0%	3.0%
Isaac Newton	5	3.0%	3.0%
Total	165	100.0%	100.0%

In this case, we can observe that figures "belonging" to Spanish history appear in the first five options, namely Christopher Columbus (23.6%), Francisco Franco (7.9%), and

Clara Campoamor (3%). The other two most-chosen people as the first option are Adolf Hitler with 22.4% and Napoleon with 4.8%, key figures in contemporary universal history.

Last, we have produced a summary with the most chosen options by the students in second, third, and fourth place. For a faster analysis, we have only included figures that were mentioned more than five times (at least 3%) in the summary table (Table 3).

Table 3. Figures most chosen in second, third, and fourth place.

Historical Figure 2				**Historical Figure 3**				**Historical Figure 4**			
Name	***n***	**%**	**V *.%**	**Name**	***n***	**%**	**V.%**	**Name**	***n***	**%**	**V.%**
Adolf Hitler	33	20.0	20.1	Christopher Columbus	19	11.5	11.9	Adolf Hitler	11	6.7	7.3
Christopher Columbus	23	13.9	14.0	Adolf Hitler	19	11.5	11.9	Marie Curie	8	4.8	5.3
Francisco Franco	8	4.8	4.9	Francisco Franco	13	7.9	8.1	Christopher Columbus	7	4.2	4.7
Isaac Newton	7	4.2	4.3	Napoleon	13	7.9	8.1	Stalin	7	4.2	4.7
Napoleon	7	4.2	4.3	Isaac Newton	6	3.6	3.8	Albert Einstein	7	4.2	4.7
Albert Einstein	6	3.6	3.7	Albert Einstein	5	3.0	3.1	Napoleon	7	4.2	4.7
Karl Marx	5	3.0	3.0	Socrates	5	3.0	3.1	Miguel de Cervantes	5	3.0	3.3
								Francisco Franco	5	3.0	3.3
								Benito Mussolini	5	3.0	3.3
Total	164	99.4	100	Total	160	97	100	Total	150	90.9	100
System losses	1	0.6			5	3			15	9.1	
Total	165	100			165	100			165	100	

* Valid percentage.

As in the previous case, the most chosen figures were Adolf Hitler and Christopher Columbus, followed by Francisco Franco, Isaac Newton, and Marie Curie. Again, except for Christopher Columbus, the students opted for figures related to the subjects they had recently studied, namely contemporary history in the last year of compulsory schooling, and scientific culture, a subject they had just begun studying when the research was conducted in their first year of the Spanish baccalaureate.

3.2. Referring to the Level of Historical Significance, Which Approach Do Students Choose to Justify Their Answer?

In the second part of this study, we wanted to analyze which approach, with reference to the level of historical significance (Hunt 2000; Lévesque 2008; Partington 1980), students select to justify their choice (Table 4).

Table 4. Historical significance chosen by the students to justify their answer.

Historical Significance Frequencies							
Category	**Answers**		**Percentage of Cases**	**Category**	**Answers**		**Percentage of Cases**
	n	%			*n*	%	
Importance	229	39.2%	156.8%	Quantity	88	15.1%	60.3%
Profundity	150	25.7%	102.7%	Durability	22	3.8%	15.1%
Relevance	95	16.3%	65.1%	TOTAL:	584	100%	400%

The first striking aspect of the overall results is the predominance of the justification focused on highlighting the historical figures' importance in their era (39.2%) with justifications such as: "he was the most important figure of his era" (Student 79, age 16) referring to Adolf Hitler, "he was the most important emperor of his time" (Student 140, age 17) referring to Napoleon or "he was a very important dictator in his country and for the whole of Europe" (Student 24, age 16) referring to Stalin.

The second type of justification most used by the students is the one in which the profundity category predominates, emphasizing that the figure in question caused a change in the society at that time (25.7%). As can be seen in Table 5, this is the reasoning most used to talk about Christopher Columbus with explanations including: "Thanks to him America was discovered and changed the map because, until then, it was believed that there was nothing between Asia and Europe" (Student 2, age 16) or "After discovering America, Spain became the first world power thanks to the gold and silver of Peru" (Student 33, age 17). It is also interesting to note how, in the field of science, arguments referring to profundity and durability are the most used with explanations including: "Thanks to his scientific theory, many others were later developed" (Student 98, age 17) to refer to Isaac Newton or "Thanks to his discovery, today we can do X-ray tests outside the hospital" (Student 105, age 16) to refer to Marie Curie.

Table 5. Most mentioned figures and the level of historical significance chosen.

Historical Figure		Categories					Total
		Importance	Profundity	Durability	Quantity	Relevance	
Adolf Hitler	Count	42	28	10	78	30	188
	%	22.5%	14.9%	5.5%	**41.5%**	15.6%	
Christopher Columbus	Count	24	42	1	5	16	88
	%	27.3%	**47.7%**	1.1%	5.7%	18.2%	
Francisco Franco	Count	13	9	0	11	5	38
	%	**35.1%**	24.3%	0%	29.7%	10.8%	
Napoleon	Count	24	3	0	3	2	32
	%	**76.5%**	8.8%	0%	8.8%	5.9%	
Isaac Newton	Count	6	8	1	0	5	20
	%	31.8%	**36.4%**	4.5%	0%	27.3%	
Albert Einstein	Count	6	3	1	2	6	18
	%	**30%**	25%	5%	10%	**30%**	
Stalin	Count	7	2	0	3	1	13
	%	**53.8%**	15.4%	0%	23.1%	7.7%	
Marie Curie	Count	3	4	1	2	2	12
	%	25%	**33.3%**	8.3%	16.7%	16.7%	
Benito Mussolini	Count	6	0	0	3	3	12
	%	**50%**	0%	0%	25%	25%	
Miguel de Cervantes	Count	4	1	1	0	4	10
	%	**40%**	10%	10%	0%	**40%**	
Socrates	Count	8	0	1	0	1	10
	%	**80%**	0%	10%	0%	10%	
Karl Marx	Count	6	0	0	0	3	9
	%	**66.7%**	0%	0%	0%	33.3%	
Clara Campoamor	Count	2	2	0	1	4	9
	%	22.2%	22.2%	0%	11.1%	**44.4%**	
Catholic Monarchs	Count	5	1	0	2	1	9
	%	**55.6%**	11.1%	0%	22.2%	11.1%	
Gandhi	Count	4	1	1	1	0	7
	%	**57.1%**	14.3%	14.3%	14.3%	0%	

Far behind these percentages, the students used the relevance category to underscore the historical figures' influence on the present (16.3%). For example, we find explanations including "it was very important because today there are still Jews looking for their relatives who were murdered during the Holocaust" (Student 29, age 17) to refer to Adolf Hitler. With similar percentages, the quantity category was also used to highlight the number of people affected by the chosen historical figure (15.1%). Finally, the least used category by the students was durability, which allows us to question whether meta-concepts or second-order concepts, such as continuity and change or historical time, are being worked on in the classroom (3.8%).

Lastly, we analyzed the approach to the level of historical significance selected by the students in the case of the most chosen historical figures (Table 5). In this case, only those figures mentioned by at least 5% of the students are shown in the summary table.

The results show the preeminence of significance referring to the figures importance in their era, above other approaches. However, in the case of the figure of Adolf Hitler, 41.5% of the responses emphasized the number of people that were affected by his actions, primarily because of the Holocaust and the casualties of the Second World War: "He was the person who did the Holocaust, he killed millions of Jews" (Student 39, age 17) or "It was because of him that the Second World War started, in which almost every country in the world except Spain took part" (Student 8, age 16).

In the case of Christopher Columbus, the approach most used was profundity (47.7%), highlighting the change the era involved and not the number of people impacted or, given that the students are Spanish, the repercussions he has had on the present, for example linguistically and culturally: "His discovery changed the map of the world to the way we see it today" (student 10, age 17) or "Thanks to him Christianity spread all over the world" (student 134, age 16). This also implies scant reflection on the consequences of colonialism, despite there being a significant percentage of students whose families come from Latin America.

In the case of political figures, such as Francisco Franco, Napoleon, Stalin, Mussolini, and the Catholic Monarchs, the predominant approach refers to the importance they had in their era (35.1%, 76.5%, 53.8%, 50% and 55.6%, respectively) with explanations including "they were the most important kings in Europe in their time" (Student 66, age 16) or "They were important because they believed in the idea of Christopher Columbus and thanks to this it was possible to discover America" (Student 32, age 16) to refer to the Catholic kings or "Among the dictators in Europe in those years, he was the most powerful, that's why he was called the Duce" (Student 113, age 17) to refer to Mussolini, and not so much their repercussion, the number of people affected or their significance in the present.

It is also interesting to note that for the scientific figures (Isaac Newton, Albert Einstein, and Marie Curie), the most-used argument refers to the change their scientific contribution brought in their era (36.4% in the case of Newton and 33.3% in the case of Marie Curie) and not so much to current scientific research, as is the case with Albert Einstein (30% of the arguments use the relevance of this figure in the present as the predominant approach).

In short, although in the figures most mentioned by students we can observe the predominant use of arguments extolling the importance of these figures in their era, the use of the other categories is significant, which enables us to emphasize that the teaching is heterogeneous as there is no single discursive or didactic approach.

4. Discussion

As mentioned in the results section, the historical figures chosen by the students have been grouped into three thematic categories: contemporary history, Spanish history, and figures belonging to the scientific field. The incipient number of relatively similar research studies on the teaching of social sciences has not only enabled us to establish some comparisons between their results but also to add new aspects to the debate.

One of the most recent studies on which figures students think are historically relevant was conducted by Ibagón-Martín and Miralles (2021). They asked a sample of over

700 Colombian students aged 15 to 17 to choose the five most important figures in Colombia's history. In summary, the results show that four out of the five most repeated figures are men belonging to the political sphere (Simón Bolívar, Jorge Eliécer Gaitán, Christopher Columbus, and Gabriel García Márquez). This choice of men partly coincides with the results of this study, in which the figure of Adolf Hitler was the most chosen, alongside others that can be classified as belonging to the political sphere, such as Francisco Franco and Napoleon. The Ibero-American cultural connection is obvious in the figure of Christopher Columbus since he appears as one of the most chosen in both studies.

Curiously, the figure appearing as the most repeated fourth choice is Policarpa Salavarrieta, considered a heroine during the independence process of Colombia due to her role as a spy. Consequently, we can speculate whether, for example, the figure of Clara Campoamor (chosen by 6% of the sample) would have been chosen by more students if we had asked about the most important figures in the history of Spain.

We also compared the results of our research with another recently published study conducted by Ibagón-Martín et al. (2021). They asked students aged 15 to 16 in Murcia (Spain) who they think are the five most relevant figures in the history of their country. Furthermore, the research sample they obtained enabled them to compare the answers given by Spanish students and students from Latin America, Asia, Africa, and other countries in Europe.

In the case of the Spanish students, the most mentioned figures were Francisco Franco, Christopher Columbus, and the Catholic Monarchs. Again, as in the research with Colombian students, the most chosen figures belong to the political sphere and are mainly male. Furthermore, out of the top twenty figures chosen, only two women are mentioned: Isabella I of Castile and Isabella II, if we view the Catholic Monarchs as an institution and not as a man and a woman. These are quite similar results to the ones obtained in our study, since the most mentioned female figures are Marie Curie and Clara Campoamor (Table 1).

The study by Ibagón-Martín et al. (2021) also highlights how immigrant students gradually transform their national "native" identity to the "converted" one as the number of years they reside in Spain increases. As a result, students from Latin America, Africa, and Eastern Europe who have lived in Spain for fewer years tend to choose figures related to their native country's independence processes, while students with more years of residence in Spain (10 to 15 years) choose figures such as Christopher Columbus, Francisco Franco, and Isabella of Castile (315–316).

Although this study uses a relatively small sample that does not lend itself to generalization (n = 56), both these results and the ones obtained in our research force us to reflect on the type of historical discourse occurring in classrooms from preschool to the end of compulsory secondary education, since the most recurrent choices are a discoverer (Christopher Columbus), a dictator (Francisco Franco), and a king and queen (the Catholic Monarchs). As far as possible, future research should focus on the type of discourse and activities Spanish teachers use when covering the history of Spain. It could prove extremely interesting to analyze their discourses using the narrative approaches proposed by Rüsen (2005), for example.

The fact that three of the four most chosen figures (Adolf Hitler, Francisco Franco, and Napoleon) are classified as belonging to the political sphere can be understood, at least in part, as a result of the design of the geography and history curricula in compulsory secondary education and also of the interpretation of them in the textbooks published by the main publishing houses, which still lean towards political and factual history. The statements in the content blocks and, above all, the assessment criteria laid down by both the state and the autonomous communities, present few examples of approaching social history (Sáiz-Serrano 2013). As mentioned above, the choices mostly refer to figures in the political–war sphere coinciding, for example, with the main topics chosen by Brazilian and Portuguese students in the research by Barca and Schmidt (2013) to address world history being the First and Second World War, totalitarian regimes, and so on.

The results are quite similar to those obtained by Apostolidou and Solé in their study with Greek and Portuguese trainee teachers, who mostly pointed to contemporary history and discoveries as the most important subjects to be imparted at school to students received as a result of migration processes (Apostolidou and Solé 2019). Also, this preponderance of characters or events of a warlike nature can be observed in the preferences when talking about Canadian history among Canadian students, who pointed to the First World War as the main event in their country's history (Lévesque 2005).

The appearance of several figures belonging to the sphere of science, such as Isaac Newton, Albert Einstein, and Marie Curie (the fifth, sixth, and eighth figures most chosen by the students), is possibly explained by the context of the students answering this questionnaire, since they were studying the subject of scientific culture when they answered the questions and, in the case of the mentioned women scientists, also because of the stimulus provided by the celebration of Women in Science Day every 11 February, which leads teachers to develop educational activities based on this celebration (Arnau et al. 2019). This is a subject in which students may experience a learning process known as romantic understanding (Egan 1997), characterized by idealizing people, and in this case, scientific figures. It turns these researchers into a kind of legend, heroes who transformed the society of their day with their skills (Hadzigeorgiou and Schulz 2019).

If we compare the results of our research with those obtained in other studies conducted in university contexts, for example the study of 660 narratives produced by university students taking a degree in preschool teacher training (Rivero and Pelegrín 2019), we can again see the predominance of Christopher Columbus (the figure chosen most frequently to write a historical account), mostly justified with arguments about how relevant the figure is today due to the transformation brought about by his arrival in America. This is reasoning in which students aged approximately 19 to 23 are capable of relating a historical figure with the present; the students aged 16 to 17 in our research sample, however, mostly identify the historical relevance of Christopher Columbus only with the change he effected during his era without considering his impact and permanence in the present.

Following the results outlined in the previous paragraph, it is striking that the topic most chosen for historical narratives produced by university students in the study (Rivero and Pelegrín 2019) covers scientific advances, inventions, and discoveries. According to these authors, this choice results from the relationship between the fascination shown for a scientist's or researcher's work and the profundity of the changes of their advances for their contemporaries and the present day (112). It can also be understood as a consequence of the content and focus of the compulsory subject of scientific culture studied in the first year of the Spanish baccalaureate in the autonomous community where both this study and the one by Rivero and Pelegrín were conducted, although as it took place at the start of the academic year, the influence of the contents covered in this subject should not yet prove significant.

Therefore, in the study by Rivero and Pelegrín (2019) the most chosen topic is scientific advances, inventions, and discoveries, with reasoning based on the profundity of the changes these advances and discoveries caused, coinciding with some of the results obtained in this research. Table 1 shows that figures including Isaac Newton, Albert Einstein, and Marie Curie are chosen by a considerable percentage of the sample (13.3%, 12%, and 8%, respectively) with a similar justification (Table 1), primarily based on the novelty value of these scientific approaches or advances during their era, without considering their long-term significance or consequences. However, there is a considerable difference between the study by Rivero and Pelegrín (2019) and this one, since the selection of figures from the world of science in the previous research is based primarily on technical and technological advances and on inventions, and, therefore, the selection of scientists focuses on scientific contributions leading to profound and prolonged changes over time (for example, electricity, aircraft, and telecommunications). In contrast, in this research, the students chose scientists based on their contribution to the history of science and, consequently, their historical relevance focuses on their own historical context and not on the change this involves for the future.

5. Conclusions

To sum up, the results of the research questions we asked—Which figures do Spanish students think are historically relevant after completing their compulsory schooling? Which approach, referring to the level of historical significance, do they use to justify their answer?—allow us to confirm, bearing in mind that we are talking about a concrete sample in a specific local context, that after completing their compulsory secondary education, students choose people they have studied a short while previously as historically relevant figures; in this case, they are figures from contemporary universal history, the history of Spain, and the history of science. Furthermore, when reasoning their response, the approach to the figure's historical significance tends to be assessed only in the context in which the person chosen lived, either due to their importance during that era or due to the change they evoked in their contemporaries. In general, reflection is still needed on change and permanence, on the consequences for society of the actions performed by historical figures considered relevant and on the change these actions or contributions introduced. The students do not understand the significance of the events or phenomena they set in motion beyond their historical moment. Their reasons do not provide a justification reflecting an analysis from a historical perspective, despite having completed their compulsory schooling. Furthermore, in the case of scientific figures, they frequently mention them without managing to explain their contribution or say what their discoveries mean for science and the lives of human beings, particularly in the cases of Newton and Einstein.

Both this research's results and the responses obtained by the other studies mentioned in this article show that, although teaching proposals and epistemological reflections on the development of historical thinking in students, teachers, and trainee teachers are becoming increasingly numerous, the responses continue to be mostly and primarily male figures related to the political sphere, who, according to the students, stand out due to their importance or the impact they had during their era. The influence of a curricular tradition with a gender bias could still be driving this difference, in the same way as the predominance of politics compared with protagonists of history that have performed an action in a more social sphere or one linked to the fight for social rights.

If the appearance in our study of female figures, such as Clara Campoamor or Marie Curie, is understood from new and necessary perspectives of approaching history, such as the conquest of political and social rights by women during the first wave of feminism or the institutionalized vindication, such as the day of women scientists, it is worth asking what new female figures and reasoning will appear if this type of initiative increases with the third wave of feminism or the visibility of women in the history of art.

In other words, the responses obtained in this research demonstrate a development of historical thinking that is still far removed from "learning to think historically" (Lévesque 2008). This not only demonstrates that second-order concepts are known, put into practice, and showcased (for instance, the use of sources, historical empathy, and continuity and change), but it also "breaks" with traditional discourses on forming national identity as a result of political milestones and the predominance of military conflicts, with a focus on history that is barely social and in which men generally take center stage. The inclusion of relevant figures in the world of science and a woman known for fighting for political rights marks an incipient change in the trend; however, critical reflection needs to be fostered in classrooms on the principle of agency in history and the continuity and change influencing the actions of historical figures and social groups over time.

6. Limits and Prospective

There is no doubt that this research, as an exploratory study, presents a certain number of limitations that must be considered, such as, for example, the fact that it is a considerable sample but not large enough to cover a greater number of different educational contexts that would allow us to obtain a more complete image. Also, although we know the official curriculum applied by teachers in the schools where the research was done, we do not know if the methodology used by the teachers may have conditioned the students' responses or if, on the contrary, the previous beliefs that emerged from their homes or peer group prevailed.

However, it is precisely because this is an exploratory study that the directions of future research or the future questions to be answered are highly attractive, for example: when choosing and arguing about the chosen character, does the student's previous beliefs or the teachers' instruction have more impact? Or, for example, does the mark a student usually obtains in the history subject have an influence on the choice and argumentation of his or her answer, and does it make a difference whether the student has more or less affinity with the subject? All these questions and other lines of research are part of the lines of work that [Research group -anonimous-] research group has been working on for some years and whose advances will be published in the future.

Author Contributions: Research designed by I.N-N., B.A. and P.R.; Conceptualization, I.N.-N. and P.R.; methodology, I.N.-N., B.A. and P.R.; formal analysis: I.N.-N.; data curation, B.A.; writing-original draft preparation, I.N.-N. and P.R.; writing review and editing, P.R. and B.A., P.R.; project administration, P.R.; and funding acquisition, P.R. All authors have read and agreed to the published version of the manuscript.

Funding: This research was funded by the project PID2020 -115288RB-I00: Competencias digitales, procesos de aprendizaje y toma de conciencia sobre el patrimonio cultural: Educación de calidad para ciudades y comunidades sostenibles MINECO/AEI-FEDER and the research group S50_20R ARGOS (Gobierno de Aragón).

Institutional Review Board Statement: Not applicable.

Informed Consent Statement: Informed consent was obtained from all subjects involved in the study.

Data Availability Statement: Not applicable.

Conflicts of Interest: The authors declare no conflict of interest.

References

Alvén, Fredrik. 2021. Immigrant Students and the Swedish National Test in History. *Frontiers in Education* 6: 459. [CrossRef]

Apostolidou, Eleni, and Gloria Solé. 2019. Different Ways to Relate with the 'Other': Prospective Teachers Ideas about Teaching History to Refugee Youth, a Comparative Study. In *Temas Controvertidos en el aula. Enseñar Historia en la era de la Posverdad*. Edited by Juan Ramón Moreno-Vera and José Monteagudo-Fernandez. Murcia: Editum, pp. 129–44.

Arias-Ferrer, Laura, Alejandro Egea-Vivancos, and Linda S. Levstik. 2019. Historical Thinking in the Early Years: The Power of Image and Narrative. In *Story in Children's Lives: Contributions of the Narrative Mode to Early Childhood Development, Literacy, and Learning*. Edited by Kelli Jo Kerry-Moran and Juli-Anna Aerila. Cham: Springer International Publishing, pp. 175–98.

Arias-Ferrer, Laura, and Alejandro Egea-Vivancos. 2019. Who changes the course of history? Historical agency in the narratives of Spanish pre-service primary teachers. *History Education Research Journal* 16: 322–39. [CrossRef]

Arnau, María José, Cristina Pertegás, Encarna Alcácer, Rebeca Valero, and José M. Rubio. 2019. La tabla periódica de las mujeres científicas: Construyendo y jgando. *Modelling in Science Education and Learning* 12: 135–52. [CrossRef]

Barca, Isabel. 2000. *O Pensamento Histórico dos Jovens*. Braga: Universidade do Minho.

Barca, Isabel. 2005. Till New Facts are Discovered: Students'Ideas about Objectivity in History. In *Understanding History: Recent Research in History Education. International Reviewof History Education*. Edited by Rosalyn Ashby, Peter Gordon and Peter Lee. Oxford: Routledge, pp. 68–82.

Barca, Isabel, and María Auxiliadora Schmidt. 2013. La consciencia histórica de los jóvenes brasileños y portugueses y su relación con la creación de identidades nacionales. *Educatio Siglo XXI* 31: 25–45.

Barton, Keith C. 2005. "Best Not to Forget Them": Adolescents' Judgments of Historical Significance in Northern Ireland. *Theory and Research in Social Education* 33: 9–44. [CrossRef]

Barton, Keith C. 2011. *Agency, Choice, and Historical Action: How History Teaching Can Help Students Think about Democratic Decision Making*. Washington, DC: American Educational Research Associations.

Barton, Keith C., and Linda S. Levstik. 2004. *Teaching History for the Common Good*. Oxford: Routledge.

Berger, Stefan. 2007. Writing national histories in Europe: Reflections on the pasts, presents, and futures of a tradition. In *Conflicted Memories: Europeanizing Contemporary Histories*. Edited by Konrad H. Jarausch and Thomas Lindenberger. Oxford: Berghahn Books, pp. 22–68.

Berger, Stefan, and Chris Lorenz, eds. 2008. *The Contested Nation: Ethnicity, Class, Religion and Gender in National Histories*. London: Palgrave Macmillan.

Calderón-López, Pablo, and Laura Arias-Ferrer. 2017. El desembarco de Colón en San Salvador. Análisis de las narraciones del alumnado español de 10–11 años en términos de perspectiva histórica. *REIRE. Revista d'Innovació i Recerca en Educació* 10: 34–51. [CrossRef]

Carretero, Mario, and Ana Atorresi. 2008. Pensamiento narrativo. In *Psicología del Pensamiento*. Edited by Mario Carretero and Mikel Asensio. Madrid: Alianza, pp. 269–89.

Carretero, Mario, Stefan Berger, and Maria Grever, eds. 2017. *Palgrave Handbook of Research in Historical Culture and Education*. London: Palgrave McMillan.

Carril-Merino, María Teresa, María Sánchez-Agustí, and Carlos Muñoz-Labraña. 2020. Empatía histórica en futuros profesores de primaria: El trabajo infantil en el siglo XIX. *Educaçao e Pesquisa: Revista da Faculdade de Educação da Universidade de São Paulo* 46: 1–18. [CrossRef]

Center of Historical Thinking Consciousness. 2016. The Historical Thinking Project. Available online: https://historicalthinking.ca/ (accessed on 23 March 2022).

Cercadillo, Lis. 2001. Significance in history: Students' ideas in England and Spain. In *Raising Standards in History Education. International Review of History Education*. Edited by Alaric K. Dickinson, Peter Gordon and Peter Lee. London: Woburn Press, vol. 3, pp. 116–45.

Cercadillo, Lis. 2002. Los alumnos de secundaria y sus ideas sobre conceptos históricos estructurales. *Iber: Didáctica de las Ciencias Sociales, Geografía e Historia* 31: 87–102.

Cercadillo, Lis. 2004. Las ideas de los alumnos sobre lo que es verdad en historia. *Enseñanza de las Ciencias Sociales: Revista de Investigación* 3: 3–14.

Cercadillo, Lis. 2006. 'Maybe they haven't decided yet what is right:' English and Spanish perspectives on teaching historical significance. *Teaching History* 125: 6–9.

Chaparro Sainz, Álvaro, María del Mar Felices de la Fuente, and Laura Triviño Cabrera. 2020. La investigación en pensamiento histórico. Un estudio a través de las tesis doctorales de Ciencias Sociales (1995–2020). *Panta Rei. Revista Digital de Historia y Didáctica de la Historia* 14: 93–147. [CrossRef]

Clark, Penney, ed. 2011. *New Possibilities for the Past. Shaping History Education in Canada*. Vancouver: UBC Press.

Colomer, Juan Carlos, Jorge Sáiz Serrano, and Rafael Valls. 2018. Competencias históricas y actividades con recursos tecnológicos en libros de texto de Historia: Nuevos materiales y viejas rutinas. *Ensayos: Revista de la Facultad de Educación de Albacete* 33: 53–64.

Cooper, Hilary, and Arthur Chapman, eds. 2009. *Constructing History*. London: SAGE Publications Ltd., pp. 11–19.

Counsell, Christine, Katharine Burn, and Arthur Chapman, eds. 2016. *MasterClass in History Education. Transforming Teaching and Learning*. London: Bloomsbury.

Egan, Kieran. 1997. *The Educated Mind: How Cognitive Tools Shape Our Understanding*. Chicago: University of Chicago Press.

Egea-Vivancos, Alejandro, and Laura Arias-Ferrer. 2018. ¿Qué es relevante históricamente? Pensamiento histórico a través de las narrativas de los estudiantes universitarios. *Educaçao e Pesquisa: Revista da Faculdade de Educação da Universidade de São Paulo* 44: 1–16. [CrossRef]

Egea-Vivancos, Alejandro, Laura Arias-Ferrer, and Juan Santacana Mestre. 2018. *Y la Arqueología Llegó al Aula. La Cultura Material y el Método Arqueológico Para la Enseñanza de la Historia y el Patrimonio*. Gijón: Trea.

Gómez-Carrasco, Cosme J., Ramón López-Facal, and Jairo Rodríguez-Medina. 2019. La investigación en Didáctica de las Ciencias Sociales en revistas españolas de Ciencias de la Educación. Un análisis bibliométrico (2007–2017). *Revista de Didáctica de las Ciencias Experimentales y Sociales* 37: 67–88. [CrossRef]

Grant, S. G. 2001. It's Just the Facts, or Is It? The Relationship between Teachers' Practices and Students' Understandings of History. *Theory and Research in Social Education* 29: 65–108. [CrossRef]

Hadzigeorgiou, Yannis, and Roland M. Schulz. 2019. Engaging Students in Science: The Potential Role of "Narrative Thinking" and "Romantic Understanding". *Frontiers in Education* 4: 14267. [CrossRef]

Hunt, Martin. 2000. Teaching historical significance. In *Issues in History Teaching*. Edited by James Arthur and Robert Phillips. Oxford: Routledge, pp. 39–53.

Ibagón-Martín, Nilson, and Pedro Miralles. 2021. Relevancia histórica desde las perspectivas de estudiantes colombianos. Permanencia y transformación de los modelos de evocación histórica. *Memorias: Revista Digital de Historia y Arqueología desde el Caribe Colombiano* 44: 37–65. [CrossRef]

Ibagón-Martín, Nilson, Javier J. Maquilón-Sánchez, and Pedro Miralles. 2021. Identidad histórica y formas de expression nacionalista en contextos escolares multiculturales. *Campo Abierto* 40: 307–22.

Lévesque, Stéphane. 2005. Teaching second-order concepts in Canadian history: The importance of "historical significance". *Canadian Social Studies* 39. Available online: https://files.eric.ed.gov/fulltext/EJ1073987.pdf (accessed on 1 April 2022).

Lévesque, Stéphane. 2008. *Thinking Historically: Educating Students for the 21st Century*. Toronto: University of Toronto Press.

López, Cesar, and Margarita G. Márquez. 2018. Aportaciones de la psicología a la comprensión de las narrativas nacionales y la identidad nacional: Pensamiento narrativo, identidades sociales y emociones colectivas. In *¿Qué Saben de su Historia Nuestros Jóvenes? Enseñanza de la Historia e Identidad Nacional*. Edited by Ander Delgado and Antonio Rivera. Comares: País, pp. 99–118.

López-Rodríguez, César. 2015. Repensando las narrativas nacionales: Un análisis del origen, transmisión e influencia en el aprendizaje histórico. *Panta Rei. Revista Digital de Ciencia y Didáctica de la Historia* 9: 77–92. [CrossRef]

López-Rodríguez, César. 2020. Las narrativas nacionales en la enseñanza y el aprendizaje de la historia: Una revisión sobre los libros de texto y las narrativas de los estudiantes. *Panta Rei. Revista Digital de Historia y Didáctica de la Historia* 14: 149–66. [CrossRef]

Lowenthal, David. 1998. *The Heritage Crusade and the Spoils of History*. Cambridge: Cambridge University Press.

Metzger, Scott Alan, and Laura McArthur Harris, eds. 2018. *The Wiley International Handbook of History Teaching and Learning*. New York: Blackwell.

Miguel-Revilla, Diego, and María Sánchez-Agustí. 2018. Modelos de conciencia histórica en el alumnado de Educación Secundaria: Tradición, simbología y contextualización en torno a los restos del franquismo. *Panta Rei: Revista de Ciencia y didáctica de la Historia* 12: 119–42. [CrossRef] [PubMed]

Molina-Puche, Sebastián, and Juan Egea. 2018. Evaluar la adquisición de competencias históricas en Secundaria: Un estudio de caso centrado en la empatía histórica. *Ensayos: Revista de la Facultad de Educación de Albacete* 33: 1–22. [CrossRef]

Monte-Sano, Chauncey. 2010. Learning to Open up History for Students: Preservice Teachers' Emerging Pedagogical Content Knowledge. *Journal of Teacher Education* 62: 260–72. [CrossRef]

Navarro-Neri, Iñaki, Pilar Rivero, Borja Aso, and Ana María Mendióroz-Lacambra. 2021. Narrativa y significación histórica en la comunicación de los museos nacionales canadienses en Facebook. *Panta Rei. Revista Digital de Historia y Didáctica de la Historia* 15: 185–207. [CrossRef]

Ortuño, Jorge, Ana Isabel Ponce, and Francisca José Serrano. 2016. La idea de causalidad en las explicaciones históricas del alumnado de educación primaria. *Revista de Educación* 371: 9–34.

Pagès, Joan, and Antoni Santisteban. 2010. La enseñanza y el aprendizaje del tiempo histórico en educación primaria. *Cadernos CEDES* 30: 278–309.

Partington, Geoffrey. 1980. *The Idea of an Historical Education*. Glasgow: NFER Publishing Company.

Peck, Carla, and Peter Seixas. 2008. Benchmarks of Historical Thinking: First Steps. *Canadian Journal of Education* 31: 1015–38.

Phillips, Robert. 2002. Historical significance–the forgotten "Key Element"? *Teaching History* 106: 14–19.

Ricoeur, Paul. 1987. *Time and Narrative III*. Chicago: The University of Chicago Press.

Rivero, Pilar, and Julián Pelegrín. 2019. What History do future teachers of early childhood education think is relevant? *Cadernos de Pesquisa* 49: 96–120. [CrossRef]

Rivero, Pilar, Olaia Fontal, Marta Martínez, and Silvia García-Ceballos. 2018. La educación patrimonial y el patrimonio arqueológico para la enseñanza de la Historia: El caso de Bílbilis. *ENSAYOS, Revista de la Facultad de Educación de Albacete* 33: 23–37. [CrossRef]

Rüsen, Jörn. 2004. Historical Consciousness: Narrative, Structure, Moral Function and Ontogenetic Development. In *Theorizing Historical Consciousness*. Edited by Peter Seixas. Toronto: University of Toronto Press, pp. 63–85.

Rüsen, Jörn. 2005. *History: Narration, Interpretation, Orientation*. New York: Berghahn Books.

Sáiz-Serrano, Jorge. 2013. Alfabetización histórica y competencias básicas en libros de texto de historia y en aprendizajes de estudiantes. *Didáctica de las Ciencias Experimentales y Sociales* 27: 43–66. [CrossRef]

Sáiz-Serrano, Jorge, and Cosme J. Gómez-Carrasco. 2016. Investigar pensamiento histórico y narrativo en la formación del profesorado: Fundamentos teóricos y metodológicos. *Revista Electrónica Interuniversitaria de Formación del Profesorado* 19: 175–90.

Sáiz-Serrano, Jorge, and Ramón López-Facal. 2014. Conciencia histórica e identidad nacional en narrativas de alumnado español de Bachillerato. In *Historia e Identidades Culturales*. Edited by Joaquín Prats, Isabe Barca and Ramón López Facal. Braga: CIED, Universidade do Minho, pp. 1187–98.

Sáiz-Serrano, Jorge, and Ramón López-Facal. 2015. Competencias y narrativas históricas: El pensamiento histórico de estudiantes y futuros profesores españoles de educación secundaria. *Revista de Estudios Sociales* 52: 87–101. [CrossRef]

Sant, Edda, Joan Pagès, Antoni Santisteban, Neus González, and Montserrat Oller. 2014. Narrativas y discursos: ¿cómo podemos analizar la competencia narrativa del alumnado en el aprendizaje de la Historia? *Clio & Asociados: La Historia Enseñada* 18–19: 166–82.

Santisteban, Antoni. 2017. Del tiempo histórico a la conciencia histórica: Cambios en la enseñanza y el aprendizaje de la historia en los últimos 25 años. *Diálogo Andino* 53: 87–99. [CrossRef]

Seixas, Peter. 1998. Student Teachers Thinking Historically. *Theory & Research in Social Education* 26: 310–41.

Seixas, Peter, ed. 2004. *Theorizing Historical Consciousness*. Toronto: Toronto University Press.

Seixas, Peter. 2011. Assessment of Historical Thinking. In *New Possibilities for the Past. Shaping History Education in Canada*. Edited by P. Clark. Vancouver: University of British Columbia Press, pp. 139–53.

Seixas, Peter, and Carla Peck. 2004. Teaching Historical Thinking. In *Challenges and Prospects for Canadian Social Studies*. Edited by Alan Sears and Ian Wright. Vancouver: Pacific Educational Press, pp. 109–17.

Seixas, Peter, and Tom Morton. 2012. *The Big Six Historical Thinking Concepts*. Nelson: Nelson College Indigenous.

Solé, Gloria, and Marília Gago. 2021. The history teacher education process in Portugal: A mixed method study about professionalism development. *Humanities Social Communications* 8: 1–12. [CrossRef]

Sung, Pei-Fen. 2020. Historical consciousness matters: National identity, historical thinking and the struggle for a democratic education in Taiwan. *Journal of Curriculum Studies* 52: 685–701. [CrossRef]

Tasshakori, Abbas, R. Burke Johnson, and Charles Teddlie. 2021. *Foundations of Mixed Methods Research. Integrating Quantitative and Qualitative Approaches in the Social and Behavioral Sciences*, 2nd ed. Thousand Oaks: SAGE Publications Ltd.

VanSledright, Bruce. 2008. Narratives of nation-state, historical knowledge, and school history education. *Review of Research in Education* 32: 109–46. [CrossRef]

VanSledright, Bruce A. 2011. *The Challenge of Rethinking History Education. On Practice, Theories, and Policy*. Oxford: Routledge.

VanSledright, Bruce A. 2014. *Assessing Historical Thinking and Understanding: Innovate Designs for New Standards*. Oxford: Routledge.

Wineburg, Sam. 2001. *Historical Thinking and Other Unnatural Acts: Charting the Future of Teaching the Past*. Philadelphia: Temple University Press.

Wineburg, Sam, Daisy Martin, and Chauncey Monte-Sano. 2013. *Reading like a Historian. Teaching Literacy in Middle & High School History Classrooms*. New York: Teachers' College Press.

social sciences

MDPI

Article

Student Teachers' Willingness to Act in the Climate Change Context

Nena Vukelić [1,*], Nena Rončević [1] and Sven Toljan [2]

[1] Department of Education, Faculty of Humanities and Social Sciences, University of Rijeka, 51000 Rijeka, Croatia; nena.roncevic@ffri.uniri.hr
[2] Independent Researcher, 47000 Karlovac, Croatia; sven.toljan@gmail.com
* Correspondence: nvukelic1@ffri.uniri.hr

Abstract: In education for sustainable development, widely regarded as a framework that offers us the opportunity to improve the ways in which we cope with climate change issues, the need for student teachers to express willingness to act in order to deal with numerous issues and challenges of sustainable development, especially climate change, is of particular importance. Therefore, the focus of this study is on the examination of predictors of student teachers' willingness to act in a climate-change mitigation and adaptation context. For the purpose of this study, measurement instruments of willingness to act in climate change mitigation and adaptation context, attitudes towards climate change, perception of action possibilities in climate change mitigation and adaptation context, interest in climate change and concern for ecological problems were validated. A total of 201 student teachers from the University of Rijeka (Croatia) participated in the study. It was determined that (I) attitudes towards climate change, (II) perception of action possibilities in climate change mitigation and adaptation context and (III) interest in climate change represent significant predictors of willingness to act in climate-change direction and mitigation contexts. Based on the results of this study, recommendations for teacher education in the climate change context have been offered.

Keywords: climate change; sustainable development goals (SDG); student teachers; education for sustainable development; action competence

Citation: Vukelić, Nena, Nena Rončević, and Sven Toljan. 2022. Student Teachers' Willingness to Act in the Climate Change Context. *Social Sciences* 11: 47. https://doi.org/10.3390/socsci11020047

Academic Editor: Alvaro Morote

Received: 23 November 2021
Accepted: 24 January 2022
Published: 27 January 2022

Publisher's Note: MDPI stays neutral with regard to jurisdictional claims in published maps and institutional affiliations.

1. Introduction

Climate change represents a complex challenge of our times. It is known that natural as well as anthropogenic factors have an influence on variations in climate, i.e., that humanity's influence on the climate started during the beginning of Industrial Revolution. Moreover, climate change represents a complex topic as well as an issue at the global level with unpredictable consequences. Although evidence, as well as a consensus, exists in the scientific community regarding the causes and consequences of climate change on the future of humankind (Masson-Delmotte et al. 2021), different attitudes towards this issue are still present (Liu et al. 2015). Consequently, this situation leads to slow changes in behavior. A study conducted by Gatersleben et al. (2010) indicated that, on the one hand, a number of people simultaneously express high level of materialism and concern for climate change. Based on that, they concluded that behavioral changes are slow, despite the fact that society began to recognize climate change as a major problem. One of the potential explanations of these findings is "aridity" of climate change, which is present even among highly motivated individuals (Howell and Allen 2019). Additionally, Morton (2013) stated that additional explanations and studies of climate change can contribute to an (un)conscious delay of activities focused on climate protection. Therefore, focus should be put on adaptation as well as climate change mitigation strategies.

The Sustainable Development Solution Network just relisted Climate Action Edition 2021, in which they conclude that "this decade will define the course of people and the planet far into the future, and it must be a Decade of Action . . . to bring about an inclusive, resilient, and net-zero world "(SDSN 2021, p. 7). Additionally, Sustainable Development Goal, Goal 13, is directly focused on Climate Action, whereby the UN calls for urgent action to combat climate change and its impacts, and Target 13.3 seeks to "Improve education, awareness-raising and human and institutional capacity on climate change mitigation, adaptation, impact reduction and early warning.". Furthermore, European Parliament resolution from March 2019 states that youth education represents one of the most efficient ways of combating climate change.

Learning how to live with climate change represents both an ontological and existential task of newly defined understandings of what life actually is, what it means to live and how to live well, leading to the creation of new, sustainable lifestyles (Verlie 2019). It is important to urge not only students, but also people in general to participate more actively in what we refer to today as a "super wicked problems" (Lazarus 2008, p. 1153) as well as the hardest moral challenge (Hudson 2017). Learning how to cohabitate with climate change represents a new discipline of pedagogy, which confirms that climate change does not only represent "a condition which we should be ashamed of" or "a problem that has to be solved" (Hulme 2009, p. 364). Despite the fact that education for climate change is not sufficient to solve this problem, it still plays a key role in the switch towards a post-carbon world. Due to this fact, it is important to educate people in line with climate change pedagogy principles in order to adopt pro-ecological attitudes. At its core, education about climate change refers to learning during the period of risk, uncertainty, and rapid changes. People have never faced a situation similar to which they have to face today in the past, which raises the question of how to educate the youth about it. It is important to understand that climate change does not only represent a scientific problem, but also a complex social problem that requires more than pure content learning (McKeown and Hopkins 2010). Teachers have to have clear ideas regarding the problem as well ways in which they can deal with it in order to implement necessary changes into school curricula. Students who demand action on climate crisis have clearly stated "Our education system failed us. We are not adequately taught about climate crisis in our classrooms. Schools do not prepare us for the world which we will enter soon, instead, they prepare us for the jobs and society that is based on a system which created the crisis. We require an urgent reform of national curriculum as well as placing ecological crisis as our top educational priority" (Irwin 2020). This message clearly indicates that students are becoming aware of the seriousness of issues regarding climate change, whose harmful consequences they will have to suffer if we do not change our current behavior immediately. They do not want to become the victims of the current adult generation's selfishness which, in its short-term policies and reflections, does not take into account those who will succeed them. Consequently, they justifiably believe that education in its current form is inadequate. Clearly, a paradigm shift in the direction of transformative education is needed. Teachers should create educational environment in which transformative, student-centered teaching and learning is encouraged and eventually achieved. This challenge to (student) teachers best outlines the significance of their role in moving toward a sustainable future.

The need for (student) teachers to show a willingness to act towards the direction of sustainable development as well as a willingness to implement education for sustainable development (e.g., Sleurs 2008; UNECE 2012; Rauch and Steiner 2013; UNESCO 2020) is particularly emphasized due to the fact that education for sustainable development represents one of the key areas within which it is possible to make progress in climate change mitigation context. Therefore, an important question is being raised regarding the factors that shape (student) teachers' willingness to act in the direction of sustainable development as well as how to encourage it through teachers' initial training.

The results of previous research on climate change from the student teachers' perspective point out the fact that student teachers are mostly aware of the sustainable development issues at both the local and national level, while frequently underestimating the importance of global ecological problems such as climate change (Spiropoulou et al. 2007). While reflecting on sustainable development as well as their role in shifting towards it, student teachers are mostly focused on its social dimension, while neglecting critical global topics as well as their role in coping with issues such as climate change (Koskela and Kärkkäinen 2021). One potential explanation for this situation can be found in the results of IEA International Civic and Citizenship Education Study from 2016 that revealed that percentages of teachers in Croatia reporting having participated in training courses on civic-related topics, namely the environment and environmental sustainability during pre-service and/or in-service training was only 28% (Schulz et al. 2018). Furthermore, Demant-Poort and Berger (2021) found that student teachers assess that they do not possess sufficient knowledge about climate change.

Although both policy documents (UNESCO 2015, 2017) and theoretical assumptions (e.g., Rauch and Steiner 2013; Sleurs 2008) put considerable emphasis in education for sustainable development on (student) teachers, who are perceived as potential agents of change needed for achieving sustainable development goals in the direction of sustainable future, it seems that they do not comprehend the importance of their own role in dealing as well as coping with sustainable development issues and problems such as climate change. Therefore, with climate change being one of the sustainable development's burning questions, the focus of this study is placed particularly on the examination of (predictors of) student teachers' willingness to act in climate change context.

1.1. (Predictors of) Willingness to Act in Climate Change Mitigation and Adaptation Context

Student teacher's willingness to act refers both to their intention and will to participate in significant and meaningful activities that bring desired changes in a sustainability context (Vukelić 2021). These activities can either specifically refer to student teacher's future professional actions such as implementation of certain education for sustainable development's aspects in their teaching or their general, daily actions.

In the context of climate change, actions one (intent to) exhibits usually fall into the following two broad categories: climate change mitigation and climate change adaptation (Chen et al. 2017). Climate change mitigation is aimed at minimizing the possible impacts of climate change, as well as tackling the causes of climate change. Climate change adaptation actions refer to altering our behavior and ways of life to protect the environment and overall wellbeing of planet and living beings. These actions are aimed at reducing the negative effects of climate change (Chen et al. 2017). In line with that, in this paper focus is placed on student teachers' willingness to act in climate change mitigation and adaptation context.

Action competence represents one of the key constructs while thinking about (teacher)'s willingness to act in climate change context or, in general, sustainable development (Breiting and Mogensen 1999; Jensen and Schnack 1997; Sass et al. 2020). At the individual level, action competence is defined as individual's latent capacity to act in the direction of sustainable development (Olsson et al. 2020). Sass et al. (2020) specify that a person possessing action competence is characterized by dedication and passion in dealing with sustainable development issues, knowledge regarding the problems in question, a critical but positive attitude towards various ways of solving the previously mentioned issues as well as trust in personal skills and capacities needed to create improved conditions. In other words, key components of action competence are willingness to act and perception of action possibilities (Breiting and Mogensen 1999; Mogensen and Schnack 2010; Sass et al. 2020).

Perception of action possibilities refers to individual's capacity to search for information and build network of knowledge based on the possibility to use personal list of activities in the direction of sustainable development, or in the context of this study, in the direction of climate change. Individuals have to understand which possible actions could be taken as well as how complex issues of sustainable development function (Kaplan and Kaplan 2009).

However, in order for student teachers to express willingness to act in climate change mitigation and adaptation context as well as switch toward sustainable future, multiple motivational factors have to be present. Additionally, the same group has to become aware of the importance of understanding the possibilities to act. On the one hand, motivational factors include attitudes and interest in issues regarding climate change, while on the other hand, they involve perceptions of the future in climate change context as well as concern for the future.

In education for sustainable development, it is assumed that student teachers have to develop both skills and knowledge for sustainable development, and also express positive attitudes towards sustainability issues in order to become prepared to integrate and facilitate this type of education (Sleurs 2008; Rauch and Steiner 2013; UNESCO 2015, 2017). The abovementioned assumption is considered to be one of the perquisites for the successful implementation of education for sustainable development (Albareda-Tiana et al. 2018; Cebrián and Junyent 2015). It is expected that attitudes towards both supporting and accepting the idea of sustainable development, including climate change, form not only part of future practice, but also a subjective evaluation of student teacher's willingness to act in the direction of sustainable development through the implementation of education for sustainable development (Vukelić 2021). Additionally, not only do previous research results mostly indicate (student) teachers' relatively positive attitudes towards ecological topics (for example, climate change) (e.g., Boon 2016; Competente 2019), but they also indicate their relatively lower level of knowledge as well as numerous misconceptions regarding climate change (Boon 2016; Nyarko and Petcovic 2021; Seroussi et al. 2019).

In situations in which the interrelation between attitudes and consequential behavior is observed, not only do numerous conclusions of previous empirical studies remain vague, but contradictory findings also often occur. For example, on the one hand, results of previous studies support the existence of connection between attitudes towards ecological matters and pro-ecological behavior (e.g., Heimlich and Ardoin 2008; Hines et al. 1987), while on the other hand, certain studies, whose findings point out the lack of connection between attitudes and behavior or behavioral intention, can also be found (e.g., Chen 2010; Chen et al. 2010; Cleveland et al. 2012). Due to the vagueness of the conclusions of previous empirical studies, it is important to examine the predictive value of attitudes towards climate change in the prediction of student teachers' willingness to act in climate change mitigation and adaptation context.

Apart from attitudes towards climate change, student teachers' interest in this phenomenon also represents an important motivational factor that has to be considered as a potential predictor of willingness to act in climate change mitigation and adaptation context. Interest is closely connected with intrinsic motivation, i.e., it is a part of its core. In this context, an intrinsic motivation refers to individual's motivation to follow personal interest as well as to put effort into searching challenges which they find significant (Ryan and Deci 2000). In an attempt to explain factors that form pro-ecological behaviors, the significance of intrinsic motivation is often singled out (Steg et al. 2016; Van der Werff et al. 2013). Specifically, individuals who express an intrinsic interest in environmental protection matters, climate change mitigation or general sustainable development issues, and feel intrinsic responsibility for solving sustainability issues, behave pro-ecologically more often (Steg et al. 2016). Based on that, in this study, interest in climate change was identified as a potential factor of student teachers' willingness to act in the climate change mitigation and adaptation context.

It is believed that encouraging views oriented towards the future represents a precondition for the creation of society based on sustainable development (UNESCO 2020). Results of previous studies point out that individuals mostly experience climate change as something which will occur in the distant future without any consequences for them, which is ultimately related with a lack of motivation to behave pro-ecologically (Coulter et al. 2019; Jones et al. 2017). Therefore, by focusing on personal future, in which climate change poses a real threat, can lead to an increasing feeling of concern for ecological issues resulting in higher level of pro-ecological behavior (Lee et al. 2020). In other words, perception of (personal) future in climate change context as well as concern for the future represent one of the key motivational factors that fuel both pro-ecological behavior and potential acting in the direction of climate change.

In research on education, studies that examine teachers' willingness to act in the climate change mitigation and adaptation context are scarce. For example, Seroussi et al. (2019) found that only a small number of participating teachers were ready to act in order to mitigate climate change. Moreover, the percentage of teachers who were willing to act is lower than the percentage of teachers who possessed sufficient knowledge regarding climate change. Additionally, Seroussi et al. (2019), found that their concern is connected with the intention to act in order to mitigate climate change. Based on that, in this study, the concern for ecological problems is identified as one of the potential predictors of willingness to act in the climate change mitigation and adaptation context.

1.2. Socio-Demographic Correlates of Willingness to Act in Climate Change Mitigation and Adaptation Context

While thinking about potential correlates of willingness to act in the climate change mitigation and adaptation context, it is necessary to consider potential effects of socio-demographic characteristics such as student teachers' gender and age. Namely, the results of previous empirical research indicate that the concern regarding the contribution to sustainable development as well as switching towards sustainable future differs depending on the individual's gender (Meinzen-Dick et al. 2014). Therefore, women express stronger positive attitudes towards sustainable development as well as higher level of willingness to act in the direction of environmental protection and general pro-ecological behavior (e.g., De Silva and Pownall 2014; Rončević and Cvetković 2016). The abovementioned differences are also present in education for sustainable development and the student teacher area where it has been continuously found that female student teachers express more positive attitudes towards environment and ecological topics, higher level of ecological literacy, clearer conceptions regarding sustainable development topics as well as a higher level of awareness about burning questions related to sustainable development such as climate change (e.g., Al-Naqbi and Alshannag 2018; Kilinc and Aydin 2013; Larijani 2010; Oerke and Bogner 2010; Tuncer et al. 2006, 2009; Vukelić 2021; Zelezny et al. 2000).

Apart from gender, student teachers' age also represents one of the potential socio-demographic correlates of their willingness to act in the climate change mitigation and adaptation context. Previous research on age differences in variables related to climate change (or any other sustainable development issue) are not only relatively scarce, but they also indicate markedly inconsistent results (Wray-Lake et al. 2010). Therefore, this study will try to answer the question of whether participants' age represents significant predictor of willingness to act in climate change mitigation and adaptation context.

2. Materials and Methods

The aim of this study was to examine predictors of student teachers' willingness to act in the climate change mitigation and adaptation context. Moreover, by reviewing relevant databases as well as studying literature on education for sustainable development, it was determined that this is a topic which is not only markedly scarce in international, but also completely neglected in national scientific discourse. A lack of focus on factors that form student teachers' perceptions regarding climate change is evident. Similarly, a lack of instruments constructed to measure the abovementioned perceptions is present. Thus, the first objective of this study was to examine the measurement characteristics of the measurement instruments used to measure various aspects of student teachers' perceptions regarding climate change. More precisely, the instruments were constructed and validated to measure (I) willingness to act in climate change mitigation and adaptation context, (II) attitudes towards climate change, (III) perception of action possibilities in climate change mitigation and adaptation context, (IV) perception of future in climate change context, (V) interest in climate change and (VI) concern for ecological problems. The second objective of this study was to examine which of the abovementioned factors represent significant predictors of student teachers' willingness to act in climate change mitigation and adaptation context.

2.1. Sample

A total of 201 student teachers from the University of Rijeka (Croatia), out of which 78.6% of them were female, participated in the study. The study used convenience sampling. Participants' average age was 22.95 years (SD = 2.57). Online surveying was used in order to collect data. Prior to surveying, the participants were familiarized with the purpose of the survey as well as the anonymity of the given data. The participants needed around 10 to 15 minutes to complete the survey.

2.2. Instruments

Data on participants' socio-demographic characteristics, i.e., their age and gender, were collected. This study used questionnaires with the purpose of measuring the aspects of student teachers' perception on climate change: (I) Willingness to act in climate change mitigation and adaptation context scale (8 items), (II) Attitudes towards climate change scale (5 items) (III) Perception of action possibilities in climate change mitigation and adaptation context scale (4 items), (IV) Perception of future in climate change scale (7 items), (V) Interest in climate change scale (5 items) and (VI) Concern for ecological problems scale (10 items). The concern for ecological problems scale was adopted from Cifrić (2005). All the other used scales, i.e., its items, were adopted from Hadžiselimović (2015). The participants expressed their agreement with all of the items on a 5-point Likert scale (1—*I completely disagree*, 5—*I completely agree*).

All measurement instruments were validated, and their internal consistency was determined. For the purpose of this study, an exploratory factor analysis was used, and Cronbach's alpha coefficient of internal consistency (α) was calculated. Additionally, descriptive data were calculated (means (M) and standard deviations (SD) for all measures. Data were analyzed by using IBM SPSS 24.

In order to obtain initial factor solution for all measurement instruments, exploratory principal component analysis was carried out. The significance of KMO test as well as Bartlett's test of sphericity was calculated. For all measurement instruments, it was determined that KMO test was not significant ($p > 0.05$), while Bartlett's test showed statistical significance. The Guttman-Kaiser as well as Scree test criterion were used to determine the number of extracted factors. According to these criteria, one factor was extracted in every measurement instrument. Not a single item had a low factor loading (<0.3). Furthermore, Cronbach's alpha (α) coefficient values for all measurement instruments used in this research ranged between $\alpha = 0.7$ and $\alpha = 0.94$.

The results of an exploratory principal component analysis for all measurement instruments are shown below. Tables 1–6 show factor (structure) matrix and factor loadings for one-factor solutions for all measurement instruments as well as their descriptive data and coefficients of reliability.

For Willingness to act in climate change mitigation and adaptation context scale one factor that explains 50.37% of variance was extracted (Table 1).

For Attitudes towards climate change scale, according to Guttman-Kaiser criterion as well as Scree test criterion, one factor that explains 71.52% of variance was extracted (Table 2).

For Perception of action possibilities in climate change mitigation and adaptation context scale one factor that explains 52.72% of variance was extracted (Table 3). Prior to the creation of a simple linear composite, first as well as second item of the questionnaire were recoded as they carried opposite meaning compared to other items (marked with (R) in Table 3).

For Perception of future in climate change scale, according to Guttman-Kaiser criterion as well as Scree test criterion, one factor that explains 70.19% of variance was extracted (Table 4).

For Interest in climate change scale according to Guttman-Kaiser criterion as well as Scree test criterion, one factor that explains 68.75% of variance was extracted (Table 5). Prior to the creation of the simple linear composite, last item of the questionnaire was recoded as it carried opposite meaning compared to other items (marked with (R) in Table 5).

For Concern for ecological problems scale, according to Guttman-Kaiser criterion as well as Scree test criterion, one factor that explains 67.15% of variance was extracted (Table 6).

Table 1. Factor (structure) matrix and factor loadings for one-factor solution, coefficient of reliability and descriptive data for Willingness to act in climate change mitigation and adaptation context scale.

Item	Factor Loadings	M (SD)
(1) I am ready to self-initiatively do what it takes to mitigate climate change.	0.683	4.0 (0.87)
(2) I am ready to do what it takes in order to mitigate climate change if somebody would require me to do it (e.g., local government)	0.742	4.13 (0.92)
(3) Protecting climate represents a more significant task compared to others.	0.690	3.23 (1.02)
(4) I am ready to limit my future travels, especially by plane.	0.673	3.27 (1.09)
(5) I will try to learn as much as possible about climate change.	0.774	3.80 (0.91)
(6) Along with previous formal education (school and faculty), I am studying about climate change informally.	0.648	3.48 (1.10)
(7) I would like to teach about climate change at my future job.	0.746	3.35 (1.19)
(8) I intent to include the issue of climate change at my future job.	0.712	3.77 (1.11)
Explained variance	50.37%	
M (SD)	29.03 (5.81)	
Cronbach's α	0.86	

Table 2. Factor (structure) matrix and factor loadings for one-factor solution, coefficient of reliability and descriptive data for Attitudes towards climate change scale.

Item	Factor Loadings	M (SD)
(1) Climate change represents a very serious problem.	0.855	4.57 (0.69)
(2) Climate change represents a threat to my future wellbeing and safety.	0.782	4.07 (0.94)
(3) Climate change represents a threat to future generation, their lives and safety.	0.892	4.51 (0.69)
(4) Climate change represents a threat to humankind on planet Earth.	0.891	4.48 (0.76)
(5) Climate change represents a threat to all living beings on the Earth (including animals and plants).	0.803	4.53 (0.74)
Explained variance	71.52%	
M (SD)	22.16 (3.2)	
Cronbach's α	0.89	

Table 3. Factor (structure) matrix and factor loadings for one-factor solution, coefficient of reliability and descriptive data for Perception of action possibilities in climate change mitigation and adaptation context scale.

Item	Factor Loadings	M (SD)
(1) What I do as an individual will not help in mitigating climate change. (R)	−0.724	2.68 (1.20)
(2) We can't do anything to stop climate change. (R)	−0.791	1.84 (0.93)
(3) What we do can stop climate change from becoming even bigger problem.	0.713	3.84 (1.01)
(4) I believe I can persuade others to put effort into mitigating climate change.	0.671	3.22 (1.00)
Explained variance	52.72%	
M (SD)	14.54 (3.01)	
Cronbach's α	0.7	

(R)—reverse item.

Table 4. Factor (structure) matrix and factor loadings for one-factor solution, coefficient of reliability and descriptive data for Perception of future in climate change scale.

Item (In the Next 50 Years, ...)	Factor Loadings	M (SD)
(1) ... there will be more heat weaves, droughts and wildfires.	0.885	4.38 (0.76)
(2) ... we are going to experience energy supply issues.	0.681	3.89 (0.98)
(3) ... streets, tunnels and roads will get more frequently flooded.	0.854	4.10 (0.85)
(4) ... we are going to experience more frequent and heavier storms.	0.886	4.17 (0.84)
(5) ... we are going to experience more frequent health issues and epidemics.	0.775	4.12 (0.99)
(6) ... (un)settled low-elevation coastal zones will get flooded due to storms and sea level rise.	0.858	4.22 (0.86)
(7) ... we are going to experience heavy rainfalls and landslides.	0.903	4.18 (0.85)
Explained variance	70.19%	
M (SD)	29.08 (5.09)	
Cronbach's α	0.92	

Table 5. Factor (structure) matrix and factor loadings for one-factor solution, coefficient of reliability and descriptive data for Interest in climate change scale.

Item	Factor Loadings	M (SD)
(1) I would like to know more about climate change.	0.879	4.02 (0.97)
(2) I would like to know what I can do on my own for environment and climate protection.	0.867	4.23 (0.88)
(3) I would like to influence the decisions regarding climate change.	0.834	3.85 (1.07)
(4) I am interested in how to influence climate protection through international democratic decision-making process.	0.864	3.89 (1.16)
(5) I am not interested in problems related to climate change. (R)	−0.687	1.84 (1.04)
Explained variance	68.75%	
M (SD)	20.15 (4.23)	
Cronbach's α	0.88	

(R)—reverse item.

Table 6. Factor (structure) matrix and factor loadings for one-factor solution, coefficient of reliability and descriptive data for Concern for ecological problems scale.

Item (I Am Concerned about …)	Factor Loadings	M (SD)
(1) … air pollution.	0.768	4.39 (0.79)
(2) … accumulation of hazardous waste.	0.883	4.38 (0.86)
(3) … the influence of industry on environment and people's health.	0.893	4.42 (0.77)
(4) … extraction, destruction, and pollution of natural resources (forests, water, soil, oil)	0.830	4.49 (0.76)
(5) … the pollution of rivers, lakes, seas, and oceans.	0.853	4.64 (0.65)
(6) … the pollution of food and drinking water (preservatives, additives, pesticides …)	0.770	4.54 (0.75)
(7) … climate change in general.	0.754	4.34 (0.81)
(8) … forest dieback.	0.830	4.38 (0.81)
(9) … inadequate disposal of municipal waste.	0.827	4.43 (0.83)
(10) … the reduction of arable land (desertification, soil erosion, urbanization and traffic, sea level rise …).	0.828	4.27 (0.89)
Explained variance	67.15%	
M (SD)	44.28 (6.48)	
Cronbach's α	0.94	

3. Results

Table 7 shows the intercorrelations (Spearman correlation coefficient) between all composite variables as well as age (in years). A statistically significant correlation was found between almost all variables. The only nonsignificant correlations were found between age and (I) perception of action possibilities in climate change mitigation and adaptation context, (II) interest in climate change, (III) concern for ecological problems as well as (IV) perception of future in climate change context.

Table 7. The correlation matrix between all composite variables and age.

Variables	Correlation Coefficients						
	1	2	3	4	5	6	7
1. Willingness to act	1	−0.11	0.61 **	0.57 **	0.67 **	0.59 **	0.45 **
2. Age (in years)		1	−0.27 **	−0.10	−0.10	−0.10	−0.09
3. Attitudes towards climate change			1	0.55 **	0.59 **	0.61 **	0.53 **
4. Perception of action possibilities				1	0.59 **	0.49 **	0.39 **
5. Interest in climate change					1	0.60 **	0.48 **
6. Concern for ecological problems						1	0.57 **
7. Perception of future							1

** $p < 0.01$.

A two-stage hierarchical multiple regression was carried out in which willingness to act in climate change mitigation and adaptation context represented dependent variable. Gender and age were implemented at stage one of the regression in order to control for socio-demographic variables. Attitudes towards climate change, interest in climate change, perception of action possibilities in climate change mitigation and adaptation context, perception of future in climate change context as well as concern for ecological problems were implemented at stage two.

The sample size of 201 student teachers was considered adequate given the seven independent variables included in the hierarchical multiple regression analysis. Due to the presence of statistically significant correlations between independent variables (Table 7), the assumption of multicollinearity was tested. The collinearity statistics (i.e., Tolerance and VIF) were all within accepted limits (Tabachnick and Fidell 2019) and therefore the assumption of multicollinearity was considered justified. The results of the hierarchical multiple regression are presented in Table 8.

Table 8. Results of hierarchical regression analysis for the estimation of willingness to act in climate change mitigation and adaptation context.

Variables	Model 1			Model 2		
	B	SE B	β	B	SE B	β
(constant)	29.27	3.86		0.81	3.52	
Gender	2.33	0.99	0.17 *	−0.59	0.69	−0.04
Age (in years)	−0.09	0.16	−0.04	0.07	0.11	0.03
Attitudes towards climate change				0.33	0.13	0.18 **
Perception of action possibilities				0.28	0.12	0.14 *
Interest in climate change				0.75	0.09	0.55 **
Concern for ecological problems				0.03	0.06	0.04
Perception of future				−0.03	0.07	−0.02
R^2	0.031			0.598		
F za R^2	$F_{(2,198)} = 3.116$ *			$F_{(7,193)} = 41.083$ **		

** $p < 0.01$; * $p < 0.05$.

The hierarchical multiple regression analysis revealed that at stage one, socio-demographic variables (gender and age) contributed significantly to the regression model ($R^2 = 0.03$, $F_{(2,198)}$

= 3.17; $p < 0.05$) and accounted for 3% of the variance in willingness to act in climate change mitigation and adaptation context. Participants' gender represents the sole statistically significant predictor. Women express higher level of willingness to act in climate change mitigation and adaptation context.

During stage two of hierarchical regression analysis, the remaining potential predictors of willingness to act in climate change mitigation and adaptation context were implemented. It was determined that predictors explain significant 59.8% of variance in willingness to act in climate change mitigation and adaptation context ($R^2 = 0.598$, $F_{(7,193)} = 41.08$; $p < 0.001$). More specifically, it was determined that predictors statistically significantly explain additional 56.8% of variance in willingness to act ($\Delta R^2 = 0.568$; $F_{(5,193)} = 54.58$; $p < 0.001$) along with variance explained by participants' gender and age. Additionally, (I) attitudes towards climate change, (II) perception of action possibilities in climate change mitigation and adaptation context as well as (III) interest in climate change represent significant predictors of willingness to act in climate change mitigation and adaptation context.

After implementing other predictors into the model in stage two of the analysis, gender no longer represented significant predictor of willingness to act in climate change mitigation and adaptation context. Therefore, regardless of their gender, student teachers who have positive attitudes towards climate change, perceive action possibilities in climate change mitigation and adaptation context (they believe that is possible to influence climate change mitigation and adaptation) as well as those who show interest in climate change, express higher levels of willingness to act in the climate change mitigation and adaptation context.

4. Discussion

The first objective of this study was to examine the instruments' measurement characteristics. By examining the measurement characteristics, very satisfactory validity and reliability measures were obtained. The exploratory factor analyses performed resulted in very satisfactory values and obtained Cronbach alpha values indicate satisfactory reliability of measurement instruments.

The second objective of this study was to examine the predictors of student teachers' willingness to act in climate change mitigation and adaptation context.

Based on the results of previous research (e.g., Al-Naqbi and Alshannag 2018; Kilinc and Aydin 2013; Tuncer et al. 2006, 2009; Vukelić 2021), participant's gender consistently represented a significant predictor of student teachers' variables related to sustainable development, or in this case, willingness to act in climate change mitigation and adaptation context. However, after implementing other predictors in the regression model, gender no longer represented a significant predictor of willingness to act in climate change mitigation and adaptation context. The absence of the contribution of gender to the explanation of variance in willingness to act in climate change mitigation and adaptation context after implementing other predictors into the model indicates the idea that effect of gender on willingness to act is completely mediated by attitudes towards climate change, perception of action possibilities in climate change mitigation and adaptation context as well as interest in climate change. Therefore, an additional examination of the previously mentioned variables' potential mediational effects on the interrelation between gender and student teachers' willingness to act in climate change mitigation and adaptation context should be conducted in future research. Other than that, it should be mentioned that the obtained sample had an uneven gender ratio, which could contribute to the lack of predictive power of gender in explaining student teachers' willingness to act. During the sampling process, it was achieved that participants' gender ratio represents ratios that exist in student teacher population, i.e., teachers. To be more specific, it is common that the teacher population is primarily female. According to the OECD data for Croatia 78% of teachers are women (OECD 2019). In line with that, in this study, 78.6% of participants were women. However, this could have had an effect on the lack of statistical significance of the gender—willingness to act relationship.

Moreover, regarding socio-demographical correlates, it was shown that age does not represent a statistically significant predictor of willingness to act in climate change context. If we consider the results of previous empirical studies, it is possible to notice that significant age differences in variables related to sustainable development are either scarcely found or often small (Wiernik et al. 2013). In line with these findings, age did not represent a significant predictor of student teachers' willingness to act in climate change mitigation and adaptation context in this study.

On the one hand, it was determined in this study that (I) attitudes towards climate change, (II) perception of action possibilities in the climate change mitigation and adaptation context as well as (III) interest in climate change represent significant individual predictors of willingness to act in climate change mitigation and adaptation context. On the other hand, perceptions of the future in climate change context as well as concern for ecological problems did not represent significant predictors of student teachers' willingness to act. One potential explanation of these results lies in the fact that participants' evaluations on both scales are quite high (means of its items are above 4) along with limited response variability. In other words, more or less all participants believe that climate change will represent ever-increasing problems in the future, and they express their concern for ecological problems. However, these beliefs do not motivate them to act (or to express willingness to act) in the climate change mitigation and adaptation context. Instead, what motivates their willingness to act in climate change mitigation and adaptation context are their attitudes towards climate change, perception of action possibilities in climate change mitigation and adaptation context as well as interest in climate change. Therefore, student teachers who have positive attitudes towards climate change, perceive action possibilities in climate change mitigation and adaptation context (they believe that it is possible to influence climate change mitigation and adaptation) as well as those who show interest in climate change, express higher levels of willingness to act in the climate change mitigation and adaptation context.

The obtained results point to the conclusion that attitudes towards climate change represent significant positive predictors of willingness to act in climate change mitigation and adaptation context, which is in accordance with insights form the sustainability sciences that often emphasize the role of attitudes in explaining various types of human behavior in the direction of sustainable development (e.g., Evans et al. 2007; Heimlich and Ardoin 2008). Therefore, a list of models that explain the interrelation between ecological knowledge, attitudes and pro-ecological behaviors exists in sustainable development ecological dimension area. For example, Kaiser et al. (2008) developed a pro-environmental competence model in which they define the interrelation between environmental knowledge, attitudes towards sustainable development ecological dimension topics as well as various types of pro-ecological behavior. Kaiser et al. (2008) emphasize that attitudes towards the environment represent a key predictor of pro-ecological behavior to a significantly larger extent compared to knowledge. In line with these facts, it was determined in this study that student teachers who express more positive attitudes towards climate change also express higher levels of willingness to act in climate change mitigation and adaptation context.

Furthermore, interest represents another motivational factor that explains student teachers' willingness to act in the climate change mitigation and adaptation context. This result is in accordance with previous empirical findings, which indicate that intrinsic interest in sustainable development ecological issues and problems leads to a higher level of various pro-ecological behavior (Steg et al. 2016; Van der Werff et al. 2013). In situations where student teachers express an interest in issues regarding climate change, and find these topics interesting and important, they express higher level of willingness to act in the climate change mitigation and adaptation context. In this context, the finding of McNeal et al. (2017), which states that interest in environmental issues, especially climate change, represents one of the key motivational factors in implementation of content about climate change in teachers' professional work, is particularly interesting.

Apart from attitudes and interest in climate change, perceptions of action possibilities in climate change mitigation and adaptation context also represents a significant predictor of willingness to act in climate change mitigation and adaptation context. Therefore, student teachers who believe that it is possible to influence climate change and act in order to mitigate them, express higher level of readiness to take action. From the theoretical perspective, it is interesting that the perception of action possibilities represents one of the key predictors of willingness to act. Namely, according to the experts who explore action competence for sustainable development (e.g., Breiting and Mogensen 1999; Mogensen and Schnack 2010; Sass et al. 2020) these two constructs represent key components of action competence. Previous theoretical notions on action competence did not encompass the examination of interrelation between abovementioned components, instead, they were based exclusively on their description (e.g., how an individual who expresses willingness to act in climate change direction functions). Due to the fact that interrelation between two key components of action competence was determined, the results of this study can offer a supplement to the construct in question. Additionally, this finding has potential implications for educational activities focused on encouraging the development of action competence. In other words, if we want the person to express willingness to act in the direction of sustainable development (or, more specifically, climate change), it is important to encourage the development of their awareness about various action possibilities in the abovementioned direction as well as empower their confidence in personal influence needed to make positive changes.

It is important to single out several key limitations of this study. First, one key limitation of this study is that its sample included exclusively student teachers studying at just one Croatian university. Second, due to abovementioned national and regional limitation of the sample, it is not possible to draw stronger conclusions regarding the predictors of student teachers' willingness to act in climate change mitigation and adaptation context. Therefore, in future research, we do not only recommend the expansion of the sample of participants to other Croatian universities, but also to include a multicultural sample in order to determine whether the obtained results explain student teachers' willingness to act in climate change mitigation and adaptation context at international level. Apart from that, willingness to act in climate change mitigation and adaptation context probably represents a dynamic construct, which changes over time depending on various factors that form it. In order to determine the development of student teachers' willingness to act in the climate change mitigation and adaptation context more clearly, it is important to conduct longitudinal studies that will encompass the phenomenon in different development phases. Finally, in this context, the experts should monitor the development of student teachers' willingness to act during their participation in teacher's initial training, particularly in programs focused on training student teachers for implementation of education for sustainable development as well as teaching about climate change.

5. Conclusions

The aim of this study was to examine predictors of student teachers' willingness to act in the climate change mitigation and adaptation context. According to the results of this study, regardless of their gender, student teachers who have positive attitudes towards climate change, perceive action possibilities in the climate change mitigation and adaptation context (they believe that is possible to influence climate change mitigation and adaptation) and those who show interest in climate change, are more willing to act in the climate change mitigation and adaptation context.

An apparent need exists for training student teachers to teach about climate change. Additionally, they should be empowered in order to shift towards climate change mitigation and adaptation.

In line with the results of this study, the recommendation is to raise student teachers' awareness regarding the action possibilities as well as influence they can have on climate change mitigation and adaptation during their initial teacher training programs. It is particularly important to show student teachers the methods and models they can use in order to cope with challenges of today's world, especially climate change. Apart from that, it is important to place emphasis on encouraging the development of student teachers' positive attitudes and interest in climate change. Additionally, through educational interventions focused on the abovementioned aspects, we can ensure that student teachers feel more prepared to act in the climate change mitigation and adaptation context, which will, consequently, have a positive effect on both their (future) students as well as lead to a general shift towards a more sustainable future by implementing education for sustainable development.

Author Contributions: Conceptualization, N.V., N.R. and S.T.; methodology, N.V., N.R. and S.T.; validation, N.V.; formal analysis, N.V.; investigation, N.V., N.R. and S.T.; data curation, N.V.; writing—original draft preparation, N.V. and N.R.; writing—review and editing, N.V. and N.R.; funding acquisition, N.V. and N.R. All authors have read and agreed to the published version of the manuscript.

Funding: This research was funded by the Croatian Science Foundation (grant number: UIP-2017-05-2031) and University of Rijeka (grant number: drustv-18-124/1263).

Institutional Review Board Statement: The study was conducted according to the guidelines of Ethics Committee at the Faculty of Humanities and Social Sciences (University of Rijeka), and ethical principle of conducting the research at the University of Rijeka.

Informed Consent Statement: Informed consent was obtained from all subjects involved in the study.

Data Availability Statement: More about the research and project on http://foror.uniri.hr/?lang=en (accessed on 30 August 2020).

Conflicts of Interest: The authors declare no conflict of interest. The funders had no role in the design of the study; in the collection, analyses, or interpretation of data; in the writing of the manuscript, or in the decision to publish the results.

References

Albareda-Tiana, Silvia, Salvador Vidal-Raméntol, Maria Pujol-Valls, and Mónica Fernández-Morilla. 2018. Holistic Approaches to Develop Sustainability and Research Competencies in Pre-Service Teacher Training. *Sustainability* 10: 3698. [CrossRef]

Al-Naqbi, Ali Khalfan, and Qasim Alshannag. 2018. The status of education for sustainable development and sustainability knowledge, attitudes, and behaviors of UAE University students. *International Journal of Sustainability in Higher Education* 19: 566–88. [CrossRef]

Boon, Helen J. 2016. Pre-service teachers and climate change: A stalemate? *Australian Journal of Teacher Education* 41: 39–63. [CrossRef]

Breiting, Søren, and Finn Mogensen. 1999. Action competence and environmental education. *Cambridge Journal of Education* 29: 349–53. [CrossRef]

Cebrián, Gisela, and Mercè Junyent. 2015. Competencies in education for sustainable development: Exploring the student teachers' views. *Sustainability* 7: 2768–86. [CrossRef]

Chen, Chung-Chiang. 2010. Spatial inequality in municipal solid waste disposal across regions in developing countries. *International Journal of Environmental Science & Technology* 7: 447–56. [CrossRef]

Chen, Ho-Wen, Ruey Fang Yu, Shu-Sheng Liaw, and Wei-Chien Huang. 2010. Information policy and management framework for environmental protection organization with ecosystem conception. *International Journal of Environmental Science & Technology* 7: 313–26. [CrossRef]

Chen, Wei-Yin, Toshio Suzuki, and Maximilian Lackner. 2017. *Handbook of Climate Change Mitigation and Adaptation*. New York: Springer.

Cifrić, Ivan. 2005. Ekološka zabrinutost—Percepcija ekoloških problema kao zabrinjavajućih. [Environmental concern—Perception of environmental problems as disturbing]. *Socijalna Ekologija* 14: 1–28.

Cleveland, Mark, Maria Kalamas, and Michel Laroche. 2012. "It's not easy being green": Exploring green creeds, green deeds, and internal environmental locus of control. *Psychology & Marketing* 29: 293–305. [CrossRef]

Competente, Ronnel Joseph Tracena. 2019. Pre-Service Teachers' Inclusion of Climate Change Education. *International Journal of Evaluation and Research in Education* 8: 119–26. [CrossRef]

Coulter, Liese, Silvia Serrao-Neumann, and Eddo Coiacetto. 2019. Climate change adaptation narratives: Linking climate knowledge and future thinking. *Futures* 111: 57–70. [CrossRef]

De Silva, Dakshina G., and Rachel A. J. Pownall. 2014. Going green: Does it depend on education, gender or income? *Applied Economics* 46: 573–86. [CrossRef]

Demant-Poort, Lars, and Paul Berger. 2021. "It is not something that has been discussed": Climate change in teacher education in Greenland and Canada. *Journal of Geoscience Education* 69: 207–19. [CrossRef]

Evans, Gary W., Gernot Brauchle, Aliya Haq, Rachel Stecker, Kimberly Wong, and Elan Shapiro. 2007. Young children's environmental attitudes and behaviors. *Environment and Behavior* 39: 635–58. [CrossRef]

Gatersleben, Birgitta, Emma White, Wokje Abrahamse, Tim Jackson, and David Uzzell. 2010. Values and sustainable lifestyles. *Architectural Science Review* 53: 37–50. [CrossRef]

Hadžiselimović, Dževdet. 2015. *Klima se Mijenja, a mi . . . [The Climate Is Changing, and We . . .]*. Pula: Istrian Psychological Society.

Heimlich, Joe E., and Nicole M. Ardoin. 2008. Understanding behavior to understand behavior change: A literature review. *Environmental Education Research* 14: 215–37. [CrossRef]

Hines, Jody M., Harold R. Hungerford, and Audrey N. Tomera. 1987. Analysis and synthesis of research on responsible environmental behavior: A meta-analysis. *The Journal of Environmental Education* 18: 1–8. [CrossRef]

Howell, Rachel A., and Simon Allen. 2019. Significant life experiences, motivations and values of climate change educators. *Environmental Education Research* 25: 813–31. [CrossRef]

Hudson, Marc. 2017. It's Ten Years Since Rudd's 'Great Moral Challenge', and We Have Failed It. *The Conversation*. March 31. Available online: https://theconversation.com/its-ten-years-since-rudds-great-moral-challenge-and-we-have-failed-it-75534 (accessed on 10 May 2021).

Hulme, Mike. 2009. *Why We Disagree about Climate Change: Understanding Controversy, Inaction and Opportunity*. Cambridge: Cambridge University Press.

Irwin, Ruth. 2020. Climate change and education. *Educational Philosophy and Theory* 52: 492–507. [CrossRef]

Jensen, Bjarne Bruun, and Karsten Schnack. 1997. The action competence approach in environmental education. *Environmental Education Research* 3: 163–78. [CrossRef]

Jones, Charlotte, Donald W. Hine, and Anthony D.G. Marks. 2017. The future is now: Reducing psychological distance to increase public engagement with climate change. *Risk Analysis* 37: 331–41. [CrossRef] [PubMed]

Kaiser, Florian G., Nina Roczen, and Franz X. Bogner. 2008. Competence formation in environmental education: Advancing ecology-specific rather than general abilities. *Umweltpsychologie* 12: 56–70. [CrossRef]

Kaplan, Stephen, and Rachel Kaplan. 2009. Creating a larger role for environmental psychology: The Reasonable Person Model as an integrative framework. *Journal of Environmental Psychology* 29: 329–39. [CrossRef]

Kilinc, Ahmet, and Abdullah Aydin. 2013. Turkish student science teachers' conceptions of sustainable development: A phenomenography. *International Journal of Science Education* 35: 731–52. [CrossRef]

Koskela, Teija, and Sirpa Kärkkäinen. 2021. Student Teachers' Change Agency in Education for Sustainable Development. *Journal of Teacher Education for Sustainability* 23: 84–98. [CrossRef]

Larijani, Maryam. 2010. Assessment of environmental awareness among higher primary school teachers. *Journal of Human Ecology* 31: 121–4. [CrossRef]

Lazarus, Richard J. 2008. Super wicked problems and climate change: Restraining the present to liberate the future. *Cornell L. Rev.* 94: 1153–234.

Lee, Pei-Shan, Yu-Hsien Sung, Chia-Chun Wu, Liang-Chu Ho, and Wen-Bin Chiou. 2020. Using episodic future thinking to pre-experience climate change increases pro-environmental behavior. *Environment and Behavior* 52: 60–81. [CrossRef]

Liu, Shiyu, Gillian Roehrig, Devarati Bhattacharya, and Keisha Varma. 2015. In-Service Teachers'Attitudes, Knowledge and Classroom Teaching of Global Climate Change. *Science Educator* 24: 12–22.

Masson-Delmotte, Valèrie, Panmao Zhai, Anna Pirani, Sarah L. Connors, Clotide Péan, Sophie Berger, Nada Caud, Yang Chen, Leah Goldfarb, Melissa I. Gomis, and et al., eds. 2021. *IPCC, 2021: Climate Change 2021: The Physical Science Basis. Contribution of Working Group I to the Sixth Assessment Report of the Intergovernmental Panel on Climate Change*. Cambridge: Cambridge University Press.

McKeown, Rosalyn, and Charles Hopkins. 2010. Rethinking climate change education. *Green Teacher* 89: 17–21. Available online: https://www.humphreyfellowship.org/system/files/Rethinking%20Climate%20Change%20Education.pdf (accessed on 10 August 2021).

McNeal, Peggy, Heather Petcovic, and Patricia Reeves. 2017. What is motivating middle-school science teachers to teach climate change? *International Journal of Science Education* 39: 1069–88. [CrossRef]

Meinzen-Dick, Ruth, Chiara Kovarik, and Agnes R. Quisumbing. 2014. Gender and sustainability. *Annual Review of Environment and Resources* 39: 29–55. [CrossRef]

Mogensen, Finn, and Karsten Schnack. 2010. The action competence approach and the 'new' discourses of education for sustainable development, competence and quality criteria. *Environmental Education Research* 16: 59–74. [CrossRef]

Morton, Timothy. 2013. *Hyperobjects: Philosophy and Ecology after the End of the World*. Minneapolis: University of Minnesota Press.

Nyarko, Samuel Cornelius, and Heather L. Petcovic. 2021. Ghanaian preservice science teachers' knowledge of ozone depletion and climate change, and sources of their knowledge. *International Journal of Science Education* 10: 1–22. [CrossRef]

OECD. 2019. *TALIS 2018 Results (Volume 1): Teachers and School Leaders as Lifelong Learners*. Paris: OECD Publishing. [CrossRef]

Oerke, Britta, and Franz X. Bogner. 2010. Gender, age and subject matter: Impact on teachers' ecological values. *The Environmentalist* 30: 111–22. [CrossRef]

Olsson, Daniel, Niklas Gericke, Wanda Sass, and Jelle Boeve-de Pauw. 2020. Self-perceived action competence for sustainability: The theoretical grounding and empirical validation of a novel research instrument. *Environmental Education Research* 26: 742–60. [CrossRef]

Rauch, Franz, and Regina Steiner. 2013. Competences for education for sustainable development in teacher education. *CEPS Journal* 3: 9–24. [CrossRef]

Rončević, Nena, and Klara Cvetković. 2016. Students' Attitudes and Behaviours in the Context of Environmental Issues. *Socijalna Ekologija* 25: 11–37. [CrossRef]

Ryan, Richard M., and Edward L. Deci. 2000. Self-determination theory and the facilitation of intrinsic motivation, social development, and well-being. *American Psychologist* 55: 68–78. [CrossRef]

Sass, Wanda, Jelle Boeve-de Pauw, Daniel Olsson, Niklas Gericke, Sven De Maeyer, and Peter Van Petegem. 2020. Redefining action competence: The case of sustainable development. *The Journal of Environmental Education* 51: 292–305. [CrossRef]

Schulz, Wolfram, John Ainley, Julian Fraillon, Bruno Losito, Gabriella Agrusti, and Tim Friedman. 2018. *Becoming Citizens in a Changing world: IEA International Civic and Citizenship Education Study 2016 International Report*. Berlin and Heidelberg: Springer. [CrossRef]

SDSN. 2021. SDG Action's Climate Action Edition. Available online: https://sdg-action.org/wp-content/uploads/2021/10/SDG-Action-Climate-2021.pdf (accessed on 10 October 2021).

Seroussi, Dominique-Esther, Nathan Rothschild, Eyal Kurzbaum, Yosi Yaffe, and Tahel Hemo. 2019. Teachers' Knowledge, Beliefs, and Attitudes about Climate Change. *International Education Studies* 12: 33–45. [CrossRef]

Sleurs, Willy. 2008. Competencies for ESD (Education for Sustainable Development) Teachers. A Framework to Integrate ESD in the Curriculum of Teacher Training Institutes. Available online: http://www.unece.org/fleadmin/DAM/env/esd/inf.meeting.docs/EGonInd/8mtg/CSCT%20Handbook_Extract.pdf (accessed on 10 May 2017).

Spiropoulou, Dimitra, Triantafyllia Antonakaki, Sophia Kontaxaki, and Sarantis Bouras. 2007. Primary teachers' literacy and attitudes on education for sustainable development. *Journal of Science Education and Technology* 16: 443–50. [CrossRef]

Steg, Linda, Siegwart Lindenberg, and Kees Keizer. 2016. Intrinsic motivation, norms and environmental behaviour: The dynamics of overarching goals. *International Review of Environmental and Resource Economics* 9: 179–207. [CrossRef]

Tabachnick, Barbara G, and Linda S. Fidell. 2019. *Using Multivariate Statistics*, 7th ed. London: Pearson.

Tuncer, Gaye, Ceren Tekkaya, and Semra Sungur. 2006. Pre-service teachers' beliefs about sustainable development: Effects of gender and enrollment to an environmental course. *Hacettepe Üniversitesi Eğitim Fakültesi Dergisi* 31: 179–87.

Tuncer, Gaye, Ceren Tekkaya, Semra Sungur, Jale Cakiroglu, Hamide Ertepinar, and Michaele Kaplowitz. 2009. Assessing pre-service teachers' environmental literacy in Turkey as a mean to develop teacher education programs. *International Journal of Educational Development* 29: 426–36. [CrossRef]

UNECE. 2012. *Learning for the Future: Competences in Education for Sustainable Development*. Geneva: UNECE, Available online: https://www.unece.org/fileadmin/DAM/env/esd/ESD_Publications/Competences_Publication.pdf (accessed on 14 May 2017).

UNESCO. 2015. *Rethinking Education*. Available online: http://unesdoc.unesco.org/images/0023/002325/232555e.pdf (accessed on 14 May 2017).

UNESCO. 2017. *Education for Sustainable Development: Learning Objectives*. Paris: UNESCO, Available online: http://unesdoc.unesco.org/images/0024/002474/247444e.pdf (accessed on 10 May 2017).

UNESCO. 2020. *Education for Sustainable Development: A Roadmap*. Paris: UNESCO, Available online: https://unesdoc.unesco.org/ark:/48223/pf0000374802 (accessed on 30 August 2020).

Van der Werff, Ellen, Linda Steg, and Kees Keizer. 2013. It is a moral issue: The relationship between environmental self-identity, obligation-based intrinsic motivation and pro-environmental behaviour. *Global Environmental Change* 23: 1258–65. [CrossRef]

Verlie, Blanche. 2019. Bearing worlds: Learning to live-with climate change. *Environmental Education Research* 25: 751–66. [CrossRef]

Vukelić, Nena. 2021. Prediktori Razine Namjere Budućih Nastavnika za Implementaciju Obrazovanja za Održivi Razvoj. [Predictors of Student Teachers' Intentions to Implement Education for Sustainable Development]. Ph.D. thesis, Faculty of Humanities and Social Sciences, University of Rijeka, Rijeka, Croatia.

Wiernik, Brenton M., Deniz S. Ones, and Stephan Dilchert. 2013. Age and environmental sustainability: A meta-analysis. *Journal of Managerial Psychology* 28: 826–56. [CrossRef]

Wray-Lake, Laura, Constance A. Flanagan, and D. Wayne Osgood. 2010. Examining trends in adolescent environmental attitudes, beliefs, and behaviors across three decades. *Environment and Behavior* 42: 61–85. [CrossRef]

Zelezny, Lynnette C., Poh-Pheng Chua, and Christina Aldrich. 2000. Elaborating on gender differences in environmentalism-statistical data included. *Journal of Social Issues* 56: 443–5. [CrossRef]

MDPI
St. Alban-Anlage 66
4052 Basel
Switzerland
www.mdpi.com

Social Sciences Editorial Office
E-mail: socsci@mdpi.com
www.mdpi.com/journal/socsci

www.ingramcontent.com/pod-product-compliance
Lightning Source LLC
Chambersburg PA
CBHW040804250926
51850CB00007BA/1080
* 9 7 8 3 0 3 6 5 9 3 5 0 0 *